T0392924

Bioprospection of Co-products and Agro-industrial Wastes

A Sustainable Biotechnological Approach

Editors

Joilane Alves Pereira Freire
Department of Nutrition
Federal University of Piaui
Picos-Piaui, Brazil

Stella Regina A. Medeiros
Department of Nutrition
Federal University of Piaui
Picos-Piaui, Brazil

Jailane de Souza Aquino
Department of Nutrition
Federal University of Paraíba
João Pessoa-Paraíba, Brazil

Chistiane M. Feitosa
Department of Chemistry
Federal University of Piaui
Teresina-Piaui, Brazil

Mahendra Rai
Department of Biotechnology
SGB Amravati University
Amravati, Maharashtra, India

CRC Press
Taylor & Francis Group
Boca Raton London New York

CRC Press is an imprint of the
Taylor & Francis Group, an **informa** business
A SCIENCE PUBLISHERS BOOK

Cover photo: Prospects for use of bacuri (*Platonia insignis*) fruit. The plant is native to the Amazon region of Brazil and Guyana, but is also found in Colombia and Paraguay. Reproduced by kind courtesy of Pedro Vitor Oliveira Silva Furtado, a master's student of the Department of Chemistry, Federal University of Piaui, Teresina, Brazil. The figure used has not been taken from any copyrighted source.

First edition published 2023
by CRC Press
6000 Broken Sound Parkway NW, Suite 300, Boca Raton, FL 33487-2742

and by CRC Press
4 Park Square, Milton Park, Abingdon, Oxon, OX14 4RN

© 2023 Joilane Alves Pereira Freire, Stella Regina A. Medeiros, Jailane de Souza Aquino, Chistiane M. Feitosa and Mahendra Rai

CRC Press is an imprint of Taylor & Francis Group, LLC

Reasonable efforts have been made to publish reliable data and information, but the author and publisher cannot assume responsibility for the validity of all materials or the consequences of their use. The authors and publishers have attempted to trace the copyright holders of all material reproduced in this publication and apologize to copyright holders if permission to publish in this form has not been obtained. If any copyright material has not been acknowledged please write and let us know so we may rectify in any future reprint.

Except as permitted under U.S. Copyright Law, no part of this book may be reprinted, reproduced, transmitted, or utilized in any form by any electronic, mechanical, or other means, now known or hereafter invented, including photocopying, microfilming, and recording, or in any information storage or retrieval system, without written permission from the publishers.

For permission to photocopy or use material electronically from this work, access www.copyright.com or contact the Copyright Clearance Center, Inc. (CCC), 222 Rosewood Drive, Danvers, MA 01923, 978-750-8400. For works that are not available on CCC please contact mpkbookspermissions@tandf.co.uk

Trademark notice: Product or corporate names may be trademarks or registered trademarks and are used only for identification and explanation without intent to infringe.

Library of Congress Cataloging-in-Publication Data (applied for)

ISBN: 978-1-032-14570-9 (hbk)
ISBN: 978-1-032-14571-6 (pbk)
ISBN: 978-1-003-23999-4 (ebk)

DOI: 10.1201/9781003239994

Typeset in Times New Roman
by Radiant Productions

Preface

The book *"Bioprospection of co-products and agro-industrial wastes: A sustainable biotechnological approach"*, seeks to reveal to the academic community the importance of the full use of food from Northeast Brazil and the world, as the waste of food and the residues resulting from this practice have become an economic as well as environmental problem. Many parts of food plants are discarded, either due to lack of information or prejudice, these parts usually disposed off, such as leaves, stems, seeds, bark and roots, have high nutritional values, many times even greater than the part usually consumed. Being able to improve the nutritional quality of food and reduce waste generated for the environment, providing sustainable consumption and low cost. The book in question addresses different aspects involved in full use, as well as the functional properties present in food, namely bioactive compounds, vitamins and minerals essential for human metabolism, with emphasis on *in vivo* studies and different clinical applications. In addition to the use of residues in the generation of renewable energies, such as carnauba wax and babassu coconut, regional products from Northeast Brazil and the world, as well as the bioprospecting of nutrients present in the unconventional parts to be consumed have been discussed in the book. Moreover, this material aims to identify and explain the functional properties of fruit and vegetable by-products, also addressing the possible toxicological components that may be present in conventional and non-conventional parts of food waste. The role of microorganisms in the full use of food was worked as an optimizing factor in this process since they are essential in the fermentation process, bio-industries and other mechanisms that govern biotechnology. A chapter with a study of great importance is presented with research carried out in the state of Piaui, which suggests that constituents of Buriti oil as inhibitors of the main peptidase of SARS-Coronavirus: an investigation by Anchoring and Molecular Dynamics. We address the technological potential of unconventional food plants (PANCs), which are not usually consumed in the diet, but have high nutritional value, easy cultivation and management. In addition, the increase in production and consumption of these plants contributes to the valorization of family farming and income generation, and to the expansion of its use in the area of food science and technology. One of the chapters aims to demonstrate, by means of technological prospecting, how the biodiversity of food plants can be used as raw material for the development of products in different industries, such as food, pharmaceutical and cosmetics. The present book will cater to the need of undergraduate and postgraduate students of food technology, plant science, phytochemistry and chemistry. It will also be useful for researchers in the related field, particularly in biotechnology.

Joilane Alves Pereira-Freire, Brazil
Stella Regina A. Medeiros, Brazil
Jailane de Souza Aquino, Brazil
Chistiane M. Feitosa, Brazil
Mahendra Rai, India

Contents

Preface iii

1. **Bioprospecting of Co-Products and Agro-industrial Wastes: Opportunities and Challenges** 1
   Joilane Alves Pereira-Freire, Stella Regina Arcanjo Medeiros, Chistiane Mendes Feitosa, Jailane de Souza Aquino and *Mahendra Rai*

2. **Technological Prospection for the Use of Food Plants with a Focus on the Food, Pharmaceutical and Cosmetics Industry** 8
   Tatiana de Oliveira Lemos

3. **Technological Potential of Unconventional Food Plants (UFP)** 20
   Tatiana de Oliveira Lemos, Anida Maria Moraes Gomes, Bianca Almeida Souza, Brenda Paiva Campi Neves, Dárcia Souza Araújo, Djany Souza Silva, Elder dos Santos Araújo, Luana Costa de Souza and *Mauriane Maciel da Silva*

4. **Biotechnological Applications of Flours and By-products of *Mauritia flexuosa* L. Fruits** 41
   Márcia Luiza dos Santos Beserra Pessoa, Ennya Cristina Pereira dos Santos Duarte, Thially Braga Gonçalves, Vivianne Rodrigues Amorim, Lívia Maria Nunes de Almeida, Carla Lorena Silva Ramos, Paulo Michel Pinheiro Ferreira and *Joilane Alves Pereira-Freire*

5. **Sustainable Use of Food Waste in the Preparation of Biodegradable Packaging for the Food Industry** 58
   Alyandra de Sousa Nascimento, Maria Crisnanda Almeida Marquesa, Patrícia e Silva Alves, Maria das Dores Alves de Oliveira and *Chistiane Mendes Feitosa*

6. **Edible Films and Coatings: Raw Material, Properties, and its Application in Food** 72
   Gabriela Almeida de Paula, Sandra Helena de Mesquita Pinheiro, Lindalva de Moura Rocha, Ronnyely Suerda Cunha Silva, Whellyda Katrynne Silva Oliveira, Francisco Eduardo Pereira Rocha, José Ribeiro dos Santos Júnior and *Erivan Santos Lima*

7. **Perspectives For Using the Fruits of Cacti from the Caatinga of Piauí For Nutritional and Biotechnological Purposes** **87**

Tamiris Ramos Silva, Tiago Soares, Mariany de Alencar, Maria Nilka de Oliveira, Boris Timah Acha, Ana Cibele Pereira Sousa, Alessandro de Lima, Joilane Alves Pereira-Freire and *Stella Regina Arcanjo Medeiros*

8. **Cashew's Industry: Products, Co-products and its Applications** **105**

Rafael Marinho Bandeira, Lindalva de Moura Rocha, Leanne Silva de Sousa, Gabriela Almeida de Paula, Josélia Borges de Moura Furtado, Paulo Afonso Furtado Filho, Raimundo Oliveira Lima Júnior, Thamires Pereira dos Santos Lima, Inês Maria de Souza Araújo and *José Ribeiro dos Santos Júnior*

9. **Chemistry, Management, Functional Properties and Biotechnological Applications of *Ocimum* Essential Oils in the Food Industry** **118**

Celma de Oliveira Barbosa, Selene Maia de Morais, Halisson Araújo de Sousa, Vitor Carvalho Martins, João Francisco Câmara Neto, Icaro Gusmão Pinto Vieira, Rita de Cássia Alves Pereira, Ana Livya Moreira Rodrigues and *José Osvaldo Beserra Carioca*

10. ***Moringa oleifera* Seed Cake: A By-product of Oil Extraction with Biotechnological Potential** **133**

Ardilles Juan Carlos Alves dos Santos, Eva Luana Almeida da Silva, Panait Abu Nidal Alves dos Santos, Hanna Gracie Inez de Freitas Lima, Welton Aaron de Almeida, Romildo de Albuquerque Nogueira and *Emmanuel Viana Pontual*

11. **The Role of Microorganisms in the Use of Food-Waste** **144**

Monalisa de Alencar Lucena, Igor Frederico S. Ramos, Valesca Lima Fernandes, Allana Costa Araújo Souza, Márcia dos Santos Rizzo, Alessandra Braga Ribeiro and *Marcília Pinheiro da Costa*

12. **General Toxicological Aspects of Toxins from Conventional and Non-Conventional Foods** **162**

Antonia Amanda Cardoso de Almeida, Ranyelison Silva Machado, Aldenora Maria Ximenes Rodrigues, Railson Pereira Souza and *Paulo Michel Pinheiro Ferreira*

13. **Annonaceae Family: Characteristics, Properties and Applications** **184**

Valéria Lima Silva Veras, Pedro Vitor Oliveira Silva Furtado, Pelrry da Silva Costa, Mahendra Rai, Chistiane Mendes Feitosa and *João Marcelo de Castro e Sousa*

14. Potential Exploitation of Residual Avocado (*Persea americana*) Seeds in the Development of Functional Foods with Glycemia- and Cholesterol-lowering Properties **201**

Mercedes V. Urquiza Martínez, José Pérez Villarreal, Zaira I. Bedolla Valdez, José Venegas González, Luz Torner, Raul Manhães de Castro, Ana E. Toscano and *Omar Guzmán Quevedo*

15. Perspectives for the Reuse of Bacuri (*Platonia insignis*) for Nutritional and Biotechnological Applications **214**

Ana Paula Rodrigues de Sousa, Pedro Vitor Oliveira Silva Furtado, Maria Geovana Santos Ribeiro, Antônio do Nascimento Cavalcante, Mahendra Rai and *Chistiane Mendes Feitosa*

16. Red Pitaya (*Hylocereus polyrhizus*) as a Functional Food: Nutritional Value, Phytochemical Content and Health-Promoting Effects **232**

Chayane Gomes Marques, Bruno Bezerra da Silva, Sandra Machado Lira, Ana Paula Dionísio, Maurício Fraga van Tilburg, Maria Izabel Florindo Guedes and *Claisa Andrea Freitas Rabelo*

17. Functional Properties of Plant By-products: Studies in Animal Models and Clinical Applications **253**

Kamila Sabino Batista, Naís Lira Soares, Juliana Gondim de Albuquerque, Renata Leite Tavares, João Victor Gama de Albuquerque, Geyse Araújo Costa and *Jailane de Souza Aquino*

18. The Benefits of Integral Use of Food in Waste Reduction, Nutrient Density and Human Health **272**

Eliakim Aureliano da Silva, Tamiris Ramos Silva, Maria Clara Feijó de Figueiredo, Beatriz Gonçalves Feitosa dos Santos, Lavínia Alves de Sousa, Rosa Maria de Oliveira Santos, Fhanuel Silva Andrade, Joilane Alves Pereira-Freire and *Stella Regina Arcanjo Medeiros*

Index **281**

Editors' Biography **283**

Chapter 1

Bioprospecting of Co-Products and Agro-industrial Wastes

Opportunities and Challenges

Joilane Alves Pereira-Freire,[1] *Stella Regina Arcanjo Medeiros,*[2] *Chistiane Mendes Feitosa,*[3] *Jailane de Sousa Aquino*[4] and *Mahendra Rai*[5,6]

Introduction

The challenges of bioprospecting in Brazil and in the world for understanding the main advances and obstacles encountered in the search for its development are immense. Conceptually, bioprospecting is to carry out a systematic search for organisms, genes, enzymes, compounds, processes and parts from living beings in general, which may have economic potential and eventually, lead to the development of a product (Saccaro-Júnior, 2011). Therefore, it is of great relevance for many sectors and activities, including biotechnology, agriculture, nutrition, pharmaceutical and cosmetics industry, bioremediation, biomonitoring, health and fuel production through biomass, among others.

The targets of bioprospecting are collectively called genetic resources. Its set forms the genetic heritage and currently, there are many challenges to improve the

[1] Research Laboratory II, Federal University of Piauí, Campus Senador Helvídio Nunes de Barros, 64607-670, Picos, Piauí, Brazil.

[2] Research Laboratory II, Federal University of Piaui, Campus Senator Helvídio Nunes de Barros - Junco, 64607-670, Picos - PI.

[3] Department of Chemistry, Federal University of Piauí, 64049-550 Teresina-PI, Brazil.

[4] Experimental Nutrition Laboratory, Department of Nutrition, Federal University of Paraíba (UFPB), Cidade Universitária, s/n-Castelo Branco III, João Pessoa 58051-900, Paraíba, Brazil.

[5] Department of Biotechnology, SGB Amravati University, Amravati-444602, Maharashtra, India.

[6] Department of Microbiology, Nicolaus Copernicus University, Torun, Poland.

* Corresponding author: joilane@ufpi.edu.br

legislation on access to genetic resources, as well as the sharing of the generated benefits; In addition to investing in regional research infrastructure and safely expanding the use of biodiversity in industrial and regional development policies, encouraging the legal participation of private companies to effectively control the illegal appropriation of information and biological material. This highlights the lack of connection between the regulatory system, expectations about the economic potential of biodiversity and the process of research and development (R&D).

Another relevant point would be the control of international bodies to regulate the patenting of products derived from biodiversity and overcoming all these challenges can mean an important contribution to economic development at the national and international level, associated with the reduction of regional inequalities and environmental preservation. So, there is an urgent need to expand research in the bioprospecting of natural products because they are relevant and essential in this process of food waste control and for their great impact on society and the environment. Losses are observed throughout the production chain, which establishes a need to use food in an integral way.

In this perspective, it is essential to carry out bioprospection and technological innovations of natural products, especially when it comes to residues and by-products, with the challenge of promoting food security, expanding the distribution of nutrient-rich foods, earlier without commercial value, promoting adaptations for consumption and facilitating access to families and assistance entities.

Historical context

Developed countries, even though they are not great holders of terrestrial biodiversity, understand the importance of molecular wealth from nature, so there is a significant number of medicines extracted and developed from natural products on the world market.

As for Brazil, which holds one of the greatest natural resources on the planet, a vigorous bioeconomy based on innovations in its rich biodiversity—a sophisticated natural factory of bioactive compounds of different classes and unusual chemical structures that would have, if well used, are an enormous potential for radical and incremental innovations for pharmaceutical, cosmetics, fragrances, agrochemicals and food supplements sectors. In the last 15 years, many discussions, both on the academic and business side, were made, converging to the same consensus: the delay in innovations from natural products and derivatives from our rich biodiversity was due, in part, to bureaucratic obstacles caused by the edition, in 2000, of the Provisional Measure (MP) 2050, reissued as MP-2186 – 16/2001, by the Federal Government, referring to access to genetic heritage, traditional knowledge, protection and benefit-sharing (Brasil, 2008).

Institutions such as the Brazilian Food Industry Association (ABIA), the Organization for Economic Cooperation and Development (OECD), the Food and Agriculture Organization of the United Nations (FAO) and the United States Department of Agriculture (USDA) have reiterated that in 10 years food production in the world needs to grow 20% to be able to feed the entire planet. And for that to happen, Brazil will have to increase its production and there is no way to feed the world without sustainability, this is a challenge that requires innovation with environmental conservation and social responsibility (CGEE, 2014).

Thus, whether in biotechnologies applied to the pharmaceutical industry or the food industry, investments in sustainability are on all fronts, to reduce the use of natural resources, professional training, social projects and technical research for the development of local communities and a strategy for waste management.

Bioprospecting of co-products and agro-industrial wastes in food

The development of new products in the food industry becomes increasingly challenging, as it seeks to meet consumer demand for products that are healthier and more attractive. As a consequence, the diet of individuals with a healthy lifestyle tends to be a desirable act that, at the same time, aims at health and well-being (Komatsu et al., 2008).

The benefits of certain types of food on health have been proven for a long time, which contribute to nutrition, being called functional foods, because they contain substances that can be considered biologically active, providing clinical or health benefits (Komatsu et al., 2008). According to Pereira-Freire et al., 2022, there are several factors affecting the quality of life, so it is necessary to make the population aware of the importance of consuming foods containing substances that help to promote health, providing an improvement in nutritional status.

In Brazil, food waste reaches up to 39,000 tons per day, putting it among the 10 countries that waste the most food in the world (Martins et al., 2022), thus it is necessary to adopt measures for the practice of conscious consumption of the population in relation to food and the full use of fruits and vegetables (pulp, peels, stalks and leaves), in the development of new products, it is a clean technological alternative that is within everyone's reach, as it can be applied both in industrial and residential environments. The use of food in a sustainable way reduces the production of organic waste, prolongs the shelf life of food, promotes food security and benefits family income (FAO/Biodiversity, 2012).

In addition, the full use of fruits and vegetables, as a way of encouraging the consumption of this food group, is a healthy eating practice and contributes to health promotion. Studies have revealed that fruits and vegetables are essential to human nutrition as they are a source of micronutrients. These foods are rich in water-soluble vitamins (vitamin C and B-complex vitamins) and pro-vitamin A, as well as minerals, and are considered good sources of essential phytochemicals in the prevention of degenerative diseases such as cancer and cardiovascular diseases (Pereira-Freire et al., 2022).

Therefore, in view of the reality of waste, the full use of food becomes a technological alternative that contributes to reducing losses in the volume of food, as well as expanding the possibilities of use and consumption, promoting a healthier and more adequate diet for the population (Martinelli et al., 2020; OECD and FAO, 2021).

For developed and developing countries, this represents the opportunities arising from this new development paradigm and assumes a government commitment to be expressed in National Policies for the Bioeconomy. To this end, a coherent alignment of priorities, investment and modernization of the regulatory framework is needed that effectively promotes and develops impact actions on science, technology and innovation and on policies to support business development in the areas covered by the so-called bioeconomy (CNI, 2013; OECD and FAO, 2021).

Bioprospecting of co-products and agro-industrial wastes for biomedical applications

In recent decades, the use of natural products in the development of new drugs has been widely discussed. From bioprospecting, new plant species are collected to be tested in screening models and later, in analytical techniques, generating several molecules whose activities can configure potential biological actions in the treatment of different diseases, such as cancer for example. Research carried out by Marques et al. (2022) indicates that although there are considerable efforts to take advantage of biodiversity in the advancement of science, they still present several difficulties, especially in Brazil, which has never established an expressive program, at the national level or contract to implement a national network of bioprospecting, in the country, as has occurred in several other nations (Brasil, 2004; Brasil, 2005; Brasil, 2007; Brasil, 2008).

Literature reports point to the 17 countries that concentrate 70% of the planet's biodiversity, being recognized by the World Center for Conservation and Monitoring as megadiverse countries, namely: Australia, Brazil, China, Colombia, the Democratic Republic of Congo (DRC) (ex-Zaire), Ecuador, India, Indonesia, Madagascar, Malaysia, Mexico, Papua New Guinea, Peru, the Philippines, South Africa, United States of America and Venezuela. This mega biodiversity determines the diversity and endemism of higher plant species in different countries.

Brazil is the main among the mega biodiverse countries that have in their territories between 15 and 20% of the planet's biodiversity distributed in an enormous natural heritage. The greatest diversity of flora in the world is found in Brazil, with more than 55 thousand described species (22% of the total in the world), as well as some of the richest ecosystems in a number of plant species (Amazon, Atlantic Forest, and Cerrado) (Marques et al., 2022).

Research results indicate that nanotechnology is one of the main resources for development and innovation in Brazil and that these companies allocate resources to research this new technological option, indicating an important option in the fight against cutaneous cell depreciation, in the identification of new trends in the cosmetics sector oriented towards innovation and sustainability.

Being considered as an organizational complex in the world of beauty, the cosmetics segment is a growing sector in the globalized economy. Some industries that are located in Brazil, namely: Boticário and LÓreal, have been using new trends in the cosmetics sector oriented towards innovation and sustainability. For L'Oréal, sustainable innovation also implies designing products that are effective, respecting ecosystems and biodiversity and providing well-being for consumers.

In this sense, the impact on the environment can be reduced through the efficient use of natural resources and the minimization of post-consumer waste. However, innovation needs to be covered in the social context, as technological innovations trigger their own needs (Feldmann, 2005; Berry et al., 2015). Regarding sustainability, advances in cosmetic research also reflect a global technological trend focused on natural products and biodiversity related to plants, essential oils, fruits and seeds as sources of raw materials and active principles for use in cosmetic products (ABDI, 2011).

There is a need to increase research in the area of innovation and sustainability, seeking greater awareness of environmental problems. Thus, there is also the necessity to intensify the forms of interaction and cooperation among companies, research institutes and universities. It appears that large companies in the sector and mainly, transnational companies assume defined strategies in relation to the development of new products by investing in their own Research and Development center (R&D) with structured knowledge management processes and technologies.

Patents, technological innovation and growth of bioprospecting research in Brazil and other countries

Bioprospection has been an activity that allows recognizing the economic value of biodiversity and has helped industries and research centers to take advantage of this wealth, in this case, the wealth of megadiverse countries, among these countries Brazil, stands out with 20% of the biodiversity of the planet distributed (Marques et al., 2022). Bioprospecting activities are based on three elements, namely—researching, transforming into products and marketing where they are inserted: macro policies; in biological diversity surveys and information management systems; technology transfer and industrial property development; and business development. All these subjects are normally developed through a national strategic plan.

Studies evaluated patents that resulted from research through bioprospecting, and provided an update on the world panorama of bioactive molecules resulting from natural products (Atanasov et al., 2021). The results point to the United States (US) as the country that deposited the most patents, with 503, followed by China (CN) with 189 patents and Australia (AU) with 162 patents, the Republic of Korea (61), Canada (56), France (54) India (42), Denmark (40), Spain (35), Germany (29), Singapore (15) and Brazil (3). These countries are usually well represented in the innovative and technological scenario through the filing of patents, granting exclusivity in the commercial exploitation of the research developed. In this context, patents are used to show the production of innovative technologies and activities capable of significantly influencing an industry, economy and society as a whole (Ryan, 2010).

In Brazil, the lack of a more adequate research infrastructure is well known due to better investment in this area, including bureaucratization, the lack of qualified personnel to absorb the knowledge made available in the patent offices and financial resources to develop from technology. In this sense, it is seen that the system of cutting-edge technologies is closed in the field of industrialized countries, but the fault lies not only with the intellectual property regime.

Conclusions

Science, technology and innovation must be stimulated in a balanced way and it is, therefore, a great challenge to bioprospecting in Brazil and in many other countries, in order to add value to products and processes, enable and develop sustainable technologies, act significantly in the support of bioprospecting activities—research, transform into products and commercialize.

In order to exploit the full potential of biodiversity in different countries, it is evident that there is a need to overcome challenges to expand and make innovative activities more efficient, aiming at the generation and transfer of technology, in close collaboration with research groups from institutes and universities, in order to encourage the creation of new businesses and technology-based food products with greater added value. The opportunities are numerous and must be taken advantage of, especially the new and existing consumer trends, such as the growing preference for healthy, better-quality foods and conscious consumption.

References

ABDI. 2011. Associação Brasileira de Desenvolvimento Industrial.

Agência Brasileira de Industrial. 2011. II Caderno de tendências: higiene pessoal, perfumaria e cosméticos. ABDI 2(2): 1–69. Available in: chrome-extension://efaidnbmnnnibpcajpcglclefindmkaj/https://www.abihpec.org.br/conteudo/caderno_tendencias.pdf.

Agência Brasileira de Desenvolvimento Industrial. 2016. Química de Renováveis: Panorama Econômico. ABDI 1–40. Available in: https://silo.tips/download/q-uimica-de-r-enovaveis.

Atanasov, A.G., Zotchev, S.B., Dirsch, Taskforce, T.N.P.S. and Supuran, C.T. 2021. Produtos naturais na descoberta de medicamentos: avanços e oportunidades. Nat. Rev. Drug. Discov. 20: 200–216. DOI: 10.1038/s41573-020-00114-z.

Berry, E.M., Dernini, S., Burlingame, B., Meybeck, A. and Conforti, P. 2015. Food security and sustainability: can one exist without the other? Public Health Nutr. 18(13): 2293–302. DOI: 10.1017/S136898001500021X.

BRASIL. Lei n. 10.973, de 2 de dezembro de. 2004. Dispõe sobre incentivos à inovação e à pesquisa científica e tecnológica no ambiente produtivo e dá outras providências. 2004. Available in: http://www.planalto.gov.br/ccivil_03/_ato2004-2006/2004/lei/l10.973.htm.

BRASIL. Lei n. 11.105, de 24 de março de. 2005. Regulamenta os incisos II, IV e V do § 1o do art. 225 da Constituição Federal, estabelece normas de segurança e mecanismos de fiscalização de atividades que envolvam organismos geneticamente modificados – OGM e seus derivados, cria o Conselho Nacional de Biossegurança – CNBS, reestrutura a Comissão Técnica Nacional de Biossegurança – CTNBio, dispõe sobre a Política Nacional de Biossegurança – PNB, e dá outras providências. 2005. Available in: http://www.planalto.gov.br/ccivil_03/_ato2004-2006/2005/lei/l11105.htm.

BRASIL. Decreto n. 6.041, de 8 de fevereiro de. 2007. Institui a Política de Desenvolvimento da Biotecnologia, cria o Comitê Nacional de Biotecnologia e dá outras providências. 2007. Available in: http://www.planalto.gov.br/ccivil_03/_ato2007-2010/2007/decreto/d6041.htm.

BRASIL. Decreto n. 6.476, de 5 de junho de. 2008. Promulga o Tratado Internacional sobre Recursos Fitogenéticos para a Alimentação e a Agricultura, aprovado em Roma, em 3 de novembro de 2001, e assinado pelo Brasil em 10 de junho de 2002. 2008. Available in: http://www.planalto.gov.br/ccivil_03/_Ato2007-2010/2008/Decreto/D6476.htm.

Centro de Gestão e Estudos Estratégicos – CGEE.; Agência Brasileira de Desenvolvimento Industrial - ABDI. 2008. Estudo prospectivo: Visão de Futuro e Agenda INI - Biotecnologia: 2008–2025. CGEE. ABDI 1–265. Available in: http://www.abdi.com.br/Estudo/Estudo%20Prospectivo%20Biotecnologia.

Centro de Gestão e Estudos Estratégicos. 2014. Sustentabilidade e sustentação da produção de alimentos no Brasil: O papel do País no cenário global. CGEE 1: 1–148. ISBN: 978-85-60755-

73-8. Available in: chrome-extension://efaidnbmnnnibpcajpcglclefindmkaj/https://www.cgee.org.br/documents/10195/734063/Papel_Brasil_Global_Vol_I_WEB_29042015_10159.pdf/3cdacbdf-64e8-49b3-bc46-e9c5e3771e35?version=1.1.

Confederação Nacional da Indústria. 2013. Bioeconomia: uma agenda para o Brasil. – Brasília: Confederação Nacional da Indústria. CNI 1–40. ISBN: 978-85-7957-101-5. Available in: chrome-extension://efaidnbmnnnibpcajpcglclefindmkaj/https://static.portaldaindustria.com.br/media/filer_public/78/86/7886aeb1-57a8-4be2-9ad9-f8f31b176a8f/bioeconomia_uma_agenda_para_brasil.pdf.

FELDMANN, F. 2005. A parte que nos cabe: consumo sustentável? In: TRIGUEIRO, André (Coord.). Meio Ambiente no Século 21 – 21 especialistas falam da questão ambiental nas suas áreas de conhecimento. 4 ed. Campinas – SP: Armazém do Ipê, p. 143–157.

Food and Agriculture Organization of the United Nations. 2012. Sustainable diets and biodiversity: directions and solutions for policy, research and action, FAO 1–309. ISBN-13: 978-92-5-107288-2. Available in: chrome-extension://efaidnbmnnnibpcajpcglclefindmkaj/https://www.fao.org/3/i3022e/i3022e.pdf.

Komatsu, T.R., Buriti, F.C.A. and Saad. S.M.I. 2008. Inovação, persistência e criatividade superando barreiras no desenvolvimento de alimentos probióticos. Braz. J. Pharm. Sci. 44(3): 329–347. DOI: 10.1590/S1516-93322008000300003.

Marques, L.G.A., Vieira Neto, J.B., Sales, S.L.A., Costa, P.M.S., Guimarães, C.J., Manso, M.P. Pereira, J.V.M. and Pessoa, C.O. 2022. The impact of bioprospecting for the discovery of new drugs. Revista Fitos. Supl. (2): 293–314. DOI: 10.32712/2446-4775.2022.1313.

Martinelli, S.S., Cavalli, S.B., Fabri, R.K., Veiros, M.B., Reis, A.B.C. and Amparo-Santos, L. 2020. Strategies for the promotion of healthy, adequate and sustainable food in Brazil in times of Covid-19. Rev. Nutr. 33: e200181. DOI: 10.1590/1678-9865202032e200181.

Martins, M.R.S.F., Viana, L.F. and Cappato, L.P. 2022. Perfil do desperdício de alimentos nas Unidades de Alimentação e Nutrição brasileiras e as ações corretivas implementadas. Food Sci. Technol. Campinas 42: e100421. DOI: https://doi.org/10.1590/fst.100421.

Organization for Economic Co-operation and Development, Food and Agriculture Organization of the United Nations. 2021. Agricultural Outlook 2021–2030. OECD Publishing 1–337. DOI: 10.1787/19428846-en.

Pereira-Freire, J.A., Aquino, J.S., Campos, A.R.N., Viana, V.G.F., Costa Junior, J.S., Silva, J.N., Moura, A.K.S., Cito, A.M.G.L., Moreira-Araujo, R.S.R., Frota, K.M.G., Medeiros, S.R.A. and Ferreira, P.M.P. 2022. Nutritional, physicochemical and structural parameters of *Mauritia flexuosa* fruits and by-products for biotechnological exploration of sustainable goods. Food Technol. Biotechnol. 60: 100–112, DOI: 10.17113/ftb.60.02.22.7106.

Ryan, M.P. 2010. Patent incentives, technology markets, and public–private bio-medical innovation networks in Brazil. World Dev. 38(8): 1082–93. DOI: 10.1016/j.worlddev.2009.12.013.

Saccaro-Júnior, N.L. 2011. TD 1569 - Desafios da Bioprospecção no Brasil. IPEA 1–38. ISSN: 1415-4765. https://www.ipea.gov.br/portal/index.php?option=com_content&view=article&id=7066.

Chapter 2

Technological Prospection for the Use of Food Plants with a Focus on the Food, Pharmaceutical and Cosmetics Industry

Tatiana de Oliveira Lemos

Introduction

The research and development activity and the general economic scenario are surrounded by several uncertainties that lead innovative companies to seek appropriate competitive strategies for their products and market. Among the strategies, one has access to technological information that also integrates the companies' competitive intelligence. Thus, in order for companies to position themselves ahead of the competition and remain competitive in the market, they seek to differentiate themselves through innovation and by trying to foresee trends and indications of change (Paranhos and Ribeiro, 2018).

In this context, informational prospecting and informational monitoring are essential steps in the process of competitive intelligence, since, by means of informational prospecting, it is possible to build a primary map containing sources of information and knowledge fundamental to the competitiveness of a specific sector. With regard to monitoring, it is necessary to carefully select, among the information, those with relevant potential, enabling the identification of new possibilities and indications of changes in the environment. With this, the technological prospection

Food Engineering Course, Federal University of Maranhão, Maranhão, Brazil.
Email: tatiana.lemos@ufma.br

deserves to be highlighted, because it works as a guide for decision-makers to elaborate innovation strategies, which can help in the mapping of scientific and technological developments and in the perception of market trends, indicating the competition, which facilitates the decision making (Paranhos and Ribeiro, 2018).

Technology prospecting is a systematic means of mapping future scientific and technological developments, with the ability to have a relevant influence on an industry, the economy or society in general, acting as a tool to guide the actions undertaken for research, development and innovation (Paranhos and Ribeiro, 2018).

Technology prospecting studies include four phases: (1) preparatory, where the objectives, scope, approach and methodology to be applied are defined; (2) pre-prospective, is the detailing of the methodology and the survey of the source of information; (3) perspective, where the collection, treatment and analysis of the information obtained in phase 2 takes place; and (4) post-prospective, which includes the communication of results, implementation of actions and monitoring. Technology prospecting methods can be classified into three categories: (1) monitoring, which deals with the systematic and uninterrupted follow-up of the evolution of facts and the identification of drivers of change; (2) forecasting, which deals with running projections based on historical data and trend modeling; and (3) vision, which is about anticipating future possibilities, with reference to unstructured interaction between experts (Amparo et al., 2012).

The alarming growth of the world's human population brings with it new challenges, which implies an increase in food production, with the consequent depletion of natural resources, making it necessary to explore alternative food sources. Thus, the transition from diets rich in animal products to diets rich in plant-based food products presents itself as a more ecological alternative, both in terms of environmental sustainability and human health (Salvi, 2016; Duarte, 2018). Food plants are those that have one or more parts or derivatives thereof, which can be used as food in the human diet. This concept includes plants for direct consumption, such as tuberous roots, tubers, bulbs, rhizomes, stalks, leaves, shoots, flowers, fruits and seeds or even latex, resins and gums. And plants for indirect consumption, such as those that provide oils, spices, seasoning and aromatic substances, those that are used as salt substitutes, as sweeteners, meat tenderizers and food coloring. But there is still a group of underutilized food plants that have received increasing attention, especially in relation to the expansion of monocultures, and have been called Unconventional Food Plants (UFP) (Leal et al., 2018; Kinupp and Lorenzi, 2014).

The research carried out in the context of Brazilian biodiversity, enables technological studies of flora as natural potentialities, allowing the emergence of potential niches for investment and market. With this, the technology prospecting of food plants can identify existing technologies, aspects of competing technologies and possible gaps to be filled (Benevides Júnior et al., 2020). Therefore, in this chapter some technological prospection studies, using the patent databases on food plants will be discussed.

Pomegranate (*Punica granatum*)

Punica granatum or pomegranate is a popular food plant with antioxidant, anti-inflammatory, anti-parasitic, antimicrobial, antidiabetic and collagen synthesis stimulating activities. In addition, it has a significant content of total phenolic compounds, such as anthocyanins and tannins. These properties inspire therapeutic and cosmetic interest in the development of *pomegranate*-based products. Thus, Coelho et al. (2017) conducted a patent search in the databases of the National Institute of Industrial Property (NIIP), European Patent Office (EPO), United States Patent and Trademark Office (USPTO) and World Intellectual Property Organization (WIPO), to verify pomegranate-derived pharmaceutical and food products through technology prospection of the topic. As a result of the research, they found that 101 patents were filed at the EPO, of which 86 were related to the development of products for therapeutic purposes and 15 to the development of food products. In WIPO 87 patents were filed, 68 with therapeutic purposes and 19 proposing the development of drinks or functional foods. At the USPTO, there were six patents filed related to the therapeutic or food use of *pomegranate*, which due to the relevance of their claims and promising market exploitation, were also filed at the EPO and/or WIPO, extending the patent protection. While in the INPI, there were two patents filed, which notify the research and development of products derived from the *pomegranate*, considering the therapeutic and cosmetic properties already proven and its ease of cultivation in Brazilian soil (Coelho et al., 2017).

Updating the data from the INPI patent database, obtained in a research conducted in February 2022 on the subject, it was possible to verify the occurrence of nine filed patents, two being patents related to food application, six with pharmacological purpose, and one with a cosmetic purpose (NIIP, 2022).

Regarding the patents with food applications filed, there are:

1. **"Application of edible parts of pomegranate (*Punica granatum*) in bread and noodle formulations"**, the invention deals with the inclusion of edible parts (peel and pulp) of pomegranate in formulations of French bread, mold, ciabatta and/or baguette, sweet and savory cookies and pasta for pasta and lasagna.
2. **"Beer and malt-based alcoholic fermentation enriched with bioactive compounds present in *pomegranate* (*Punica granatum*)"**, the invention relates to the preparation of a fermented beer-type beverage based on aqueous extract (tea) and/or methanolic extract and/or ethanolic extract, and/or leaves of pomegranate, *Punica granatum* L., both in dry and fresh form.

Considering the patents filed for pharmacological purposes, there are:

1. **"Formulations containing fractions and compounds isolated from the leaves of *Punica granatum* L.: process for obtaining them and their uses as anti-inflammatory, antimicrobial and healing agents"**, the invention deals with obtaining a formulation for oral use and the therapeutic use of the compositions in the treatment of inflammatory processes as well as in the treatment of other pathologies such as those related to infections of the upper and lower respiratory tract. The formulation is produced from fractions or purified compounds

of freeze-dried hydroalcoholic extracts obtained from the leaves of *Punica granatum* L.

2. **"Use of a combination of *Curcuma longa, Punica granatum* and *Zingiber officinale* extracts in the prevention and/or treatment of mucosal lesions"**, the invention relates to formulations containing a combination of extracts of *Curcuma longa, Punica granatum* and *Zingiber officinale*, useful for the treatment and prevention of mucosal lesions, in particular those caused by the herpes virus or Behcet's syndrome.
3. **"Antitumor formulations with apoptosis-inducing potential containing *Punica granatum*"**, the invention deals with a formulation with antitumor effect and apoptosis induction containing *Punica granatum*, as well as the biological effect of this product.
4. **"Pharmaceutical formulations containing the freeze-dried hydroalcoholic extract of the leaves of *Punica granatum* L., and its use in inflammatory processes"**, the invention deals with obtaining a pharmaceutical formulation and its therapeutic use in the treatment of acute inflammatory processes, assisting in the prevention of complications of chronic inflammatory processes.
5. **"Oral formulation containing *Punica granatum* extract for the treatment of stomatitis"**, the invention relates to an oral formulation (gel or tablet) for the treatment of candidiasis associated with prosthetic stomatitis or of other natures, whose bases are extracts, fractions and substances isolated from the plant species *Punica granatum*.
6. **"Pharmaceutical compositions based on a crude extract of the pericarp of Punica granatum fruits and its use as antimicrobial, antiseptic, astringent and healing agent"**, the invention deals with pharmaceutical compositions based on a crude extract of the pericarp of *Punica granatum* fruits, with antimicrobial properties for dermatological and oral antisepsis and dental and veterinary use.

Finally, as a patent deposited for cosmetic purposes we have:

1. **"Using an extract of *Punica granatum* to fight baldness"**, the invention relates to the use of at least one plant of the species *Punica granatum* or one of its extracts as an agent to diminish or prevent whitening of hair. The extract is used orally. This also refers to a cosmetic treatment process by administration of *Punica granatum*.

Soursop (*Annona muricata* Linnaeus)

The *soursop* (*Annona muricata* Linneus) is a fruiting food plant, which has large, greenish, edible fruits, whose pulp is white, fibrous, soft and with pleasant sensory attributes, and which is used as a raw material for the fruit pulp industry. Besides the fruit, the unconventional parts of the *soursop*, such as leaves, seeds, stem and bark, have bioactive compounds with health-promoting properties, such as antioxidant, anti-inflammatory, anticarcinogenic, gastro-protective, anti-diabetic, antirheumatic, antiparasitic and antineuralgic. However, research shows that most of the bioactive compounds present in the fruit are concentrated in its peels, seeds and pomace, which become waste products when processing *soursop* pulp. Considering the bioactive potential of *soursop* and its industrial residues, Arôxa et al. (2020) carried out a mapping of technological prospection regarding the industrial use of these residues to evaluate the panorama of patent filing in the databases of the Derwent Innovations Index (DII) and the National Institute of Industrial Property (INPI), in order to subsidize future strategic applications in the areas of science, technology and innovation of this productive chain.

In the mapping of the industrial potential of *soursop* and its residues, a total of 90 patents filed were identified, with 14 patents filed at the NIIP and 76 patents filed at the DII. The patent applications of *soursop* and its residues were divided into classes: medical science with 39 patents; food products with 34 patents; agriculture with 12 patents; chemical with two patents; electricity with one patent; household appliance with one patent; paper production with one patent. The main patent application is related to medical science, more precisely the production of drugs, probably due to the influence of the benefits of bioactive compounds, present in *soursop*, for health, and as the second-largest application the production of food (Aroxa et al., 2020).

A different picture was found in a search conducted in the NIIP database in February 2022, using as keywords the terms "*graviola*" and "*Annona muricata*", where it was verified that seven patents had been filed, three of which were for food use, one for pharmacological purposes, and one for food/pharmacological purposes. No patent application was verified for cosmetic purposes, an application field that can be explored.

Of the patents filed at the NIIP with food purposes there are:

1. **"Production of a freeze-dried juice based on *soursop* and soy milk"**, the invention deals with the production of a freeze-dried juice based on *soursop* and soy milk.
2. **"Production process of beer with *soursop* added"**, the invention deals with the production of beer with added *soursop*, exploring the form and quantity of fruit or derivatives added, and the timing of the addition, which gives the product distinct sensory aspects, such as in aromas, flavors, residual sweetness and mild acidity, besides functional aspects.
3. **"Process for obtaining mixed natural drink, a mixed natural drink containing *taperebá*, *camu-camu* and *soursop*"**, the invention relates to formulations of mixed natural drink containing *taperebá*, *soursop* and *camu-camu* as well as the process for obtaining these formulations with the field of application being the food processing industry.

The patent filed at the INPI for pharmacological purposes is as follows:

1. **"Extraction process of the *soursop* seed oil (*Annona muricata* L.) and the product obtained"**, the invention relates to an extraction process for obtaining crude oil using *soursop* seeds (*Annona muricata* L.) as raw material, with potential pharmacological and biotechnological uses and applications.

The patent filed at the INPI with food/pharmacological purpose there are:

1. **"Dried *soursop* leaves in a foam layer"**, the invention deals with the production of *soursop* leaf foam (*Annona muricata* Linn) and then its drying in a circulating air oven, to obtain the powder. The powder obtained can be a strong ally in the treatment of several diseases, and can be used in pharmaceutical products as well as in the food industry, and can be stored at room temperature, thus facilitating access to different places.

Umbu (Spondia tuberosa)

Spondia tuberosa also known as umbunzeiro, imbuzeiro, ombuzeiro, umbu, imbu, ambu and giqui, is a UFP belonging to the *Anacardiaceae* family, whose fresh fruits, the watery tubers and the young leaves are edible, and with great economic possibilities. The fruit in nature has, in 100 g: 89% moisture; 37 kcal energetic value; 9,0 g carbohydrates; 1,0 g proteins; 2,0 g dietary fiber; 0,5 g minerals; 12,0 mg calcium; 11,0 mg magnesium; 13,0 mg phosphorus; 0,1 mg iron; 152,0 mg potassium; 0,4 mg zinc; 24,0 mg vitamin C (Kinupp and Lorenzi, 2014). In addition, UFP has antioxidant activity, containing bioactive substances such as phenolic compounds and flavonoids (Oliveira et al., 2020).

Based on the economic and medicinal potential of UFP, Oliveira et al. (2020) conducted a technological prospection study of *Spondia tuberosa* in the patent databases European Patent Office (EPO), World Intellectual Property Organization (WIPO), United States Patent and Trademark Office (USPTO) and the National Institute of Industrial Property (NIIP). As a result of the study, they obtained a total of 20 patents filed in the consulted databases, eight patents being filed in WIPO, six patents filed in NIIP, four patents filed in USPTO, and two patents filed in EPO. Analyzing the patent applications they found that: 23.40% of the patents were in the class of preparations for medical, dental or hygienic purposes; 19.14% of the patents were in the class of acyclic or cyclic organic compounds; 19.14% of the patents were in the class of food or food products; 14.89% of the patents were in the class of specific therapeutic activity of chemical compounds or medicinal preparations (Oliveira et al., 2020). In the study, the main patent application is related to medical science, and the second to food production application.

In a search in the NIIP database in February 2022, using the terms "*umbu*" and "*Spondia tuberosa*" as keywords, a total of 13 patents filed were obtained, 11 with food purposes and one with food/cosmetic purposes. There have been no patent filings for pharmacological purposes, a field of application to be investigated.

Considering patents with food purposes there are:

1. **"Extra jelly of sapoti (Manilkara zapota) and *umbu cajá* (*Spondia tuberosa* X *Spondia mombin*)"**, the invention deals with the production of extra jelly from fresh sapoti and *umbu* cajá fruits in a suitable stage of maturation for processing.
2. **"*Umbu-cajá* jelly formulated with *okara* powder as a substitute for commercial pectin"**, the invention relates to the preparation of *umbu-cajá* jelly

formulated with *okara* powder. Produced from the pulp of whole *umbu-cajá* and okara powder obtained by freeze-drying, as a substitute for commercial pectin.

3. **"*Umbu-cajá* pulp in powder with 10% maltodextrin"**, the invention deals with the production of *umbu-cajá* powder, from *umbu-cajá* pulp, added to maltodextrin and submitted to a freeze-drying process to obtain the powder.
4. **"Elaboration and processing of *Umbu* flatbread with added flour from the *umbu* peel and seed (*Spondias tuberosa* Arruda Câmara)"**, the invention relates to the preparation of a loaf of bread with the addition of flour from the peel and seed of *umbu*, fresh, from the exotic species *Spondias tuberosa* Arruda Câmara found in northeastern Brazil and cultivated without the use of pesticides and pesticides.
5. **"Elaboration and processing of flour from *umbu* peel and seed (*Spondias tuberosa* Arruda Câmara)"**, the invention deals with the elaboration and processing of flour obtained from the mixture of the peel and the seed of the *umbu* fruit, in the ripe stage, of the exotic species *Spondias tuberosa* Arruda Câmara, found in the northeast region of Brazil.
6. **"Mixed jelly with *umbu* (*Spondias tuberosa* Arruda Câmara) and *mangaba* (*Hancornia speciosa* Gomes) added to fructooligosaccharide"**, the invention deals with two formulations of mixed jelly, which are differentiated by the addition of dietary fibers (fructooligosaccharide) in amounts above the minimum (5 g) according to the Brazilian legislation, to be classified as a functional jelly.
7. **"Freeze-drying of *umbu* for juice production and use as ingredients in the formulation of food products"**, the invention is about a food intended for human feeding with the addition of maltodextrin (0 to 30%) as an ingredient of the formulation to obtain the *umbu* powder by the freeze-drying process, this product is used for consumption in the form of juice after the dilution of the powder in drinkable water and/or in the preparation of food products, due to its nutritional composition and, also, to the process used that allows its storage.
8. **"*Peanut* and *umbu* powder extract mix"**, the invention deals with the production process of a mix based on peanut powder extract with the addition of *umbu* powder, *peanut* powder obtained from the preparation of the peanut base without the skin and mineral water, with application in the preparation of beverages for human consumption, aiming to increase the nutritional and technological utilization of the two vegetable species.
9. **"Process for obtaining *umbu* powder and ready-mix *umbu* powder"**, the invention deals with the process of obtaining a ready mix of umbuzada powder, whose steps involve the selection of fruits at the turning stage of maturity, followed by washing and sanitization in chlorinated water and heat treatment, is submerged in heated water for a period of 10 to 20 min. After this time, the fruits are drained and pulped. The obtained pulp is acidified with 0.01 to 0.05% citric acid, and a 2% solution of carboxymethylcellulose (CMC) is also added. The dehydrated material is ground to be finally mixed with powdered milk, refined sugar, anti-humectant (Tricalcium Phosphate - INS 341iii) at a concentration of

0.05 to 0.1%, and an emulsifier (Soy Lecithin - INS 322i) at a concentration of 0.1 to 0.3%, to obtain the mixture ready for umbuzada.

10. **"Process for the preparation of pickled *umbu* (*Spondia tuberosa*)"**, the invention relates to the processing of the raw material *umbu* or imbu (*Spondia tuberosa*), the canned fruit of the umbuzeiro.
11. **"Process of extraction and isolation of active substances present in *umbu* pulp, active substances, nutraceutical and/or functional foods comprising said active substances and their use"**, the invention relates to the process of extraction and isolation of active substances present in the *umbu* pulp (*Spondias tuberosa* Arr. Câmara). Where the fractionated dichloromethane extract and the actives obtained by this proposed process can be applied in nutraceutical and/or functional foods.

Regarding the patent for food/cosmetic purposes there are:

1. **"Process of extraction and isolation of active substances present in *umbu* pulp, nutraceutical and/or functional foods and cosmetics comprising these active substances and their uses"**, the invention refers to the extraction and isolation process of active substances present in the *umbu* (*Spondias tuberosa* Arr. Câmara) pulp, where the fractionated extract and the actives obtained by the proposed process can be applied in nutraceutical and/or functional foods or also in cosmetics. As food, the actives are used in the treatment of neurodegenerative diseases. In the form of cosmetics, the actives are used against aging.

Moringa (*Moringa oleifera* Lam.)

A *Moringa oleifera* Lam. is a cruciferous plant belonging to the Moringaceae family, native to the southern Himalayan Mountains in northern India and cultivated in backyards, mainly in the Northeast region of Brazil. *Moringa* is considered a UFP and is consumed not only for its nutritional value but also for its health benefits (Kinupp and Lorenzi, 2014; Kou et al., 2018). *Moringa* leaves are high in carotenoids, excellent sources of pro-vitamin A (up to 44,817 IU), rich in vitamin C, vitamin E and polyphenols, and are good sources of natural antioxidants. Green fruits are good sources of calcium and vitamin C. In addition, the presence of phytochemicals such as flavonoids, saponins, anthraquinones and alkaloids makes this UFP a good medicinal agent (Kinupp and Lorenzi, 2014; Kou et al., 2018; Fink et al., 2018).

Considering the relevance of the use and potential of *moringa*, Barreto et al. (2016) conducted a search in the patent databases of the Brazilian National Institute of Industrial Property (NIIP), European Patent Office (Espacenet - Worldwide) and World Intellectual Property Organization (WIPO), on the use and applications of the plant in the area of food science and technology. For this, the keywords applied in the search were initially "*moringa*" in the title and title and/or abstract fields. Then the terms "food" and "nutrition" were added to the search, as well as their respective translations when searched in the national NIIP database. As result, Barreto et al. (2016) obtained in the WIPO database for the application of *moringa* referring to the food area 11 patent filings between the years 2000 and 2012, using in the search the keywords "*moringa*" and "food" and 16 filings applying the keywords "*moringa*" and "nutrition". In both searches, China and the United Kingdom led the ranking of filing countries. In the ESPACENET database, the results obtained were 15 patents filed between the years 1995 and 2013, using the keywords "*moringa*" and "food" and 13 patents filed between the years 2003 and 2013, using the keywords "*moringa*" and "nutrition". In the ESPACENET search, China, the Republic of Korea, Japan and the United States appeared as depositing countries (Barreto et al., 2016). Checking the NIIP database, no applications were found when combining the keywords "*moringa*" and "food" or "*moringa*" and "nutrition".

Almeida et al. (2017) conducted a technological prospection for the use and application of *Moringa oleifera* in the cosmetics sector. For this, searches were conducted in the patent databases of the National Institute of Industrial Property (INPI) of Brazil, the European Patent Office (Espacenet - Worldwide) and the Derwent Innovations Index. Regarding the terms used in the search, initially, the keywords "*Moringa*" and "*Moringa oleifera*" were chosen, followed by the term cosmetic.

The results obtained by Almeida et al. (2017) were two patents filed at NIIP, using in the search the keywords "*Moringa*" and "cosmetic"/"*Moringa*" and "*oleifera*" and "cosmetic". In the ESPACENET patent base, 42 patents filed using the keywords "*Moringa*" and "cosmetic" were verified, and 11 patents filed using the keywords "*Moringa*" and "*oleifera*" and "cosmetic" in the search. In the Derwent patent base, 103 patents were verified as having been filed using the keywords "*Moringa*" and "cosmetic", and 34 patents were filed using the keywords "*Moringa*" and "*oleifera*" and "cosmetic".

The prospective studies by Barreto et al. (2016) and Almeida et al. (2017) demonstrated the difference in the existing quantitative data between the protected moringa technological products in Brazil and in the other countries of the world. With this, innovative research regarding *moringa* products still needs recognition for intellectual protection (Almeida et al., 2017).

Research conducted in the patent base of the INPI in February 2022, using as keywords the terms "*Moringa*" and "*Moringa oleifera*", we obtained as a result of a total of 28 patents filed, one patent with pharmaceutical/food/cosmetic purpose, two patents with pharmaceutical purpose, seven patents with food purpose and one patent with food/cosmetic purpose (NIIP, 2022).

Considering patents with pharmaceutical/food/cosmetic purposes there are:

1. **"Bioactive extracts from fermented *Moringa oleifera* seeds",** the invention relates to obtaining bioactive extracts from *moringa* seed flour. These extracts have shown potential for future application as sources of bioactive compounds of interest to the pharmaceutical, cosmetic and food industries.

Regarding patents for food purpose there are:

1. **"Preparation and processing of chocolate brownie enriched with *moringa* leaf flour",** the invention refers to the preparation of chocolate brownie enriched with moringa leaf flour, produced from *moringa* leaf flour, nutritionally enriching the food. This invention stands out for not using dyes and flavorings.

2. **"Elaboration of a biodegradable plastic film based on extracts of Syrah grape (*Vitis vinifera*) and *Moringa oleifera* residue for food packaging and increasing shelf life"**, the invention refers to the process of elaboration of a biodegradable plastic film containing acidified crude ethanolic extract of *Moringa oleifera* leaves and residue of *Syrah grape* pomace (*Vitis vinifera*) for food packaging, also aiming at its application as active packaging with antimicrobial and antioxidant action for perishable products, seeking to increase the shelf life.
3. **"Elaboration of the farinaceous product from the seed of *Moringa oleifera* Lam."**, the invention relates to the elaboration of a floury product from the seed of *Moringa oleifera* Lam. by microwave drying.
4. **"Bioactive food compound: anti-hyperglycemic effect isolated or associated with *Moringa oleifera* L. (*Moringaceae*)"**, the invention refers to a food product with functional properties on Non-transmissible Chronic Diseases (NCDs), especially regarding cardiovascular (CVD) and respiratory diseases, Diabetes Mellitus (DM), obesity and cancer. Its formulation is composed of oat bran (*Avena sativa* L.), brown flaxseed (*Linum usitatissimum* L.), Textured Soy Protein (TSP) (*Glycine max* L.) and powder from the leaves of the plant *Moringa oleifera* Lam. (*Moringaceae*).
5. **"Food in the form of a spreader, food preparation process in the form of a spreader and the use of a mono- or diester diglycerol and *moringa* oil"**.
6. **"Food product in the form of a spreadable process for the preparation of a food product in the form of a spreadable, use of a mono- or diester of glycerol and *moringa* oil for the preparation or stabilization of a spreadable, process and use"**.
7. **"Manufacturing process of liposoluble and water-soluble nutritional products for animals and humans extracted from *moringa* leaves in its various species consisting of six operational phases of extraction, which are, preparation, water-soluble extraction fat-soluble extraction, water-soluble evaporative extraction and fat-soluble evaporative extraction and drying, generating five products rich in proteins, vitamins A, vitamins C, minerals, insoluble fibers, chlorophylls and carbohydrates applied to human and animal nutritional additives"**.

The patents with pharmaceutical purposes include:

1. **"Process of extraction, fractionation and bioguided isolation of antioxidant substances from *Moringa oleifera* leaves"**, the invention refers to the process of extraction, purification and bioguided isolation of substances with the high antioxidant potential present in the leaves of *M. oleifera*. The actives with multifunctional character, obtained from the described process, can be applied in nutraceutical and/or functional foods, in formulations to combat anti-inflammatory activity, high blood pressure, hyperglycemia, hyperlipidemia and digestive problems, besides being used to combat aging.

2. **"*Moringa oleifera* for inhibition of intimal layer hyperplasia"**, the invention relates to *Moringa oleifera* extract that will be employed to inhibit arterial intimal hyperplasia and prevent atherosclerotic disease and its complications, such as arterial stenosis and occlusion.

Considering patents with food/pharmaceutical purposes there are:

1. **"Extraction and manufacturing process of natural products from Moringa oleifera seeds for the industries of animal feed, fuel, food, cosmetics, sanitizing and in sugar and ethanol plants, consisting of the stages of extraction/extrusion of the *Moringa* seed coat, cold extraction and instantaneous filtration of the vegetable oil, grinding, pulverization, amide drying and the flocculant and clarifying enrichment phase of the coagulant bioactive".**

Conclusions

Based on technological prospection studies using patent databases on unconventional food plants pomegranate, soursop, umbu and moringa, it was possible to identify the existing and protected technologies with these UFP and possible gaps to be filled. In this context, there were no records of patent deposits for the invention of products and/or processes with soursop applied to the cosmetic industry or with umbu applied to the pharmaceutical industry, fields of application to be explored.

References

Almeida, C.B.L., Sá, C.C., Carvalho, R.C.D. and Almeida, E.S. 2017. Estudo prospectivo da moringa na indústria de cosméticos. Caderno de Prospecção 10(4): 905–918, out./dez.

Amparo, K.K.S., Ribeiro, M.C.O. and Guarieiro, L.L.N. 2012. Estudo de caso utilizando mapeamento de prospecção tecnológica como principal ferramenta de busca científica. Perspectivas em Ciência da Informação 17(4): 195–209.

Aroxa, C.N.F., Silva, G.F., Santos, J.A.B., Moreira, J.J.S. and Barretto, L.C.O. 2020. Prospecção tecnológica sobre o potencial industrial do resíduo da graviola (*Annona muricata* L.). Revista INGI 4(1): 664–673.

Barreto, L.C.O., Gois, J.M.V., Oliveira, D.S., Gama, G.J., Silva, G.F. and Santos, J.A.B. 2016. Prospecção tecnológica sobre Moringa oleifera aplicada a ciência e tecnologia de alimentos. Caderno de Prospecção 9(2): 219–229, abr./jun.

Benevides Júnior, A.Y., Gama, B.T.S., Bezerra, T.T.C., Silva, D.A. and Kieling, A.C. 2020. Prospecção tecnológica do cumaru (*Dipteryx odorata*). Cadernos de Prospecção 13(4): 1103–1121.

Coelho, A.G., Feitosa, T.C.B., Campos, R.C., Vilarinho, A.C.S.G., Cunha, F.V.M. and Nunes, L.C.C. 2017. Prospecção tecnológica: aplicação de *Punica granatum* (*Punicaceae*) em produtos medicamentosos e alimentícios. Revista GEINTEC 7(4): 4100–4111.

Duarte, M.C.R. 2018. Perspectivas futuras para a sustentabilidade alimentar novas fontes de proteína na alimentação dos portugueses. 89 f. Dissertação (Mestrado em Biotecnologia e Inovação) – Escola Superior de Biotecnologia, Universidade Católica Portuguesa.

Fink, S.R., Konzen, R.E., Vieira, S.E., Ordonez, A.M. and Nascimento, C.R.B. 2018. Benefícios das Plantas Alimentícias Não Convencionais - PANCs: Caruru (*Amatanthus viridis*), *Moringa oleifera* Lam. e Ora-pro-nóbis (*Pereskia aculeata* mill). Pleiade 12(S1): 39–44, Set.

Kinupp, V.F. and Lorenzi, H. 2014. Plantas alimentícias não convencionais (PANC) no Brasil: guia de identificação, aspectos nutricionais e receitas ilustradas. São Paulo: Instituto Plantarum de Estudos da Flora.

Kou, X., Li, B., Olayanju, J.B., Drake, J.M. and Chen, N. 2018. Nutraceutical or pharmacological potential of *Moringa oleifera* Lam. Nutrients 10: 343.

Leal, M.L., Alves, R.P. and Hanazaki, N. 2018. Knowledge, use, and disuse of unconventional food plants. Journal of Ethnobiology and Ethnomedicine 14: 6.

National Institute of Intelectual Production (NIIP). 2022. Disponível em: https://busca.inpi.gov.br/pePI/jsp/patentes/PatenteSearchBasico.jsp. Acesso em: Fev./2022.

Oliveira, F.G.S., Araújo, A.D., Almeida, J.R.D.S., Silva, M.V. and Correia, M.T.S. 2020. Prospecção tecnológica de *Spondias tuberosa* (*Anacardiaceae*). Revista GEINTEC 10(3): 5546–5552.

Paranhos, R.C.S. and Ribeiro, N.M. 2018. Importância da prospecção tecnológica em base de patentes e seus objetivos da busca. Cadernos de Prospecção 11(5): 1274–1292.

Salvi, J. and Katewa, S.S. 2016. Review: Underutilized wild edible plants as a potential source of alternative nutrition. International Journal of Botany Studies 1(4): 32–36.

Chapter 3

Technological Potential of Unconventional Food Plants (UFP)

Tatiana de Oliveira Lemos,[1,*] *Anida Maria Moraes Gomes,*[2] *Bianca Almeida Souza,*[1] *Brenda Paiva Campi Neves,*[1] *Dárcia Souza Araújo,*[1] *Djany Souza Silva,*[1] *Elder dos Santos Araújo,*[1] *Luana Costa de Souza*[1] and *Mauriane Maciel da Silva*[1]

Introduction

The knowledge and use of plants have been driven by practical needs and cultural preferences and always have been part of human history. Among the widely distributed and commonly used plants, are food plants. Food plants have one or more parts that can be used as food in the human diet. This definition includes plants for direct or indirect consumption, such as oilseeds, spices, flavoring and aromatic substances, and those that are used as substitutes for salt, sweeteners, meat tenderizers and food colorings. However, there is still a group of underutilized food plants that have received increasing attention, especially in reaction to the expansion of monocultures, and which received the name Unconventional Food Plants (UFP) (Leal et al., 2018).

UFP are plant species that have one or more parts with food potential and unusual use, which have unconventional processing methods and generally have no market value or are marketed only on small scales. Thus, native, exotic, wild, cultivated and spontaneous plants are classified as UFP. Furthermore, they are cultures that are not organized in the production chain, not arousing the interest of companies in the sector (Brasil, 2010). Many of these species are undergoing genetic erosion, with

[1] Food Engeneering Course, Federal University of Maranhão, Maranhão, Brazil.
[2] College Laboro, Maranhão, Brazil.
* Corresponding author: tatiana.lemos@ufma.br

a consequent loss of genetic diversity and associated knowledge about them (Leal et al., 2018).

In this context, the UFP is a sustainable alternative for biodiversity and food security to meet the demand for food and the changes in consumer behavior. As they are easy to grow species and are available in vacant lots, gardens, backyards and parks, among others, the UFPs are popularly known as crabgrass or weed and, despite being little known in cuisine by the general population, recently they have gained emphasis on being an alternative for preserving biodiversity in human food, revealing its nutritional potential (Liberalesso, 2019).

The nutritional quality of the UFP, depending on the species, is related to significant levels of nutrients, such as mineral salts, vitamins, fibers, carbohydrates and proteins, in addition to the recognized functional effect. As an example of functionality and nutritional value, there is yam (*Dioscorea*), recognized as a blood purifier indicated for strengthening the immune system or the *ora-pro-nobis*, known as "poor meat" for its high levels of proteins. Thus, the recovery and valorization of the UFP in food represent important gains from the cultural, economic, social and nutritional point of view, considering the tradition in cultivation, by various communities, and its contribution in terms of nutrition. Therefore, stimulating the production and consumption of the UFP is a matter of food security and sovereignty, given their nutraceutical characteristics and their rusticity of cultivation (Brasil, 2010).

In this chapter, the nutritional and functional quality of some UPF and the potential for their use in obtaining products in the food industry will be described.

Arowroot (*Maranta arundinacea* L.)

Arrowroot is a perennial herbaceous plant originating from tropical regions of South America, including central Brazil, with a wide range of distribution, from the Northeast to the South. It belongs to the *Marantaceae* family and can reach up to 1.5–1.8 m in height. The rhizomatous stalks are rich in starch, which is particularly valued and sought after for its characteristics as excellent digestibility, which can be used by people with dietary restrictions on gluten and used in the production of creams, porridges, bread, biscuits, cheese bread and as a thickener for sauces (Brasil, 2010; Kinupp and Lorenzi, 2014). The most frequent varieties of arrowroot are Common and Creole. "*Common*" produces light rhizomes, while "*Creole*" has darker ones (Melo, 2019).

Regarding nutritional quality, it is known that rhizomes, on a dry basis, have: 68.2% moisture; 24.2% starch; 1.44% fibers; 1.34% protein, 1.83% ash; 0.19% lipids; 1.08% total soluble sugars; data from plants grown in Porto Alegre/RS (Kinupp and Lorenzi, 2014). The analysis of arrowroot rhizomes cultivated in Venezuela presented the following composition: 79.9% moisture; 5.5% protein; 6.0% lipids; 2.8% ash; 7.5% fiber; 78.2% of total carbohydrates; expressed on a dry basis (Melo, 2019). Considering the mineral content of plants cultivated in Porto Alegre/RS, rhizomes (dry) have: 0.37% phosphorus; 2.0% potassium; 0.04% calcium; 0.15% magnesium; 0.18% sulfur; 5 mg/kg copper; 18 mg/kg zinc; 18 mg/kg iron; 7 mg/kg manganese; 65 mg/kg of sodium (Kinupp and Lorenzi, 2014).

Arrowroot has nutritional, functional and technological potential that has been neglected. In this context, research that validates its use in the preparation of food products should be encouraged (Santos, 2017).

Starch extracted from native arrowroot has a low amylose content and a low tendency to retrograde with consequent loss of water by the final product, which favors its use as an additive in food production (Papalia, 2017). Papalia (2017) used arrowroot starch in combination with rice starch in the production of frozen cheese bread and concluded that it was possible to obtain cheese bread with substitution of up to 90% cassava starch, using 72% arrowroot starch and 18% rice starch.

Arrowroot flour has favorable characteristics for application in the food industry: high fiber content, high gelling capacity and significant resistant starch content, which makes it possible to improve sensory quality (texture) and survival of probiotics in yogurt, as well as in extruded products, which showed an improvement in their physical and functional properties due to the amount of resistant starch in this flour (Santos, 2017). The study by Santos (2017) indicated the great potential for the application of arrowroot flour for the formulation of dairy desserts, as the flour presented a high yield of resistant starch, in addition to having a low glycemic index and prebiotic potential.

Arrowroot derivatives are also used in the production of active packages, packages that interact with the food, prolonging its shelf life. Silva (2018) used arrowroot starch and thyme essential oil to produce an active biofilm and concluded that the formulated films had great potential to be used as antimicrobial-active packaging. Arrowroot starch can also be used in the production of an edible coating for fruit. Repolho et al. (2019) evaluated the application of the edible coating on the postharvest quality of acerola (*Malpighia emarginata*), Costa Rica during refrigerated storage (T = 10°C) and concluded that the edible coating based on arrowroot starch was an effective alternative in preserving the quality of acerola during refrigerated storage for 12 days.

Yam (*Dioscorea* spp. L.)

Yam (*Dioscorea* spp.) of the *Dioscoreaceae* family is a multispecies crop that generally produces large, starchy tubers. The genus includes more than 600 different species especially cultivated in tropical and subtropical countries and used as a popular food in Africa and Asia, playing an important role in food security, nutrition and income generation for more than 60 million people. The main yam species cultivated globally are *D. alata* L., *D. bulbifera* L., *D. cayenensis* Lam., *D. esculenta, D. opposita-japonica, D. numularia* Lam., *D. pentaphylla* L., *D. rotundata* and *D. trifida* L. The species *D. cayenensis* Lam. and *D. rotundata* are indigenous to West Africa, and are the two most important yam species in terms of production. On the other hand, *D. alata* L., of Asian origin, is the most widespread yam species globally and one of the oldest yam species cultivated in the world (Bomer et al., 2019). Annual global production is estimated at 73.01 million tons, with Nigeria currently being the largest producer representing 67% of global production (Silva, 2019). Furthermore, in some tropical countries, it has become the third most important tropical root crop, after cassava (*Manihot esculenta* Crantz.) and sweet potato (*Ipomoea potatoes* L. Lam.) (Wu et al., 2016).

In addition to economic values, yam ownership and cultivation have cultural and social meanings for various specific ethnic groups (Obidiegwu et al., 2020). In Brazil, yam is also known as "Cará", and has great socio-economic importance, with the Northeast region being the largest producer, with emphasis on the states of Sergipe, Paraíba, Pernambuco, Alagoas, Bahia and Maranhão, in which the main species cultivated are yellow yam (*D. cayennensis* Lam.) cultivar "yam-da-Costa" and water yam (*D. alata* L.) or "yam São Tomé" (Lovera et al., 2020; Brito et al., 2011). In practice, they are consumed in different ways, cooked, fried and baked, but can also be processed into flours, flakes, chips and dry-roasted slices by the food industries, serving as a good source of nutritional supply and staple foods throughout the world (Wu et al., 2016; Padhan et al., 2020).

In nutritional terms, according to the Brazilian Food Composition Table (TACO) estimates that raw yam has: 2.0 g of protein, 23.0 g of carbohydrate, 1.7 g of dietary fiber, 1.2 g of ash, 12.0 mg calcium, 29.0 mg magnesium, 0.2 mg manganese, 65.0 mg phosphorus, 0.4 mg iron, 0.4 mg sodium, 568.0 mg potassium, 0.17 mg of copper, 0.3 mg of zinc and 6.0 mg of vitamin C for every 100 grams of edible portion. In addition to nutritional value, yam roots are also known to be an important source of steroid saponins and other special bioactive constituents, which have important pharmacological effects, including promoting cell proliferation, decreasing plasma glucose, antifungal, antithrombotic, anticancer and hepatoprotective properties (Wu et al., 2016; Chen et al., 2017).

Despite its wide distribution and importance as a pharmacological agent and food source, yam is referred to as an underutilized species. The family is a source of native starches whose functional, physicochemical and nutritional properties can generate applications interesting in the food and non-food industry as exploited property (Lovera et al., 2020).

Liu et al. (2019) investigated the effects of partial replacement of wheat flour by purple yam flour (*Dioscorea alata* L.) on the *in vitro* starch digestibility of bread. Their research showed that the resistant starch content increased from 34.9% (bread made with 100% wheat flour) to 41.3% (bread made with 70% wheat flour and 30% red yam flour) and that the content of rapidly digestible starch and slowly digestible starch decreased with the addition of red yam flour to bread, thus concluding that red yam has potential for use as a functional ingredient, reducing starch digestibility in ready-to-eat food products and bread, which would be of great importance for the food industry.

The processing of yam into flour has been an alternative of great advantage as it does not contain gluten and has a lower number of calories than other flours. The trend today is the development of new gluten-free products, which would serve the population hypersensitive to gluten consumption (Dias et al., 2020). Dias et al. (2020) used yam flour to prepare cereal bars and obtained an item with high content of protein and dietary fiber and low lipid content.

Research in this field has also helped in the development of yam as a sustainable crop. Costa et al. (2020) developed biodegradable films based on purple yam starch, chitosan and glycerol for food application. Purple yam starch has shown good physical properties to be used as a matrix for edible films. Its characteristics indicated that it could be a viable coating for food products, contributing to the conservation of the environment, reducing the use of non-biodegradable polymers and improving food industries.

Jacatupé (*Pachyrhizus tuberosus* (L.) Urb.)

Pachyrhizus tuberosus is world-renowned as monkey bean, potato bean, jack bean, vegetable sausage, jicama, yam bean, yeticopé-yacatupé (Tupi) and potato-Cochon (French). In the Brazilian Southeast, it is called jacatupé (Silva, 2015). Belonging to the Fabaceae family, it is a legume species from the headwaters of the Amazon River, whose tuberous root is consumed by the indigenous populations of the western Amazon, in fresh or cooked form, and such as salads, cakes and puree (Silva, 2015; Cardoso, 1997). There are reports of its cultivation and consumption in Minas Gerais and Goiás, among other states in Brazil and Mexico, and Peru (Madeira et al., 2013; Silva, 2015). Green pods, leaves and seeds of jacatupé are consumed only after cooking due to being toxic (Madeira et al., 2013).

Jacatupé is a UFP perennial herbaceous vine that can reach up to 3.0 m in height when trussed. The leaves are alternate, composed and trifoliate, and the inflorescence is racemose. It produces tuberous roots, generally in small numbers with brown bark and white pulp, reaching four or five as in good fertility and management conditions (Brasil, 2010). It develops fully in Brazilian humid tropical regions but adapts to summer crops in other regions of Brazil (Madeira et al., 2013). It presents some characteristics to the Amazon region, being extremely rustic, undemanding in soil fertility and tolerant to high temperatures and drought (Silva, 2015).

The interest in jacatupé as food is due to the protein and starch content of the tuberous roots, which can reach 9.4% of proteins and 52.3% of starch, on a dry basis. Its nutritional composition contains: 1.12 to 9.46% of proteins; 5.6 to 45.6% sugars; 0.78% dietary fiber; 10.0 to 14.5% starch (Kinupp and Lorenzi, 2014; Souza et al., 2018).

The productive potential of the jacatupé has been little explored and incipient. Although, its tuberous roots can reach an average production that varies from 28 t ha^{-1} to 50 t ha^{-1} or 108 t ha^{-1}. From the roots, it is possible to produce flour and extract starch. In research carried out in the state of Amazonas, it was possible to obtain flour and starch from the roots of jacatupé, using the artisanal processing practiced in flour houses, with the same technology used in cassava, obtaining a yield for flour and starch, respectively, of 8 and 10% (Souza et al., 2018). Jacatupé tuberous root flour for human consumption can contribute to the use and conservation of its nutritional characteristics, a longer period of life for the flour. Starch from *P. tuberosus*, on the other hand, has low retrogradation comparable to waxy starches, which indicates the possibility of using it as a food additive, in its gelling function, in certain food products with undesirable retrogradation (Silva, 2019). In addition, the extracted starch, together with the flour, can be used in the production of bakery and confectionery products, as a total or partial replacement for wheat flour (Silva, 2015).

Tuberous roots can also be used whole or grated in regional cuisine to prepare recipes such as raw salad, cake and puree (Kinupp and Lorenzi, 2014).

Mangarito (*Xanthossoma mafaffa* Schott)

Mangarito (*Xanthosoma mafaffa* Schott), an herbaceous plant without an aerial stem, presents a main underground rhizome (mother or primary) with lateral shoots (child or secondary rhizomes) and several large leaves sprouting from the main rhizome. Belonging to the Araceae family, it originates from the Central American region that encompasses Central and South America. Mangarito is also known as tannia, tiquisque, malangay and in Brazil, as mangará, Portuguese taioba and mangareto, and called tayao by the Guarani population. It is one of the starchy tuberous species with the highest energy value, consisting primarily of starch and with a nutritional value comparable to potatoes. The offspring rhizomes, with small dimensions, are unattractive for culinary use and, consequently, have low commercial value (Leite et al., 2007; Costa et al., 2008).

Regarding the nutritional quality of the UFP, the approximate composition of the raw mangarito rhizomes, according to a survey conducted by Yoshime (2007), was: 62.96 to 78.25% moisture; 95.12 to 143.60 kcal/100 g of energy value; 0.61 to 3.53% protein; 15.11 to 31.83% carbohydrates; 0.12 to 0.24% lipids; 0.28 to 1.44% ash; 3.50% dietary fiber; 290.60 mg.100 g^{-1} calcium; 3.90 mg.100 g^{-1} of iron; 1408, 70 mg.100 g^{-1} potassium; 4.2 mg.100 g^{-1} of zinc. Furthermore, the presence in rhizomes of α-tocopherol (72.04 ¼g/g), α-carotene (1.26 ¼g/g) and β-carotene (3.41 ¼g/g) were also verified. Ávila (2011) in his research, obtained the following composition for the rhizomes of mangarito, on a wet basis: 73.51% moisture; 100.56 kcal/100 g of energy value; 2.15% protein; 0.12% lipids; 22.72% carbohydrates; 0.19% reducing sugars; 0.12% non-reducing sugars; 0.31% of total sugars; 2.39% crude fiber; 1.50% ash. This research also quantified, on a dry basis, the starch present in the rhizomes, and obtained 91.35 g/100 g.

The species of the starchy tuberose family, in which starch predominates as a component, such as mangarito, except for cassava, are little explored, resulting in the need to study the obtainment of this starch and its technological application, boosting research involving its use. In this context, Ávila (2011) in his research, extracted and characterized mangarito starch and used it as a basis for the elaboration of biodegradable films, keeping the amount of starch constant (2.0 g), varying the mass of the filmogenic solution and the concentration of glycerol (plasticizing agent). In the research, it was found that mangarito starch has a high pasting temperature, stability under heat under mechanical agitation and a marked tendency to retrograde, which indicates the possibility for use in the packaging and food industry. Furthermore, it was also found that films produced with 10% glycerol and 24.73 g of film-forming solution mass would be those that would meet the desired parameters of low water vapor permeability and thickness and high solubility.

From the extraction of the mangarito starch, a residue is generated, considered a problem for the environment. Seeking the use of this residue, Ávila et al. (2010) produced flour from this and verified its proximate composition and functional properties to apply it in the preparation of cookies. Mangarito flour had a considerable nutritive value, on a wet basis: 358.43 kcal/100 g; 78.8 g/100 g of carbohydrates; 0.87 g/100 g of total sugars; 0.44 g/100 g of reducing sugars; 0.42 g/100 g of non-reducing sugars; 9.15 g/100 g protein; 0.71 g/100 g lipids; 2.62 g/100 g crude fiber; 3.56 g/100 g of minerals. And include functional properties (swelling power, water absorption index, oil absorption index, solubility index and swelling volume) favorable to the use of flour as an ingredient in the production of cookies.

Taro (*Colocasia esculenta* (L.) Schott)

Taro is an ancient crop grown in all tropical, subtropical and warm temperate regions. Also known as "yam" in South-central Brazil, taro (*Colocasia esculenta* Schott) is an herbaceous plant belonging to the Araceae family, characterized by its dark green leaves and its tuberous rhizome, which forms corms with a scaly appearance and of variable diameter (Morishita, 1988; Silva, 2011; Vidigal et al., 2016). The crop also presents good rusticity characteristics to environmental and biological factors, in addition to a high yield per unit of the planted area, which makes it an attractive alternative crop for family-based agriculture (Monhol et al., 2018). In 2018, countries like Nigeria, China and Cameroon stood out worldwide in taro production. On the other hand, in Brazil, the crop still does not appear among the 20 main commodities produced in the country, but it is observed that production has been growing in recent years (FAOSTAT, 2020). Until then, taro was the main crop produced in Espírito Santo, reaching 3,200 ha and generating 90 thousand tons of product. This cultivar is produced in 36 municipalities in the state of Espírito Santo, with Alfredo Chaves, Laranja da Terra and Marechal Floriano being the main producers (Balbino et al., 2018).

Nutritionally, taro has a value similar to that of English potato and has great versatility in the form of consumption in human food. It has high levels of starch, 50% of which is composed of readily digestible starch fractions, and it also has a medium glycemic index, thus being a good alternative to carbohydrates in the diet of diabetic people. The crop also has a good content of minerals and reasonable amounts of B-complex vitamins (Brasil, 2010; Simsek and El, 2015). Among some industrial applications, taro can be used as a raw material in the gluten-free dextrin, gum and bread industry (Calle et al., 2020). Regarding proximate content, tuberose evaluated by Rabiou et al. (2019) had a moisture content of 81.74%, 1.17% of proteins, 0.95% of lipids, 1.00% of ash, 21.38% of carbohydrates and energy value of 98.83 kcal/100 g of food for the variety sold in Niamey city (Nigeria). For the variety "monkey", grown in the municipality of Miranda-RS (Brazil), values of 78.92% (moisture), 2.19% (proteins), 0.12% (lipids) were found; 1.70% (fibers); 14.46% (starch) and 67.68 kcal/100 g (energy value) (Ramos Filho et al., 1997). This indicates the variation in nutrient content according to species and crop management.

Regarding digestion, Simsek and El (2015) elucidated that the starch present in taro is more resistant to digestion compared to potato starch, and as a result, tends to have a lower Glycemic Index (GI). Some GI values are found in literature: 63.1 (Simsek and El, 2015), 72 (Bahado-Singh et al., 2006), and 79 (Foster-Powel et al., 2002), such variations may be justified by the variety and methodology of analysis of tuberose. However, these data show that taro could turn out to be an excellent alternative from a GI perspective for people suffering from type II diabetes.

When fresh, the tuber has a bitter and unpleasant taste, due to the presence of acidity, which is associated with calcium oxalate in the form of raphid crystals (Desai et al., 2017). Values between 433, 8 to 856.1 mg of calcium oxalate may be present in 100 g of taro wet matter, but this content varies according to cultivar (Du Thanh et al., 2017). The acidity is so pronounced that, when consumed raw, the corms cause swelling of the lips, mouth and throat, in addition to the astringent taste and scratches in the mouth and throat (Owusu-Darko et al., 2014). Considered as an anti-nutrient, calcium oxalate can also cause severe damage to the gastrointestinal system when the tuber is eaten undercooked (Kumoro et al., 2014). According to literature, this irritating agent, probably a protease, can be reduced by cooking the parts used or also with fermentation (Balbino et al., 2018). For this reason, taro is usually consumed after prolonged cooking. Since, calcium oxalate is removed by leaching or may undergo a decomposition reaction of calcium oxalate to calcium oxide at elevated temperatures (Brasil, 2010; Lima and Krupek, 2016; Sabir et al., 2017).

Although it has the high nutritional quality and is highly appreciated in Brazilian cuisine, tuberose has undergone little scientific investigation. There are still few studies on management techniques, genetic improvement, characterization, nutritional and industrial potential, as well as not many reports of drying and storage studies (Castro et al., 2017). Nevertheless, taro has great potential for export and industrial processing. This is because the crop demonstrates greater starch production potential per area depending on its agricultural yield, showing great potential for industrial starch production (Anjos, 2012; Leonel and Cereda, 2002). In addition, it has an excellent percentage of starch in its constitution, which can be used in several industries, such as paper, textile and food industries, where it is used as an ingredient in processed foods, which is one of its main areas of application (Almeida, 2012; Sit et al., 2015).

There are also other problems faced during crop processing, such as variation in shape and size, internal coloration and crop color changes, in addition to the problem of high corm acidity. As a result, the processing possibilities are significantly reduced. Thus, new technologies complementary to those currently available are needed to improve the productivity and quality of taro starch, increasing the bioavailability of nutrients and enabling the offer of a product that meets market requirements and ensures consumer food safety (Sharma et al., 2016). Likewise, the need for advances in productivity and incentives for research, related to agricultural practices and product development, should also be considered, keeping in mind the tuberose varieties with their own characteristics for specific purposes. There is also a need to improve industrial technologies already applied to other raw materials (Balbino et al., 2018).

For use in industrialization, Almeida (2012) highlights that the native taro starch has a low content of substances considered contaminants, which do not influence the functional properties, and also that after modification and combination processes, the starch is shown to be in conformity with the requirements of the Brazilian legislation, it is possible to use the processes to improve the functional properties for use in the food industry in different ways. Modifications can be made for use in quick or instant preparation foods such as dehydrated soups, pasta and desserts, among others, also for application in frozen products for bakery and embedded meat products industry, or in the dairy beverage industry and in the manufacture of cookies, candies, jellies, candies in syrups, fillings and toppings for cakes (Alcantra et al., 2013; Almeida et al., 2013; Krisnaningsih et al., 2019).

The flour obtained from taro, the main by-product, is very easy to handle and store, as it obtains good microbiological stability compared to the natural product, in addition to maintaining its nutritional characteristics (Castro et al., 2017; Sá et al., 2018). According to Deo et al. (2009), taro flour is a by-product usually used in the diet of people allergic to some cereals and can be added in the formulation of canned baby foods, especially those products aimed at children with sensitivity to milk.

Flour can be used in the bakery industry. In a study carried out by Calle et al. (2020), the flour showed good characteristics for application in the preparation of gluten-free bread, which includes good fiber content, low-fat content and good technological properties, which together with additives bakery products can expand the functionality of the by-product, increasing the technological quality of bread. Thus, according to the authors, the flour has the potential to be used in the preparation of "gluten-free" bread with technological quality parameters similar to those obtained by other gluten-free flours. A similar result was also found by Arici et al. (2020), when evaluating different bread formulations with the addition of taro flour. The authors observed positive aspects with the replacement of up to 18.70%, such as the browning effect in bread and the increase in dietary fiber content with the taro flour content in the formulation. Thus, taro flour presents itself as a valuable ingredient in commercial gluten-free bread mixes.

For the manufacture of biscuits, Hegazy (2019) evaluated that the incorporation of 10, 20 and 30% of taro flour in the formulation considerably improves the nutritional value and sensory quality of the biscuits, in addition to not promoting significant differences in the parameters of cookies, such as diameter, thickness, specific volume and density of the dough, in relation to the control sample (wheat flour). A similar result was also observed by Alflen et al. (2016) when using the same concentrations to replace wheat flour with taro flour in the production of biscuits, which increased the nutritional value of the product without significant changes in its physical properties. In this way, it is noted that taro flour presents possibilities for application in the biscuit industry, improving nutritional quality, and also aiming to meet the current demand for more nutritious and healthier products. In addition to being able to be used with gluten-free flours for the preparation of biscuits for the food-restricted public, for example, celiacs (Giri and Sajeev, 2020).

As an additive, taro flour can also act as a stabilizer in the production of edible ice creams. Hong and Nip (1990), in their research, used taro flour as a stabilizing agent and, when comparing them with commercial additives, they observed that they could be replaced by flour without significant changes in the functional properties of the sorbet. Dardiry et al. (2018), ice cream made with the addition of 20% of cooked taro, reducing the lipid content and commercial emulsifier, proved to be a good strategy for improving the antioxidant content and nutritional value of edible ice cream without the sensory properties being impaired. In making puddings, when applying taro flour as a hydrocolloid, both the nutritional composition and rheological properties showed positive results with the flour content in the formulation. Thus, the flour has excellent emulsifying activity and potential for use in food products, which additionally improves the nutritional value of the products (Ertop et al., 2018).

Among other applications of taro flour, because it contains a good fiber content, tuberose offers potential as a prebiotic in the diet through the presence of resistant starch (RA). This is because the AR present in taro helps intestinal bacteria in the production of short-chain fatty acids. Especially for the production of health-promoting butyric acid, there is a strong positive correlation with the concentration of taro-resistant starch (Saxby et al., 2020). There is also the use of starch extracted from taro flour for packaging development. In a study carried out by Wulandari and Warkoyo (2018), the film made with taro starch and chitosan plasticizer did not obtain good water vapor permeability properties, which could be improved with the addition of hydrophobic components to the formulation and the increase the thickness of the material (Wulandari and Warkoyo, 2018). In the study by Siskawardani et al. (2020), the mechanical properties of the edible film of taro starch (3%) and glycerol plasticizer (25%), obtained good results for flexibility, reduction of vapor permeability and transparency, significantly improving the applicability of the films in foods.

Cubiu (*Solanum sessiliflorum* Dunal)

Cubiu (*Solanum sessiliflorum* Dunal) is an important genetic resource native to the Amazon, belonging to the Solanaceae family and widely distributed in equatorial regions of Brazil, Colombia and Peru. UFP is a semi-perennial, erect and branched herbaceous plant, which grows from 1 to 2 m in height, with simple leaves, a broad-oval blade and irregularly serrated margins, having pubescence on both sides, with a length of 30 to 60 cm and with the petiole of 10 to 14 cm. The flowers have no petiole, they are yellowish-green, gathered in bunches with five to eight flowers. Its fruits are fleshy berry-like, non-climacteric, with variations in size, color and shape, its juicy pulp is edible, with a characteristic acidic taste, which has already been described as similar to that of a lemon and a tomato, and of small seeds. The fruits are used in fresh form or as raw material for food, pharmaceutical and cosmetic purposes. Cubiu has economic potential, due to its rusticity, high fruit yield, precocity, easy cultivation, nutritional value and diversity of uses (Brasil, 2010; Kinupp and Lorenzi, 2014; Colodel et al., 2017; Sprey et al., 2019).

Regarding the information on the proximate composition of the cubiu pulp, there is the following range of variation, considering the analysis of 8 cubiu ethnovarities: 88.4–92.1% moisture; 24.7–43.8 kcal/100 g of energy value; 4.5–7.1% carbohydrates; 0.4–0.7% protein; 0.3–1.8% lipids; 0.5–0.7% ash; 0.9–2.2% total fiber; 0.0–0.5% soluble fiber; 0.9–1.8% insoluble fiber (Yuyama et al., 2007). In addition, cubiu has a content of total phenolic compounds in the range of 149.54–157.76 mg EAG.100 g^{-1} (Ferreira, 2020).

The use of UFP as cubiu in the diet, increases the options in food diversification, enabling the creation of new products. Therefore, to preserve the characteristics of the fruit and its functional properties, vary its usability and facilitate its acceptance by the consumer, it is necessary to apply processing technologies (Ribeiro and Durigan, 2018). In this context, Furnaleto (2015) in his study, developed a light cubiu jelly containing 60% pulp, 28% sucrose and 12% sucralose, with good sensory acceptance, total phenolic compounds content of 38.3 mg of acid gallic.100 g^{-1} and antioxidant activity of 93.9% (DPPH method). While Fares (2010) in his research, produced, characterized and verified the shelf life of cubiu mass candy, called by him cubilada, in the traditional and diet versions. The traditional version had the following proximate composition: 22.11% moisture; 310.0 kcal/100 g of energy value; 76.84% carbohydrates; 19.48% of total sugars; 13.79% reducing sugars; 5.69% non-reducing sugars; 0.56% protein; 0.04% lipids; 1.99% dietary fiber; 0.44% ash. The diet version had the following composition: 48.22% moisture; 205.64 kcal/100 g of energy value; 50.35% carbohydrates; 2.38% of total sugars; 1.72% reducing sugars; 0.66% of non-reducing sugars; 0.52% protein; 0.02% lipids; 1.71% dietary fiber; 0.39% ash. The cubilada in the traditional and diet versions had good sensory acceptance and shelf life of 90 days stored, respectively, at room temperature (T: ± 28°C) and under refrigeration (T: ± 6°C).

Another product that can be obtained from cubiu is flour. Maia et al. (2015), produced from the fruits of cubiu flour, with the following composition: 8.22% moisture; 76.55% carbohydrates; 6.99% protein; 1.49% lipids; 6.75% ash; 13.30% soluble fiber; 25.94% insoluble fiber. In the study, the authors also confirmed the popular use of the fruit as a hypocholesterolemic agent.

From cubiu, a food additive can also be obtained, such as pectin, which can act as a gelling agent, stabilizer and emulsifier. Colodel and Petkowicz (2019) extracted pectin from cubiu bark, which is composed of a highly methoxylated homogalacturonan with a low degree of acetylation. The extracted pectin formed gels in an acidic medium (pH = 1.5–2.5) and with 60% sucrose, indicating the possibility of its use as a food additive in acidic products with a high content of total soluble solids.

Maxixe (*Cucumis anguria* L.)

A plant from East Africa, *C. anguria* L. also popularly known as gherkin, wild gherkin or prickly cucumber, is a species of the *Cucurbitaceae* family. Annual climbing herbaceous, its fruit is light green, with length ranging from 5 to 7 cm, diameter from 3 to 4 cm, the cycle of 70 days, has no bitter taste and has variations in size and presence of spicules. With economic, social, and cultural importance for the Northeast, the gherkin is among the most consumed vegetables in the region. And is slso part of the popular cuisine of the north and southeast. Regarding the form of consumption, gherkin is consumed cooked, natural and preserved, in the form of pickles (Brasil, 2010; Kinupp and Lorenzi, 2014; Matos, 2016).

Regarding the composition of gherkin (raw) per 100 g of the edible part, it contains 95.1% moisture; 1.4 g of protein; 0.1 g of lipids; 2.7 g of carbohydrates; 2.2 g of dietary fiber; 0.7 g of ash; 21 mg of calcium; 10 mg of magnesium; 0.07 mg of manganese; 25 mg of phosphorus; 0.4 mg iron; 11 mg sodium; 328 mg of potassium; 0.02 mg of copper; 0.2 mg zinc; 0.06 mg vitamin B1; 0.02 mg vitamin B2; 0.04 mg vitamin B6; 9.6 mg of vitamin C. Considering the presence of bioactive compounds in gherkin, researchers reported the following average contents: 136, 21 ± 3.69 mg of gallic acid.100 g^{-1} (total phenolic compounds); 16.7 ± 3.2 μg.100 g^{-1} (carotenoids). Furthermore, research on the antioxidant activity of gherkin reported values ranging from 3.1 to 6.1 mM of Trolox.100 g^{-1} and 8.0 to 18.0 mM of Trolox.100 g^{-1} samples for aqueous and ethanolic, respectively, by the ABTS radical scavenging method (NEPA, 2011; Silva, 2017).

Gherkin is a seasonal and regionalized crop, which has high perishability, as with other vegetables, requiring the use of traditional or emerging technologies for food preservation, which assist in the transformation of this raw material into a product that preserves its medicinal properties and nutritional value of this vegetable, making it available for consumption in the off-season periods and enabling its geographic distribution, stimulating its consumption. There are few reports on the use of this vegetable in human food other than the traditional way. However, it is possible to add value to this NCFP, with a possible increase in agricultural production and in the productive sector, through the generation of new products using gherkin or incorporating it in the formulation of other products (Matos, 2016).

Matos (2016) in his research to obtain gherkin fruit flour, and from this to elaborate a product, found that gherkin flour presented significant levels of nutrients when compared to the fresh fruit, highlighting the nutrients, carbohydrates and ash, the latter indicating a high content of minerals, and flour can be a considerable source of these micronutrients. In addition, it was also found in the research, the feasibility of partially replacing wheat flour with gherkin flour in the preparation of cakes, which presented an increase in dietary fiber and mineral contents, and a reduction in terms of energy, configuring a less caloric product (Matos, 2016).

According to the research conducted by Abrantes (2019) on the development of gherkin mass jam with coconut added with different types of sugars, it was possible to conclude that gherkin jam with coconut is a viable alternative for the conservation and better use of gherkin fruit, since all formulations obtained good sensory acceptance and a positive purchase intention, in addition to satisfactory physicochemical characteristics, regardless of the different types of sugar added to the product.

A potential for the use of gherkin by the food industry was also reported in the research developed by Dantas (2019) when proposing the elaboration of cocada added with gherkin and flavored with fruit pulp. With the research, Dantas (2019) concluded that it was possible to produce cocadas with gherkin flavored with pineapple, soursop and passion fruit pulps, with characteristics of identity and quality suitable for consumption, good sensory acceptance and positive purchase intention.

Chicory (*Cichorium intybus* L.)

Chicory (*Cichorium intybus* L.) is an herbaceous plant, with an annual cycle, widely distributed from north to south of Brazil. Also known as chicory, belonging to the Asteraceae family. It has lanceolate, lobed or peaked leaves, smooth light green or with purple veins with a characteristic bitter taste (MAPA, 2010; Barreira et al., 2015). Plants of this genus are known for their therapeutic, medicinal and inulin extraction properties. They are usually consumed in salads, but their leaves and roots can also be roasted and used as a coffee substitute or additive (Aisa, 2020). The most common traditional medicinal preparations made from fresh or dried *C. intybus* L. are decoctions, infusions and herbal teas (Dalar and Konczak, 2014). Chicory has a rich nutritional composition, which makes it a potential source of bioactive substances for the fortification of the human diet, such as inulin, sesquiterpene lactones, caffeic acid derivatives, fats, proteins, hydroxycoumarins, flavonoids, alkaloids, steroids, terpenoids, oils, volatile compounds, vitamins (α-tocopherol, γ-tocopherol), β-carotene, zeaxanthin and minerals. In addition to an important nutritional profile, chicory also has biological activity: hepatoprotective, anti-inflammatory, antioxidant, lipid-lowering, anti-diabetic, anti-cancer, gastro-protective and anti-microbial, among others (Perovic et al., 2020).

Regarding the nutritional profile, the following nutritional composition for fresh raw chicory per 100 g by weight: 90.52 g of water; 17.0 kcal of energy value; 4.0 g of carbohydrates; 0.9 g protein; 0.1 g of lipids; 3.1 g of dietary fiber; 0.47 g of minerals. Regarding its composition in minerals and vitamins, it contains: 19.0 mg.100 g^{-1} of calcium; 0.24 mg.100 g^{-1} of iron; 10.0 mg.100 g^{-1} of magnesium; 26.0 mg.100 g^{-1} of phosphorus; 211.0 mg.100 g^{-1} potassium; 2.0 mg.100 g^{-1} sodium; 0.16 mg.100 g^{-1} of zinc; 0.051 mg.100 g^{-1} copper; 0.1 mg.100 g^{-1} of manganese; 0.2 g.100 g^{-1} of selenium; 2.8 mg.100 g^{-1} of vitamin C; 0.062 mg.100 g^{-1} of thiamine; 0.027 mg.100 g^{-1} riboflavin; 0.16 mg.100 g^{-1} of niacin; 0.145 mg.100 g^{-1} of pantothenic acid; 0.042 mg.100 g^{-1} vitamin B6; 37.0 μg.100 g^{-1} total folate; 29.0 IU.100 g^{-1} of vitamin A (Perovic et al., 2020).

Considering the nutritional composition of the chicory parts, the following composition for the roots (g/100 g): 58.0–87.57 moisture; 70.43–90.77 carbohydrates; 11.06–12.33 total soluble sugars; 44.69 of inulin; 3.83–5.54 of proteins; 0.95–3.01 lipids; 31.15 of dietary fiber; 30.73 of insoluble dietary fiber; 0.42 soluble dietary fiber; 8.12 of minerals. For leaves it contains (g/100 g): 22.64–85.77 of moisture; 38.2–70.71 carbohydrates; 7.8 total soluble sugars; 10.95 of inulin; 14.0–14.7 proteins; 1.9 lipids; 15.13–18.65 of minerals. And for seeds (g/100 g) 4.24–6.40 moisture; 31.66–34.72 of carbohydrates; 1.79 of inulin; 10.70–19.57 of proteins; 14.40–22.89 of lipids; 6.91–17.19 of minerals (Perovic et al., 2020).

Chicory is the most used plant source for inulin extraction on an industrial scale. Inulin extracted from chicory rhizomes has a wide range of technological applications. In the food industry, it can be used as a fat substitute, stabilizer, thickener, gelling agent and prebiotic (El-kholy, 2020). In this context, Teixeira (2017) in his study, developed a rennet cheese added with inulin, acting as a prebiotic, as a potential functional food matrix, considering the inulin content per portion of the product was in accordance with what is established in the legislation relevant. Gomes (2018), in his research, produced hamburgers with a 25% reduction in fat, containing 5 and 10% of inulin as a fat substitute and found that the use of inulin in the preparation of the product is viable. Solagna et al. (2016), elaborated on pretzels with integral inulin and concluded that an additional level of up to 4% of inulin in pretzels was well accepted by children, obtaining sensory acceptance comparable to the standard product and with positive marketing expectations.

Caruru (*Amaranthus* spp.)

Considered as foods of the future, due to their nutritional and functional properties, plant species classified in the genus Amaranthus can be an alternative to diversify the human diet (Xavier et al., 2019). *Amaranthus* spp. is an annual, dicotyledonous, erect or ascending herbaceous plant, native to the Americas, also called pigweed or bramble, which has about 60 species, whose leaves and seeds (grains) are consumed as food in various regions of the world. The seeds of the species *Amaranthus deflexus* L. have the potential as pseudocereal (Brasil, 2010; Kinupp and Lorenzi, 2014; Amaya-Farfan et al., 2005). The UFP has a smooth stem, greenish or red in color, due to anthocyanins, as well as the roots. Alternate or differing, the leaves are intensely green, with dark or violet spots and very soft consistency. The flowers can occur in the leaf axils as pale green spikes. Plants are heat and drought-tolerant (Brasil, 2010).

Although it has cereal characteristics, amaranth (*Amaranthus* spp.) cannot be classified as such, due to the botanical aspects, such as the presence of panicle-like inflorescence and being a dicotyledon and nutritional properties, such as high protein and sulfurated amino acid content and lysine, which classifies it as pseudo-cereal (Bartz and Cabral, 2017).

The nutritional quality of *Amaranthus hybridus* and *Amaranthus viridis* species was verified in the research by Lima e Silva et al. (2019). Considering the *Amaranthus hydridus* species, the researchers obtained the following results for leaves and grains or seeds: moisture (74,25% leaves; 13,34% seeds); energy (47,53 kcal/100 g leaves; 358,06 kcal/100 g seeds) carbohydrates (5,28% leaves; 61,23% seeds); proteins (4,42% leaves; 13,75% seeds); lipids (0,97% leaves; 6,46% seeds); fibers (1,81% leaves; 2,99% seeds); ash (6,79% leaves; 2,23% seeds); phosphor (0,35% leaves; 0,46% seeds); potassium (1,80% leaves; 0,45% seeds); calcium (0,25% leaves; 0,37% seeds); magnesium (0,09% leaves; 0,35% seeds); iron (151,50 ppm leaves; 457,52 ppm seeds); manganese (47,90 ppm leaves; 27,26 ppm seeds); zinc (50,80 ppm leaves; 29,71 ppm seeds). For the *Amaranthus viridis* species, the following results were obtained for leaves and grains or seeds: moisture (77,79% leaves; 14,23% seeds); energy (48,80 kcal/100 g leaves; 347,26 kcal/100 g seeds); carbohydrates (3,80% leaves; 61,66% seeds); proteins (5,79% leaves; 12,60% seeds); lipids (1,16% leaves; 5,58% seeds); fibers (1,88% leaves; 2,73% seeds); ash (4,64% leaves; 3,21% seeds); phosphor (0,53% leaves; 0,42% seeds); potassium (2,15% leaves; 0,52% seeds); calcium (0,46% leaves; 0,49% seeds); magnesium (0,09% leaves; 0,32% seeds); iron (184,10 ppm leaves; 137,65 ppm seeds); manganese (42,50 ppm leaves; 20,90 ppm seeds); zinc (105,70 ppm leaves; 26,75 ppm seeds).

Analyzes carried out on leaves of the species *Amaranthus deflexus* L. contained the following composition: 88% moisture; 34 kcal/100 g of energy; 6.0 g.100 g^{-1} of carbohydrates; 3.0 g.100 g^{-1} of proteins; 1.0 g.100 g^{-1} of lipids; 4.5 g.100 g^{-1} dietary fiber; 2,6 g.100 g^{-1} cinzas; 455 mg.100 g^{-1} of calcium; 197 mg.100 g^{-1} of magnesium; 0,9 mg.100 g^{-1} of manganese; 77 mg.100 g^{-1} of phosphor; 4,5 mg.100 g^{-1} of iron; 14 mg.100 g^{-1} of sodium; 279 mg.100 g^{-1} of potassium; 6,0 mg.100 g^{-1} of zinc; 0,10 mg.100 g^{-1} of riboflavin; 0,11 mg.100 g^{-1} pyridoxine; 5,0 mg.100 g^{-1} vitamin C. The seeds had 17.2% protein and 6.5% lipids. The composition of leaves of the species *Amaranthus spinosus* L. was as follows: 91% moisture; 27 kcal/100 g;

4.3 g.100 g^{-1} of carbohydrates; 4,0 g.100 g^{-1} of proteins; 0,6 g.100 g^{-1} of lipids; 2,48 g.100 g^{-1} of dietary fiber; 2,76 g.100 g^{-1} of ash; 3.931 mg.100 g^{-1} of calcium; 629 mg.100 g^{-1} of phosphor; 32 mg.100 g^{-1} of iron; 1.166 mg.100 g^{-1} of magnesium; 3,0 mg.100 g^{-1} of manganese; 393 mg.100 g^{-1} of sodium; 15 mg/kg of zinc (Kinupp and Lorenzi, 2014).

Amaranth is also an excellent and unique source of antioxidant leaf pigments, such as β-cyanine, β-xanthine, betalain and other pigments such as carotenoids, anthocyanin, chlorophylls and amaranth, in addition to having antioxidant phytochemicals such as β-carotene, vitamin C, phenolic compounds and flavonoids. Most of these compounds are natural antioxidants and detoxify the human body, therefore, of significant importance to the food industry (Sarker and Oba, 2019).

The incorporation of amaranth (*Amaranthus* spp.) into the human diet depends on the creation of products that are acceptable to consumers. Among the products, there is amaranth flour, which is obtained by grinding amaranth seeds or grains, and which manages to maintain its high protein and amino acid content, even after the refining operation (Bartz and Cabral, 2017). Gluten-free flours, such as amaranth flour, and with specific texture characteristics, can be an alternative for the formulation of products, from mixtures or combinations between flours, which contribute to the quality of the final product, highlighted for the bakery section (Pinelli, 2018). Miranda-Ramos et al. (2019) in their research, used the whole flour of *Amaranthus spinosus* and *Amaranthus hypochondriacus* for bread production and concluded that it was possible to replace 25% of wheat flour by whole wheat flour of amaranth, in the bread formulation, with considerable improvement in nutritional quality, technological and sensory acceptance of the product.

As a product derived from amaranth, is amaranth oil, with interesting biological activities due to its components, especially tocopherols and tocotrienols known for their antioxidant activity, phytosterols that reduce serum cholesterol and squalene with anticancer and hypocholesterolemic activity. The oil can be obtained from the amaranth grain using solvent extraction (hexane or methanol) and supercritical fluids. Montes (2012) in his research found that crude amaranth oil showed good stability for 42 days at 60°C, better than other commercial oils, such as corn.

Amaranth grain is a gluten-free food alternative, in the form of amaranth popcorn, obtained by expanding the grains at a temperature of 100°C. In the research by Bindermann et al. (2020), amaranth popcorn maintained the nutritional quality of the amaranth grain, being rich in fiber, proteins and low in lipids, with a significant loss of calcium when compared to the raw grain.

Amaranth leaves and leaf branches are used in gastronomy, in the preparation of dishes such as pigweed cake, pigweed with meat, pigweed souffle and pigweed with angu (Kinupp and Lorenzi, 2014). In this context, Hiscock et al. (2020) added amaranth leaves to a cooked culinary preparation that is part of the diet of local rural consumers in order to assess consumers' hedonic responses regarding the overall acceptability of 13 cooked amaranth genotypes (leaves). As a result of the study, the researchers found that a cooking method, using potatoes, tomatoes and onions as ingredients, was able to mask the bitterness of the amaranth genotypes' leaves, increasing their acceptability. Thus, research on methods that mask the bitterness of the leaves is relevant, as they can lead to commercially viable and acceptable products for the target audience, which can help to promote the consumption of amaranth leaves (Hiscock et al., 2020).

Conclusion

Based on the information presented about the UFP, it was possible to verify their nutritional and functional potential, and the possibility of using them as raw material for food production, thus diversifying their use and the forms of insertion in the human diet.

References

Abrantes, I.F.R. 2019. Desenvolvimento de doce em massa de maxixe com coco adicionado de diferentes tipos de açúcares. 43 f. Monografia (Bacharelado em Nutrição)—Centro de Educação e Saúde, Universidade Federal de Campina Grande, Cuité.

Aisa, H.A., Xin, X. and Tang, D. 2020. Chemical constituents and their pharmacological activities of plants from Cichorium genus. Chin. Herb. Med.

Alcantara, R.M., Hurtada, W.A. and Dizo, E.I. 2013. The nutritional value and phytochemical components of taro [*Colocasia esculenta* (L.) Schott] powder and its selected processed foods. Journ. of Nutrit. & Food Scienc. [S. l.] 3(3).

Alflen, T., Quast, E., Bertan, L. and Bainy, E. 2016. Partial substitution of wheat flour with taro (*Colocasia esculenta*) flour on cookie quality. Rev. Ciênc. Exat. e Natur. [S. l.] 18.

Almeida, E.C. 2012. Amido modificado de taro (*Colocasia esculenta* L. Schott): propriedades funcionais. 2012. Universidade Federal da Paraíba [S. l.].

Almeida, E.C., Bora, P.S. and Zárate, N.A.H. 2013. Amido modificado de taro (*Colocasia esculenta* L. Schott): propriedades funcionais. Bolet. Cent. de Pesq. de Proc. de Alim., [S. l.] 31(1): 67–82.

Amaya-Farfan, J., Marceilio, R. and Spehar, C.R. 2005. Deveria o Brasil investir em novos grãos para a sua alimentação? A proposta do amaranto (*Amaranthus* sp.). •• Segurança Alimentar e Nutricional, Campinas 12(1): 47–56.

Anjos, B.B. 2012. Manejo sustentável da cultura do taro. *In*: Dossiê Técnico. Serviço Brasileiro de Respostas Técnicas.

Arici, M., Özullku, G., Kahraman, B., Yildirim, R.M. and Toker, Ö.S. 2020. Taro flour usage in wheat flour bread and gluten-free bread: Evaluation of rheological, technological and some nutritional properties. Journ. of Food Proces. Engin., n. e13454.

Avila, R., Reis, R.C., Barbosa, L.S., Rezende, F.L. and Ascheri, D.P.R. 2010. Avaliação da composição centesimal e utilização na elaboração de biscoitos tipo cookie do resíduo da extração da fécula do mangarito. Rev. Agrotec. 1(1): 21–32.

Avila, R. 2011. Caracterização dos rizomas filhos e da fécula do mangarito (*Xanthosoma mafaffa* Schot) e elaboração de filmes biodegradáveis. 76 f. Dissertação (Mestrado em Engenharia Agrícola)—Programa de Pós-graduação em Engenharia Agricola, Universidade Estadual de Goiás, Anápolis.

Azevedo, T.D. 2018. Propriedades nutricionais, antioxidantes, antimicrobianas e toxicidade preliminar do peixinho da horta (*Stachys byzantina* K. Koch). 89 f. Dissertação (Mestrado em Alimentação e Nutrição)—Departamento de Nutrição, Universidade Federal do Paraná, Curitiba.

Bahado-Singh, P.S., Wheatley, A.O., Ahmad, M.H., Morrison, E.Y. St. A. and Asemota, H.N. 2006. Food processing methods influence the glycaemic indices of some commonlyeaten West Indian carbohydrate-rich foods. British Journal of Nutrition 96: 476–481.

Balbino, J.M.S., Carmo, C.A.S., Puiatti, M., Favarato, L.F., Ramos, J.P. and Krohlin, C.A. 2018. Taro (Inhame) Boas práticas de colheita e de pós-colheita. Vitória: Incaper. Disponível em: https://biblioteca.incaper.es.gov.br/digital/bitstream/123456789/3088/1/BRT-Livro-Taro-PDF.pdf.

Barreira, T.F., Paula Filho, G.X., Rodrigues, V.C.C., Andrade, F.M.C., Santos, R.H.S., Priore, S.E. and Pinheiro-Santa'na, H.M. 2015. Diversidade e equitabilidade de plantas alimentícias não convencionais na zona rural de Viçosa, Minas Gerais, Brasil. Rev. Bras. de Plant. Med. 17(4): 964–974. https://doi.org/10.1590/1983-084X/14_100.

Bartz, C.G. and Cabral, D.D. 2017. Efeito da farinha de amaranto na redução de colesterol total e triglicerídeos em indivíduos hipercolesterolêmicos e hipertrigliceridêmicos. Nutri. Bras. 16(3): 127–134.

Bindermann, D., Teixeira, L.C., Santos, M.R., Maciel, S.O. and Balbi, M.E. 2020. Características físico-químicas, nutricionais e funcionais da pipoca de amaranto (*Amaranthus* spp., *Amaranthaceae*). Vis. Acad. 21(1).

Bomer, M., Rathanayake, A.I., Visendi, P., Sewe, S.O., Sicat, J.P.A., Silva, G., Kumar, P.L. and Seal, S.L. 2019. Tissue culture and next-generation sequencing: A combined approach for detecting yam (*Dioscorea* spp.) viruses. Physiological and Molecular Plant Pathology 105: 54–66.

Brasil. 2010. Ministério da Agricultura, Pecuária e Abastecimento. Manual de hortaliças não-convencionais. Secretaria de Desenvolvimento Agropecuário e Cooperativismo. Brasília: mapa/ACS.

Brito, T.T., Soares, L.S., Furtado, M.C., Castro, A.A. and Carnelossi, M.A.G. 2011. Composição centesimal de inhame (*Dioscorea* sp.) in natura e minimamente processado. Scientia Plena, 7: 011502.

Calle, J., Benavent-Gil, Y. and Rosell, C.M. 2020. Development of gluten free breads from Colocasia esculenta flour blended with hydrocolloids and enzymes. Food Hydroc. 98: 105243.

Cardoso, M.O. 1997. Hortaliças Não–Convencionais da Amazônia. Brasília: EMBRAPA SPI: Manaus: EMBRAPA CPAA.

Chen, X., Li, X., Huang, H., Wang, T., Qu, Z., Miao, J. and Gao, W. 2017. Effects of drying processes on starch-related physicochemical properties, bioactive components and antioxidant properties of yam flours. Food Chemistry 224: 224–232.

Costa, J.C.M., Miki, k.S.L., Ramos, A.S. and Costa, B.E.T. 2020. Development of biodegradable films based on purple yam starch/chitosan for food application. Helyon, 6, 2020. https://doi.org/10.1016/j.heliyon.2020.e03718.

Castro, D.S., Oliveira, T.K.B., Lemos, D.M., Rocha, A.P.T. and Almeida, R.D. 2017. Efeito da temperatura sobre a composição físico-química e compostos bioativos de farinha de taro obtida em leito de jorro. Brazil. Journ. of Food Techn. 20, n. Epub May 18.

Colodel, C., Bagatin, R.M.G., Tavares, T.M. and Petkowicz, C.L.O. 2017. Cell wall polysaccharides from pulp and peel of cubiu: A pectin-rich fruit. Carbohydr. Polym. 174: 226–234.

Colodel, C. and Petkowicz, C.L.O. 2019. Acid extraction and physicochemical characterizarion of pectin from cubiu (*Solanum sessiliflorum* D.) fruit peel. Food Hydrocolloids 86: 193–200.

Costa, C.A., Ramos, S.J., Alves, D.S., Fernandes, L.A., Sampaio, R.A. and Martins, E.R. 2008. Nutrição mineral do mangarito num Latossolo Vermelho Amarelo. Horticultura Brasileira, Brasília 26(1): 102–106.

Dalar, A. and Konczak, I. 2014. Cichorium intybus from Eastern Anatolia: Phenolic composition, antioxidant and enzyme inhibitory activities. Industrial Crops and Products 60: 79–85.

Dantas, E.N.A. 2019. Elaboração e caracterização física, físico-quimica, microbiológica e sensorial de cocada adicionada de maxixe (*Cucumis anguria* L.) saborizada com polpa de frutas. 42 f. Monografia (Bacharelado em Nutrição)—Centro de Educação e Saúde, Universidade Federal de Campina Grande, Cuité.

Dardiry, A.I.E., Ewis, A.M. and Abo-Srea, M.M. 2018. Impact of taro corms on functional low fat ice cream properties. Journ of Food and Dair. Scienc. [S. l.] 9(12): 399–402.

Deo, P.C., Tyagi, A.P., Taylor, M., Becker, D.K. and Harding, R.M. 2009. Improving taro (*Colocasia esculenta* var. esculenta) production using biotechnological approaches. The South Pacific Journal of Natural and Applied Sciences [S. l.] 27(1): 6–13.

Desai, D.O., Vortsman, D.O., Rdms, N. and Chary, M. 2017. Case studies in toxicology: always cook your boba. Emerg. Med. [S. l.] 49.

Dias, J.S.R., Mendes, F.Z.C., Nolasco, M.V.F.M. and Bogo, D. 2020. Obtenção de farinha de inhame para elaboração de barra de cereal como suplemento alimentar e funcional. Braz. J. of Develop., Curitiba, v. 6, n. 3, p.15716-15735,mar. 2020.

Du Thanh, H., Phan Vu, H., Vu Van, H., Le Duc, N., Le Minh, T. and Savage, G. 2017. Oxalate content of taro leaves grown in Central Vietnam. Foods (Basel, Switzerland) 6(1): 2.

El-Kholy, W.M., Aamer, R.A. and Ali, A.N.A. 2020. Utilization of inulin extracted from chicory (*Cichorium intybus* L.) roots to improve the properties of low-fat synbiotic yoghurt. Ann. of Agric. Scienc. 65(1): 59–67.

Ertop, M.H., Atasoy, R. and Akin, Ş.S. 2018. Evaluation of taro [*Colocasia esculenta* (L.) Schott] flour as a hydrocolloid on the physicochemical, rheological, and sensorial properties of milk pudding. Journ. of Food Process. and Preserv. 43.

Fares, L.C.P. 2010. Elaboração, caracterização físico-química, química, microbiológica e avaliação sensorial de doce em massa tradicional e diet de cubiu (*Solanum sessiliflorum* Dunal). 73 f. Dissertação (Mestrado em Ciências dos Alimentos)—Programa de Pós-graduação em Ciências dos Alimentos, Universidade Federal do Amazonas, Manaus.

Ferreira, F.B. 2020. Extração e purificação de compostos bioativos de cubiu ("*Solanum sessiliflorum*"), camu-camu ("*Myciaria dúbia*") e araçá-boi ("*Psidium myrtoides*") utilizando filtração por membranas com e sem promotores de turbulência. 133f. Tese (Doutorado em Engenharia Química)—Programa de Pós-graduação em Engenharia Química, Universidade Federal de Uberlândia, Uberlândia.

FAOSTAT. FAO Statistical Database. 2020. https://faostat.fao.org.

Foster-Powell, K., Holt, S.H. and Brand-Miller, J.C. 2002. International table of glycemic index and glycemic load values: 2002. Americ. Societ. for Clin. Nutri. 76(Especial): 5–56.

Furnaleto, K.A. 2015. Qualidade nutricional e aceitabilidade da geleia convencional e light de maná cubiu. 69 f. Dissertação (Mestrado em Agronomia)—Faculdade de Ciências Agronômicas, Universidade Estadual Paulista, Botucatu.

Giri, N.A. and Sajeev, M.S. 2020. Physico-mechanical and nutritional evaluation of taro (*Colocasia esculenta*) flour-based gluten-free cookies. Agricultural Research 9(1): 125–131.

Gomes, B.O. 2018. Redução do teor de gordura em hambúrguer pelo uso de prebiótico. 35 f. Monografia (Bacharelado em Engenharia de Alimentos)—Curso de Engenharia de Alimentos, Universidade Federal do Maranhão, Imperatriz.

Hegazy, A. 2019. Using of taro (*Colocasia esculent*) flour as a partial substitute of wheat flour in biscuit making. J. Biol. Chem. Environ. Sci. 14(1): 235–246.

Hiscock, L., Bothma, C., Hugo, A., Biljon, A.V. and Rensburg, W.S.J.V. 2020. Hedonic evaluation and check-all-that-apply (CATA) question for sensory characterisation of stewed vegetable *Amaranthus*. J. Food Sci. Technol. 57(2): 454–462.

Hong, G.P. and Nip, W.K. 1990. Functional properties of precooked taro flour in sorbets. Food Chemistry 36(4): 261–270.

Kinupp, V.F. and Lorenzi, H. 2014. Plantas alimentícias não convencionais (NCFP) no Brasil: guia de identificação, aspectos nutricionais e receitas ilustradas. São Paulo: Instituto Plantarum de Estudos da Flora.

Krisnaningsih, A.T., Radiati, L.E., Purwadi, H.E. and Rosyidi, D. 2019. The Effect of incubation time to the physicochemical and microbial properties of yoghurt with local Taro (*Colocasia Esculenta* (L.) Schott) starch as stabilizer. Curr. Res. in Nutr. and Food Scien. 7(2): 547–554.

Kumoro, A.C., Putri, R.D.A., Budiyati, C.S., Retnowati, D.S. and Ratnawati, D.S. 2014. Kinetics of calcium oxalate reduction in taro (*Colocasia Esculenta*) corm chips during treatments using baking soda solution. Procedia Chemistry 9: 102–112.

Leal, M.L., Alves, R.P. and Hanazaki, N. 2018. Knowledge, use, and disuse of unconventional food plants. Journal of Ethnobiology and Ethnomedicine 14: 6.

Leite, G.L.D., Silva, F.W.S., Jesus, F.M., Costa, C.A., Guanabens, R.E.M. and Gusmão, C.A.G. 2007. Efeito da adubação orgânica, espaçamento e tamanho de rizoma-semente sobre artrópodes em mangarito *Xanthosoma mafaffa* Schott. Arq. do Inst. Biol., São Paulo 74(4): 343–348.

Leonel, M. and Cereda, M.P. 2002. Caracterização físico-química de algumas tuberosas amiláceas. Ciênc. e Tecn. de Alim. 22(1): 65–69.

Liberalesso, A.M. 2019. O futuro da alimentação está nas Plantas Alimentícias Não Convencionais (NCFP)? 77 f. Dissertação (Mestrado em Agronegócios)—Programa de Pós-Graduação em Agronegócios do Centro de Estudos e Pesquisa em Agronegócios, Universidade Federal do Rio Grande do Sul, Porto Alegre.

Lima e Silva, L.F., Souza, D.C., Xavier, J.B., Samartini, C.Q. and Resende, L.V. 2019. Avaliação nutricional de caruru (*Amaranthus* spp.). Rev. Agrar. 12(45): 411–417.

Lima, A.S.G. and Krupek, R.A. 2016. Caracterização morfológica, anatômica, e toxinas endógenas em *Colocasia esculenta* (L.) Schott e *Xanthosoma sagittifolium* (L.) Schott. Lumin. 18(1): 31–40.

Liu, Y.H., Lin, Y.S., Liu, D.Z., Han, C.H., Chen, C.T., Fan, M. and Han, W.C. 2019. Effects of different types of yam (Dioscorea alata) products on the blood pressure of spontaneously hypertensive rats. Biosci. Biotechnol. Biochem. 73: 1371–1376. https://doi.org/10.1271/bbb.90022.

Lovera, M., Castro, G.M.C., Pires, N.R., Bastos, M.S.R., Araújo, M.L.H., Laurentin, A., Moreira, R.A. and Oliveira, H.D. 2020. Pyrodextrinization of yam (*Dioscorea* sp.) starch isolated from tubers grown in Brazil and physicochemical characterization of yellow pyrodextrins. Carbohydrate Polymers, 242.

Madeira, N.R., Silva, P.C., Botrel, N., Mendonça, J.L., Silveira, G.S.R. and Pedrosa, M.W. 2013. Manual de Produção de Hortaliças Tradicionais. Brasília, DF: EMBRAPA.

Maia, J.R.P., Schwertz, M.C., Sousa, R.F.S., Aguiar, J.P.L. and Lima, E.S. 2015. Efeito hipolipemiante da suplementação dietética com a farinha de cubiu (Solanum sessiliflorum Dunal) em ratos hipercolesterolêmicos. Rev. Bras. Pl. Med. 17(1): 112–119.

Matos, M.A. 2016. Bioprospecção do maxixe (*Cucumis anguria* L.). 168 f. Dissertação (Mestrado em Ciências Naturais e Biotecnologia)—Universidade Federal de Campina Grande, Cuité.

Melo, M.S.D.B. 2019. Mandioca, batata-doce e araruta como matérias-primas para obtenção de amidos fosfatados. 85 f. Tese (Doutorado em Agronomia)—Faculdade de Ciências Agronômicas, Universidade Estadual Paulista, Botucatu.

Miranda-Ramos, K.C., Sanz-Ponce, N. and Haros, C.M. 2019. Evaluation of technological and nutritional quality of bread enriched with amaranth flour. LWT – Food Scien. and Techn. 114.

Monhol, C., Costa, A.F., Galeano, E.A.V., Costa, H., Balbino, J.M.S., Rossi, D.A., Carvalho, D.R. and Piassi, M. 2018. Análise de custos da cultura do taro (*Colocasia esculenta* (L.) Schott) na região das montanhas capixabas: Estudo de caso. Intelletto 3(Especial): 43–50.

Montes, A.C.G. 2012. Evaluación fisicoquímica y capacidade antioxidante del aceite de amaranto (*Amaranthus hypochondriacus*) y estabilidade oxidativa de diferentes sistemas de encapsulación. 109 f. Tesis. (Maestro en Ciencia y Tecnología de Alimentos)—Facultad de Química, Universidad Autónoma de Querétaro, Santiago de Querétaro.

Morishita, M. 1988. Taro (Colocasia esculenta Schott.). pp. 322–338. *In*: BAJAJ, Y.P.S. (org.). Crops II. Berlin, Heidelberg: Springer Berlin Heidelberg.

NEPA - Núcleo de Estudos e Pesquisa em Alimentação. 2011. Tabela brasileira de composição de alimentos. 4ª. ed. rev. e amp. Campinas: NEPA – UNICAMP, 161 p.

Obidiegwu, J.E., Lyons, J.B. and Chilaca, C.A. 2020. The Dioscorea Genus (Yam) - An Appraisal of Nutritional and Therapeutic Potentials. Foods 9: 1304. https://doi.org/10.3390/foods9091304.

Owusu-Darko, P.G., Paterson, A. and Omenyo, E.L. 2014. Cocoyam (corms and cormels)—An underexploited food and feed resource. Journ. of Agric. Chem. and Environm. 3(1): 22–29.

Padhan, B., Mukherjee, A.K., Mohanty, S.K., Lenka, S.K. and Panda, D. 2019. Genetic variability and inter species relationship between wild and cultivated yams (*Dioscorea* spp.) from Koraput, India based on molecular and morphological markers. Physiol. Mol. Biol. Plants 25(5): 1225–1233.

Papalia, I.S. 2017. Substituição de amido de mandioca por amido de araruta e de arroz em pão de queijo congelado. 69 f. Dissertação (Mestrado Profissional em Tecnologia de Alimentos)—Programa de Pós-graduação em Tecnologia de Alimentos, Universidade Tecnológica Federal do Paraná, Londrina.

Perovic, J., Saponjac, V.T., Kojic, J., Krulj, J., Moreno, D.A., Garcia-Viguera, C., Bodroza-Solaroy, M. and Ilic, N. 2020. Chicory (*Cichorium intybus* L.) as a food ingredient—Nutritional composition, bioactivity, safety, and health claims: A review. Food Chem. 336.

Pinelli, L.C.D.F. 2018. Análise de dados multi-tabelas (CONDIM) para a avaliação de farinhas comestíveis por espectroscopia NIR e texturômetro. 44 f. Dissertação (Mestrado Profissional em Inovações Tecnológicas)—Programa de Pós-graduação em Inovações Tecnológicas, Universidade Tecnológica Federal do Paraná, Campo Mourão.

Rabiou, M.M., Tchicama, M.M., Sabo, H., Maazou, M.S. and Sadou, H. 2019. Chemical composition of taro and analysis of the level of its consumption in the city of Niamey, Niger: Case of Madina and Koira Tégui district. Food Scien. and Nutr. Res. 2(1).

Ramos Filho, M.M., Ramos, M.I.L. and Hiane, P.A. 1997. Avaliação química do inhame (*Colocasia esculenta* L. schott) cultivado em solo alagadiço da região pantaneira de Mato Grosso do Sul. Bol. Cent. de Pesq. de Process. de Alim. 15(2): 175–186.

Repolho, R.P.J., Oliveira, W.C., Carvalho, A.P., Sanches, A.G. and Sousa, J.T.R. 2019. Application of edible coatings in conservation of acerola. Appl. Res. & Agrotec., Guarapuava-PR 12(2): 59–69, May–Aug.

Ribeiro, T.P.S. and Durigan, M.F.B. 2018. Produtos alimentícios a base de cubiu (*Solanum sessiliflorum* Dunal) como oportunidade para a agroindústria. Rev. Amb.: Gest. e Desenv. 11(1): 241–250.

Sá, A.R.A., Lima, M.B., Silva, E.I.G., Mendes, M.L.M. and Messias, C.M.B.O. 2018. Caracterização físico-química e nutricional de farinhas obtidas de inhame (Dioscorea spp.) e taro (Colocasia esculenta) comercializados em Petrolina-PE. Rev. Saúde (Sta. Maria) 44(3).

Sabir, A., Reddy, D.K., Waghray, D.K., Bhaskar, V. and Maloo, S. 2017. Extending the use of a highly nutritious underutilised taro (*Colocasia esculenta*) tuber by development of tutti fruity. World Journ. of Engin. Resear. and Techn. 3(4): 503–516.

Santos, R.O. 2017. Efeito da adição de farinha de araruta (*Maranta arundinacea* L.), nas propriedades físico-químicas, reológicas e funcionais de sobremesa láctea sabor baunilha. 75 f. Dissertação (Mestrado em Ciências)—Programa de Pós-graduação em Ciência e Tecnologia de Alimentos, Seropédica.

Sarker, U. and Oba, S. 2019. Nutraceuticals, antioxidant pigments, and phytochemicals in the leaves of *Amaranthus spinosus* and *Amaranthus viridis* weedy species. Nature, Sci. Rep. 9: 20413.

Saxby, S., Tipton, L., Lee, C., Wang, L., Zhang, H., Jia, W., Boushey, C. and Li, Y. 2020. Prebiotic potential of taro (*Colocasia esculenta*) to modulate gut bacteria composition and short chain fatty acid production. Current Developments in Nutrition 4(Supplement 2): 1582.

Sharma, H.K., Kaushal, P. and Singh, B. 2016. Taro: technological interventions. pp. 325–413. *In*: Sharma, H.K., Njintang, N.Y., Singhal, R.S. and Kaushal, P. (org.). Tropical Roots and Tubers: Production, Processing and Technology.

Silva, A.J.P. 2019. Desenvolvimento de negócio a partir do inhame como matéria-prima e suas diversas possibilidades de produto final. 83 p. Dissertação (Mestrado em Administração de Negócios) - Universidade Presbiteriana Mackenzie, São Paulo.

Silva, E.S. 2019. Interação genótipo x ambiente de componentes agronômicos e físico-químicos de feijão-macuco (*Pachyrhizus* spp.) em terra firme e várzea na região metropolitana de Manaus-AM. 81 f. Tese (Doutorado em Agronomia Tropical)—Faculdade de Ciências Agrárias, Universidade Federal do Amazonas, Manaus.

Silva, F.T.S. 2018. Desenvolvimento e caracterização de biofilmes ativos de amido de araruta aditivados com óleo essencial de *Thymus vulgaris*. 72 f. Dissertação (Mestrado em Ciências dos Materiais)—Programa de Pós-graduação em Ciências dos Materiais, Universidade Federal de Pernambuco, Recife.

Silva, R.A.M. 2017. Diversidade de acesso de maxixe (*Cucumis anguria* L.) do Norte-Nordeste brasileiro. 105 f. Tese (Doutorado em Fitotecnia)—Programa Pós-graduação em Fitotecnia, Universidade Federal Rural do Semi Árido. Mossoró.

Silva, E.S. 2015. Avaliação de progênies de feijão-macuco (*Pachyrhizus tuberosus* (lam.) *Spreng.*), Fabaceae, com base na nodulação natural, caracteres associados e processamento de raízes tuberosas. 82 f. Dissertação (Mestrado)—Instituto Nacional de Pesquisas da Amazônia – INPA, Manaus.

Silva, E.E. 2011. A Cultura do Taro - Inhame (*Colocasia esculenta* (L.) Schott): Alternativa para o Estado de Roraima. *In*: Documentos 51. Embrapa Roraima.

Simsek, S. and El, S.N. 2015. *In vitro* starch digestibility, estimated glycemic index and antioxidant potential of taro (*Colocasia esculenta* L. Schott) corm. Food Chem. 168: 257–261.

Siskawardani, D.D., Warkoyo, Hidayat, R. and Sukardi. 2020. Physic-mechanical properties of edible film based on taro starch (*Colocasia esculenta* L. Schoott) with glycerol addition. IOP Conference Series: Earth and Environmental Science 458: 12039.

Sit, N., Deka, S.C. and Misra, S. 2015. Optimization of starch isolation from taro using combination of enzymes and comparison of properties of starches isolated by enzymatic and conventional methods. Journ. of Food Scien. and Techn. 52(7): 4324–4332.

Solagna, T., Rodrigues, B.M., Santos, N.M., Candido, C.J., Santos, E.F. and Novello, D. 2016. Pretzels integrais adicionados de inulina: Avaliação físico-química e sensorial entre crianças. Rev. da Univers. Vale do Rio Verde 14(1): 212–226.

Souza, L.A.G., Silva Filho, D.F., Benavente, C.A.T. and Noda, H. 2018. Ciência e tecnologia aplicada aos agroecossistemas da Amazônia Central. Manaus: Editora INPA, 283 p.

Sprey, L.M., Ferreira, S.A.N. and Sprey, M.M. 2019. Physiological quality of pelleted Cubiu (*Solanum sessiliflotum* Dunal) seeds. Rev. Bras. de Frutic. 41(1).

Teixeira, V.M. 2017. Desenvolvimento de queijo de coalho com potencial prebiótico e simulação do processo industrial. 105 f. Dissertação (Mestrado em Engenharia Química)—Programa de Pós-graduação em Engenharia Química, Universidade Federal de Pernambuco, Recife.

Teixeira, B.A. 2018. Bioprodução de fitoquímicos em plantas alimentícias não convencionais (NCFP) nas quatro estações do ano. 50 f. Dissertação (Mestrado em Ciências Agrárias)—Programa de Pós-graduação em Ciências Agrárias, Universidade Federal de São João Del-Rei, Sete Lagoas.

Vidigal, S.M., Lopes, I.P.C., Puiatti, M., Sediyama, M.A.N. and Ribeiro, M.R.F. 2016. Yield performance of taro (*Colocasia esculenta* L.) cultivated with topdressing nitrogen rates at the Zona da Mata region of Minas Gerais. Rev. Ceres 63(6): 887–892.

Wu, Z.G., Jiang, W., Nitin, M., Bao, X.Q., Chen, S.L. and Tao, Z.M. 2016. Characterizing diversity based on nutritional and bioactive compositions of yam germplasm (*Dioscorea* spp.) commonly cultivated in China. Journal of Food and Drug Analysis 24: 367–375.

Wulandari, Y. and Warkoyo, N.H. 2018. Characterization of edible film from starch of taro (*Colocasia esculenta* (L.) Schott) with addition of chitosan on dodol substituted seaweed (*Eucheuma cottonii* L.). Food Techn. and Halal Scien. Journ. 1(1): 22–32.

Xavier, J.B., Andrade, D.B., Souza, D.C., Guimarães, G.C., Resende, L.V. and Guimarães, R.M. 2019. Morphological, chemical and physiological characterization of *Amaranthus* spp. seeds. Journ. of Seed Scien. 41(4): 478–487.

Yoshime, L.T. 2007. Caracterização nutricional do mangarito (*Xanthosma mafaffa* Schott). 90 f. Dissertação (Mestrado em Ciências Farmacêuticas)—Faculdade de Ciências Farmacêuticas, Universidade Federal de São Paulo, São Paulo.

Yuyama, L.K.O., Macedo, S.H.M., Aguiar, J.P.L., Filho, D.S., Yuyama, K., Favaro, D.L.T. and Vasconcellos, M.B.A. 2007. Quantificação de macro e micro nutrientes em algumas etnovariedades de cubiu (*Solanum sessiliflorum* Dunal). Acta Amazôn. 37(3): 425–430.

Chapter 4

Biotechnological Applications of Flours and By-products of *Mauritia flexuosa* L. Fruits

Márcia Luiza dos Santos Beserra Pessoa,[1]
Ennya Cristina Pereira dos Santos Duarte,[1]
Thially Braga Gonçalves,[1] *Vivianne Rodrigues Amorim,*[2]
Lívia Maria Nunes de Almeida,[2] *Carla Lorena Silva Ramos,*[2]
Paulo Michel Pinheiro Ferreira[2] and *Joilane Alves Pereira-Freire*[1,*]

Introduction

Dyslipidemia, nowadays, is a disease that is also occurring in young people and, consequently, many other chronic diseases, due to changes in lifestyle, including physical inactivity and inadequate eating habits, in addition to other genetic and environmental factors (Ramirez-Moreno, 2022).

It is known that diets with low antioxidant content and high consumption of ultra-processed foods, sedentary lifestyle and increased oxidative stress are related to the increase of chronic diseases, as well as inflammatory processes, obesity, dyslipidemias and cancer (Amorim and Ferreira, 2020; Amorim et al., 2021). In this context, the products of plant origin derived from fruits emerge in current scientific research as relevant therapeutic resources. Thus, the interest and consumption of products containing Biologically Active Compounds (BACs) with functional properties is increasing (Carmona-Hernandez et al., 2021).

[1] Research Laboratory II, Federal University of Piauí, *Campus* Senador Helvídio Nunes de Barros, 64607-670, Picos, Piauí, Brazil.

[2] Laboratory of Experimental Cancerology, Department of Biophysics and Pharmacology, Federal University of Piauí, *Campus* Ministro Petrônio Portella, 64049-550, Teresina, Piauí, Brazil.

* Corresponding author: joilane@ufpi.edu.br

Among the Brazilian native plants containing BACs, *Mauritia flexuosa* (Arecaceae), popularly known as buriti, stands out. Its fruit has promising nutritional uses and is considered a functional food because of its nutritional composition, such as carotenoids, vitamin A, C and E, phenolic compounds, iron, calcium, essential fatty acids and fiber (Igbal, 2014; Siqueira et al., 2014; Aquino et al., 2015; Pereira-Freire et al., 2016, 2018).

Due to nutrients and other bioactive compounds identified in the buriti fruit, it has high potential for the development of food products with high added value, functional properties, pharmacological and relevant characteristics biotechnology.

Considering the nutritional value of buriti, the objective of this chapter is to focus on the biotechnological and pharmacological potential of nutrients and bioactive compounds identified in the buriti fruit (*Mauritia flexuosa* L.). This chapter discusses how these compounds and properties are beneficial to health and the future prospects of biotechnological applications of the fruit.

1. Buriti (*Mauritia flexuosa* L.)

Mauritia flexuosa (Fig. 4.1A) is a palm tree that occurs naturally in areas where the soil remains humid during all seasons of the year, such as river margins, streams, lakes and pathways. It is known by several popular names, such as "buriti", "miriti", "muriti", "palmeira-dos-brejo", "moriche", "carangucha", and "aguaje", and it

Figure 4.1. (A) Adult tree, (B) Whole fruit (C) and Parts of the fruit: epicarp (peel), mesocarp (pulp), endocarp and seed of *Mauritia flexuosa* L.

changes locally in states and/or geographic Brazilian regions (Sampaio, 2011; Martins et al., 2012; Pereira-Freire et al., 2016).

The fruit (Figs. 4.1B and 4.1C) has an elliptical oval structure, averaging 5 to 7 cm in length and 4 cm in circumference, and is surrounded by an epicarp (or peel), composed of reddish-brown triangular scales (Gazel Filho and Lima, 2001; Martins et al., 2012). It contains on average 9 g of fresh pulp, 12 g of scales, 21 g of kernels and 10 g of endocarp (Cymerys et al., 2005). Its mesocarp (pulp) varies from yellow to orange, is fleshy and oily (Gazel Filho and Lima, 2001; Martins et al., 2012), and rich in carotenoids, vitamins, unsaturated fatty acids and fiber (Cruz et al., 2020).

Buriti (*Mauritia flexuosa* L.), is considered the most abundant palm tree in Brazil, with wide distribution in the Amazon and Cerrado biomes (Pezoti et al., 2014; Resende et al., 2019; Rudke et al., 2019). According to Sampaio (2012), besides the Cerrado (savanna from Central Plateau) and Amazon (tropical forest) biomes, it is also present in the Caatinga (semiarid) and Pantanal (marshlands), with more abundant occurrence in the Brazilian Northeast and Central-west regions.

According to Pezoti et al. (2014), Resende et al. (2019) and Rudke et al. (2019), the fruits produced are coated with a thin layer of yellow pulp, which can be consumed as juices, sweets, ice cream, popsicles, jams and wines (Cândido et al., 2015; Milanez et al., 2018; Cruz et al., 2020).

Up to now, a diversity of compounds have been identified in the buriti fruit, including phenols and flavonoids, such as ferulic acid ($C_{10}H_{10}O_4$), caffeic acid ($C_9H_8O_4$), protocatechuic acid ($C_7H_6O_4$) chlorogenic acid ($C_{16}H_{18}O_9$) and quinic acid ($C_7H_{12}O_6$) and flavonoids, such as (+)-catechin ($C_{15}H_{14}O_6$), (-)-epicatechin ($C_{15}H_{14}O_6$) apigenin ($C_{15}H_{10}O_5$), luteolin ($C_{15}H_{10}O_6$), myricetin ($C_{15}H_{10}O_8$), kaempferol ($C_{15}H_{10}O_6$) and quercetin ($C_{15}H_{10}O_7$) (Bataglion et al., 2014; Pereira-Freire, 2016; Pereira-Freire et al., 2018).

According to Pereira-Freire et al. (2018), the pulp (mesocarp), peel (epicarp) and endocarp can be used to produce flours (Fig. 4.2), and have been good options for new food ingredients in the production of new food and pharmaceutical materials. These flours have shown to have several pharmacological and biological activities, including antioxidant properties. Gomes et al. (2020) and Pereira-Freire et al., 2022, demonstrated in their study with buriti flour that it can be considered an alternative for food consumption and preparation of by-products. From the buriti fruit, there are also the scrappings of buriti (obtained by solar drying the scrapped pulp of the fruit) and the peanut of buriti, when the scrapings are mixed with some manioc flour and brown sugar.

Sampaio (2011) described this oil as edible and traditionally used by Brazilians to help heal wounds and burns, relieve pain from insect and snake bites and to ease respiratory problems. Sousa et al. (2010) and Cândido et al. (2015) reported that buriti oil has high levels of monounsaturated fatty acids, tocopherols that protect against the harmful effects of ultraviolet radiation, carotenoids (precursors of vitamin A) and vitamin E, fatty acids with predominance of oleic and palmitic and antioxidants and high antioxidant capacity. Such a composition suggests a good perspective in the use of this product as a therapeutic and cosmetic alternative (Rosso and Mercadante, 2007) and physiological properties (Aquino et al., 2015).

Figure 4.2. Illustration of buriti fruits after the drying process and production of flour from the pulp, peel and endocarp. Source: Pereira-Freire et al., 2018.

2. *Mauritia flexuosa* and its health benefits and dyslipidemia reduction

Studies developed in recent years have identified the existence of a direct association between hyperlipidemia and the occurrence of oxidative liver damage, indicating that hyperlipidemia negatively affects the antioxidant defense systems (Wong et al., 2012; Dahecha et al., 2013; Pereira-Freire et al., 2018; Melo-Cavalcante et al., 2019; Santos et al., 2019).

Among the most accepted and effective treatments for reducing dyslipidemia is the use of 3-hydroxy-3-methylglutaryl coenzyme A reductase (HMG-CoA reductase) inhibitors and statins. However, these classes of drugs often cause muscle damage within weeks or even years after the start of treatment (Faludi et al., 2017; Baigent et al., 2010; Bonfim et al., 2015). In addition to the pharmacological approach to control the lipid profile, dietary changes and ingestion of low-fat foods, continuous physical exercise with body weight control, restriction of alcohol, salt consumption and discontinuation of tobacco use are also recommended (Vaz, 2016).

In phytotherapy, compounds from plants are used as therapeutic ingredients, with applications in health promotion, disease prevention, diagnosis and treatment, encompassing dietary advice and guidance on lifestyles. It is known that increased lipids, poor diet, sedentary lifestyle and increased oxidative stress are related to the increase of chronic pathologies (cancer, dyslipidemias, inflammatory states) (Nogueira and Quintans, 2019). In this context, for the treatment of dyslipidemia it is very common to make therapeutic use of products of plant origin that contain Biologically Active Compounds (BACs) with functional properties (Nonato et al., 2015).

Research carried out with parts of the fruit of *M. flexuosa* has shown to reduce blood concentrations of LDL, triglycerides, total cholesterol and also indirectly reduce the atherogenic index (Pereira-Freire et al., 2018; Amorim et al., 2021), which is very important for reducing the risk factors for cardiovascular diseases, since studies show that reducing blood concentrations of LDL decreases the chance of cardiovascular events both in the primary and secondary prevention stage. The reduction of risk factors in association with the increase in HDL-C levels establishes ideal conditions for the inhibition of the atherosclerotic process (Silverman et al., 2016; Pereira-Freire et al., 2018; Amorim and Ferreira, 2020; Noce et al., 2021).

Other studies have shown that fatty acids present in the buriti fruit are typically unsaturated, which can favorably reduce serum lipids (Manhães and Sabba Srur, 2011; Aquino et al., 2012; Pereira-Freire et al., 2022), and demonstrating that buriti oil has high amounts of oleic acid (n-9) and approximately four times more linolenic acid (n-3) than the olive oil (Manhães and Sabaa-Srur, 2011). In addition, young rats supplemented with refined buriti oil showed a reduction in total cholesterol, LDL, triglycerides and TGO levels compared to those fed with a diet containing crude oil (Aquino et al., 2012; Milanez et al., 2016; Pereira-Freire et al., 2018), which indicates the BACs in buriti oil are often associated with the reduction of serum cholesterol and oxidative stress (Fig. 4.3).

An extensive research developed by Pereira-Freire et al. (2018) showed that pulp, pell and endocarp have significant amounts of phenolic compounds, especially chlorogenic and protocatechuic acids (Fig. 4.4). Both compounds are able to reduce fatty acid synthesis by inhibiting the activity of HMG-CoA reductase, which impacts on the lipid metabolism and levels of obesity-related hormones such as leptin and adiponectin (Amorim et al., 2021; Cho et al., 2010). Such phytochemical characterization of methanolic extracts of pulp, peel and endocarp by high performance liquid chromatography described in Fig. 4.4 (A, B and C) and Table 4.1 also found other phenolic compounds, as quercetin, apigenin, catechin and epicatechin) with the following retention times at 230 nm: 16.3, 33.6, 41.7, 53.6 and 49.3 min, respectively (Table 4.1).

Earlier investigations also described that natural sources rich in β-carotene (precursor of vitamin A), flavonoids, tocopherols and ascorbic acid such as *Mauritia flexuosa* (Romero et al., 2015; Pereira-Freire et al., 2016) are promising options in

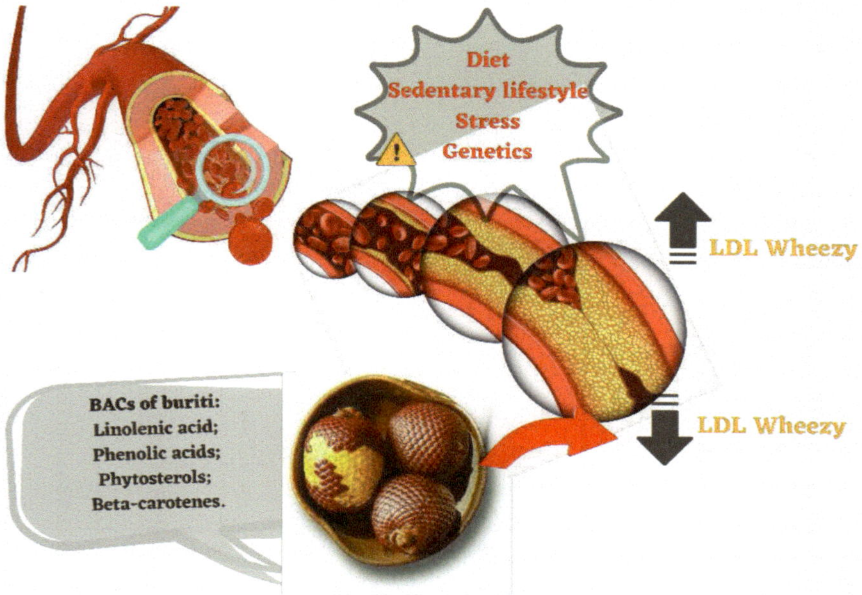

Figure 4.3. Illustration of BACs present in buriti fruit associated with reduction of serum cholesterol and oxidative stress. Source: self-authored.

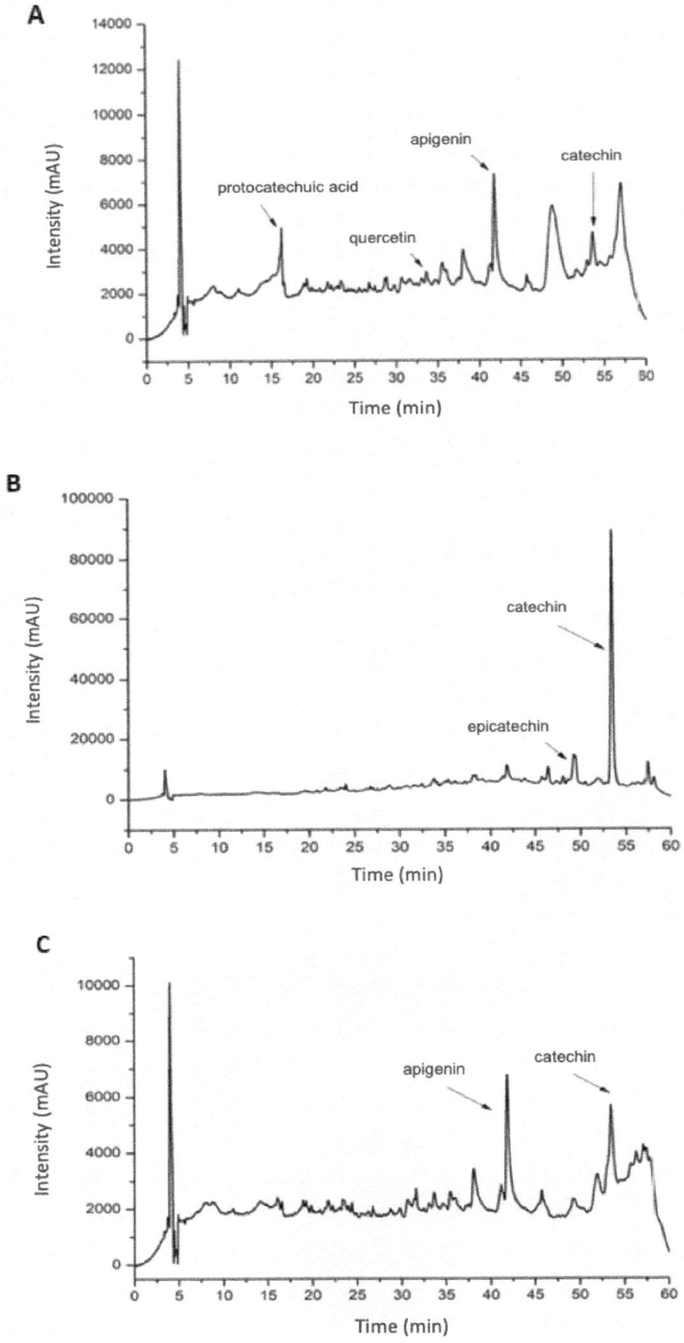

Figure 4.4. Chromatographic profile obtained by CLAE-DAD of the methanolic extract of (A) pulp, (B) peel and (C) endocarp (230 nm).

Table 4.1. Identification of compounds by HPLC in samples of the buriti fruit. Source: Pereira-Freire et al., 2018.

IUPAC name	Chemical structure	Class	Retention time (min)	Sample
3,4-Dihydroxybenzoic acid (Protocatechuic acid)	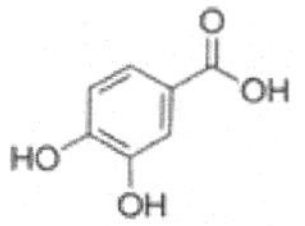	Phenol	16,3	Pulp
2-(3,4-dihydroxyphenyl)-3,5,7-trihydroxyichromen-4-one (quercetin)	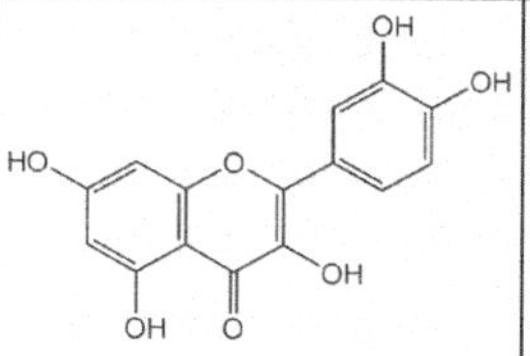	Flavonoid	33,6	Pulp
4′,5,7-trihydroxyflavone (*apigenin*)	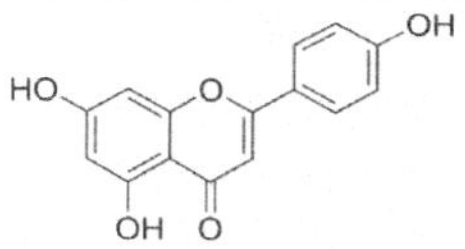	Flavonoid	41,7	Pulp Endocarp
(-)-trans-3,3′,4′,5,7-pentahydroxyflavone, (2S,3R)-2-(3,4-dihydroxyphenyl)-3,4-dihydro-1(2H)-benzopyran-3,5,7-triol (catechin)	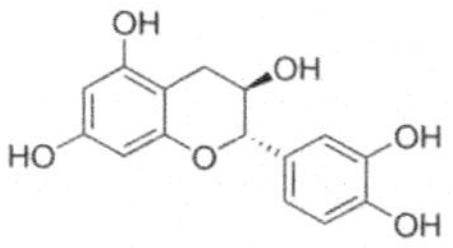	Condensed Tannin	53,6	Endocarp Bark Pulp
(-)-cis-3,3′,4′,5,7-pentahydroxyflavone, (2R,3R)-2-(3,4-dihydroxyphenyl)-3,4-dihydro-1(2H)-benzopyran-3,5,7-triol (epicatechin)	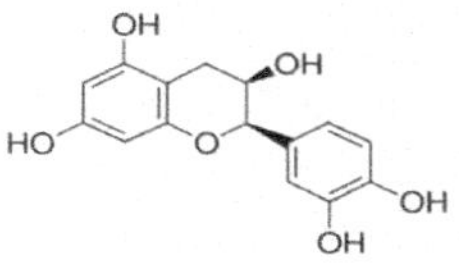	Condensed tannin	48,3	Bark

restoring health by improving antioxidant enzyme profile, lipid metabolism, blood pressure and organics biomarkers (liver, kidneys and heart), as well as reducing dyslipidemia.

Finally, it should be noted that the consumption of foods rich in soluble fibers can help to diminish total cholesterol (Mudgil and Barak, 2013) and buriti bioproducts, especially the pulp, have shown relevant amounts of soluble fibers (Carneiro, 2016; Pereira-Freire et al., 2022), which provides viscosity and ability to form gels that act as emulsifiers, improving the lipid profile.

3. Future perspectives and biotechnological applications of the fruit of *Mauritia flexuosa* L.

Taking into account the composition of the buriti fruit and its possible applications, researchers have been expanding the studies that aim to evaluate the possibility of using this fruit and by-products as an alternative local non-expensive source

(Camelo-Silva et al., 2020). So, we performed a (bio)technological prospection at the European Patent Office (EPO), the World Intellectual Property Organization (WIPO), the United States Patent and Trademark Office (USPTO), the National Institute of Industrial Property (INPI), the Canadian Intellectual Property Office (CIPO) as well as at the Latin American Patent Bank (LATIPAT), using the descriptors 'buriti' and '*Mauritia flexuosa*'. Only filed patents published until August 2021 mentioning these terms in their abstract or title and related to the therapeutic and nutritional application of buriti were reported, excluding duplicates ones (Table 4.2).

Of the 643 patents found in databases, only 16 were related to the therapeutic and/or nutritional application of buriti as defined in this study, from the year 1999 until the present, with low annual records of patent deposits, but with an increase in the number of these applications from the year 2014 (Table 4.2).

According to the International Patent Classification (IPC), 75% of the selected patents are classified in the human needs session (A61K/Q), which considers medical, dental and cosmetic compositions or similar preparations for personal hygiene purposes, like the patent BRPI0303404-6 B1, US20120251602 and BR102018 0743244A2, which use the oil extracted from the buriti pulp (*M. flexuosa* and/or *Mauritia vinifera*) for preparation of cosmetic, hair and pharmaceutical formulations, as a sun protection enhancer, as well as a source of carotenoids (pro-vitamin A) and vitamin E (tocopherols-natural antioxidants) (Nonato et al., 2021; Chaves et al., 2018). The patent DE102008044701 refers to various cosmetic compositions such as lip gloss, eyeshadow, among others (Table 4.3) (Leitner et al., 2010).

Many patents were classified in session A23L/D, related to food or food products and their processing. The patent BRPI9904761-6 deals with a nutritional complement composed of a balanced and complete mix of vitamins, mineral salts, enzymes, coenzymes, catalysts and other trace elements found exclusively in natural sources such as the mesocarp of buriti (Oliveira et al., 2001); the patent BRPI0800470, in which a food product was formulated based on ingredients selected according to their nutritional value and functional characteristics and contains the dehydrated buriti pulp flour to incorporate healthier habits, suggesting functional benefits to prevent non-transmissible chronic diseases (Table 4.3) (Rodrigues et al., 2021).

A smaller number of patents were classified in session C11D, which refers to detergent and soap compositions derived from vegetable or animal oils. One patent at session C08 - BRPI0403407-4 B1—deals with chemical preparations or processes of organic macromolecular compounds (Table 4.3) (Murta et al., 2006).

In recent years, researchers have also been expanding studies to evaluate the uses of the buriti fruit and its by-products as an alternative for emerging issues (Table 4.3), considering recycling waste that would be discarded, generating and utilizing food products rich in nutrients with great positive impacts on the environment (Soares et al., 2011; Pereira-Freire et al., 2018; Amorim et al., 2021).

In this perspective, the method for elaboration of flours from the buriti peel can be mentioned. In a study that aimed to characterize the composition, total energy value, content of phenolic compounds, vitamin C, total carotenoids and anthocyanins of the buriti peel *in natura* and dried in an oven at 55°C for the preparation of flours, it was noted that this process was efficient and did not denature the analyzed constituents

Table 4.2. Search results regarding patent deposits using the descriptors buriti and *Mauritia flexuosa*. Source: Self-Authored.

	Patent	Class*	Country	Year	Title	Author
1	**PI 9904761-6 A2**	A23L 33/00	BRA	1999	Natural nutritional supplement based on alternative foods and process of its preparation	Oliveira et al., 2001
2	**PI 0301845-8 A2**	A61K 7/047	BRA	2003	Enamel removal composition for nails	Alvarenga, 2004
3	**PI 0303404-6 A2**	A62K 7/42	BRA	2003	Use of Mauritia palm pulp oil as a sun protection potential and as a source of carotenoids (pro-vitamin a) and vitamin E (tocopherols) natural antioxidants in cosmetics, hair and pharmaceutical products	Barrera-Arellano et al., 2005
4	**PI 0500512-4 A2**	A23L 2/04	BRA	2005	Preservation of oleaginous fruit juice	Armando, 2006
5	**PI 0800470-6 A2**	A23L21/12 A23L35/00 A23L7/10	BRA	2008	Cereal bar with banana and dehydrated buriti pulp	Srur and Manhães, 2009
6	**BRPI0905343A2**	A23G9/04 A23G9/36 A23G9/42	BRA	2009	Manufacturing process of symbiotic, fruit-based, non-dairy edible ice cream, edible ice products and their use	Saad and Vasconce-los, 2011
7	**BRPI1000759A2**	A23D9/007 A23D9/013	BRA	2010	Vegetable-based preparation used as a nutritional supplement	Vasconcelos, 2010a
8	**WO201111 3128**	A23D 9/00 A61K 31/20 A61K 31/201	USA	2011	Fish- and vegetable oil-based preparations used as supplements	Vasconcelos, 2010b
9	**BR102013 033412 0 A2**	A61K 9/51 A61K 31/567 A61K 47/44 A61K 47/14 B82Y 5/00	BRA	2013	Lipid carrier nanoparticle and its use	Caballero et al., 2015
10	**BR102014 020266 8 A2**	C12P 7/64 C12N 9/20 A61P 31/00 A23D 9/00 A61K 8/00	BRA	2014	Enzymatic interesterification process, interesterified lip composition, its uses, pharmaceutical composition, food product and cosmetic composition	Macedo et al., 2016
11	**BR102016 0034850 A2**	A23L 7/126	BRA	2016	Cereal bar and manufacturing method of cereal bar made from buriti mesocarp and pumpkin seeds	Queiroz et al., 2017

Table 4.2 contd. ...

...Table 4.2 contd.

	Patent	Class*	Country	Year	Title	Author
12	**BR102016 015728 5 A2**	A61K 8/92 A61K 8/34 A61K 8/36 A61Q 19/10	BRA	2016	Antimicrobial liquid soap based on buriti oil	Soares et al., 2018
13	**BR102016 020796 7 A2**	A23G 3/48	BRA	2016	Fruit pulp bullets and their production process	Boeira, 2018
14	**BR102017021942**	A23G 3/00	BRA	2017	Jelly candies without sugar and with fruit pulp	Boeira, 2019
15	**BR102017 013407 5 A2**	A61K 36/889 A61K 38/28 A61K 47/10 A61P 3/10	BRA	2017	Poloxamer 407, vegetable oil and insulin, thermal-eversible micella system, production process, pharmaceutical composition and uses	Barichello et al., 2019
16	**BR102017015017A2**	A21D13/043A21D13/06 A21D2/36-	BRA	2017	Cakes enriched with buriti	Silva et al, 2019
17	**BR132018 003837 0 E2**	A61K 36/63 A61K 36/185 A61K 36/88	BRA	2018	Manufacturing process of formulation of mineral, animals and various ozonized vegetable oils and product obtained	Oliveira, 2019
18	**WO2019164651A1**	A23C20/00 A23D7/005 A23D7	USA	2019	Edible product comprising plant oils or creams and cooked vegetables	Altman and Mcclure, 2019
19	**BR102018 005239 0**	A61K 36/889	BRA	2019	Medicinal, cosmetic and food formulations from the epicarp of *Mauritia flexuosa*	Ferreira et al. 2019
20	**BR102018 016091 5 A2**	A61K 36/889 A61K 127/00 A61P 31/04 A61P 31/10	BRA	2020	Antimicrobial herbal medicine obtained from the extract of the leaves of *Mauritia flexuosa*	Rocha et al., 2020
21	**BR102018 074324 4 A2**	A61K 8/97 A61K 8/92 A61K 8/02 A61Q 17/00	BRA	2020	Edible sunscreens (bars and capsules) able to promote the increase of the SPF of the skin	Longo, 2020
22	**BR102020 003381 6 A2**	A23P 20/00	BRA	2020	Active packaging based on gelatin with buriti oil (mauritia flexuosa l.f.) and the process to prepare such packaging	Franceschi et al., 2021
23	**BR102020 006660 9 A2**	A23B 7/04 F24S 20/67	BRA	2020	Raft and process for production of frozen fruit pulp	Bertolini, 2020

Table 4.3. Patents related to the therapeutic or nutritional properties of *Mauritia flexuosa*. Source: Self-Authored.

	Bases					
Descriptors	**INPI**	**USPTO**	**WIPO**	**EPO**	**LATIPAT**	**CIPO**
Buriti	35	47	63	184	36	40
Mauritia flexuosa	13	33	41	119	15	18
Total	48	80	104	303	51	58

and promoted a significant increase in carbohydrates, total energy value, vitamin C, carotenoids and total phenolic compounds of the evaluated samples. Thus, it could be a potential for the production of new foods, especially for bakery products (Morais et al., 2019; Chu, 2022).

Another study aimed to prepare and characterize a bread enriched with *Mauritia flexuosa* flour. This study proposes that it could be added to the daily diet since its by-products present adequate values of lipids and proteins, providing a high caloric intake (Gomes et al., 2020). In addition, recent studies have also indicated that burti by-products have promising applications for the treatment of diabetes and cholesterol, as a cicatrizant, photoprotective and anti-inflammatory agent, in the production of biodiesel and activated charcoal, besides being considered a potential bactericidal agent against some pathogenic bacteria (Pereira-Freire et al., 2016; Camelo-Silva et al., 2020; Amorim et al., 2021). Table 4.4 shows some of the most recent scientific research and biotechnological uses.

Thus, it is essential to emphasize that *M. flexuosa* fruits have an amazing marketing potential from culinary to the medical field. However, there is much to be investigated about it and many questions must be answered to make available buriti biotechnological products to the final consumer.

Conclusions

Fruits from *Mauritia flexuosa* has shown several properties: *in vitro* and *in vivo* cytotoxic studies tested as chemotherapeutic agents; bakery products for enrichment and functional potential; antioxidant action and photoprotective effects or as new pharmaceutical formulation with antimicrobial, healing and anti-inflammatory option; as an alternative to other oils for the preparation of foods. However, *in vivo* studies are necessary to confirm this hypocholesterolemic and hypoglycemic effects. Consequently, it is extremely important to explore the nutritional, functional, physicochemical and phytochemical features of the fruits for human/livestock foods and to install biofriendly techniques and sustainable biotechnology handling of natural resources. For Brazilian local communities, it is important to reuse residues, especially for people from poor regions as a way to give better opportunities and improve the quality of life.

Table 4.4. Recent biotechnological applications and therapeutic or nutritional activities of *Mauritia flexuosa*. Source: Self-Authored.

Aims	**Results**	**Reference**
To evaluate the influence of adding crude or refined buriti oil in the diets of young rats on biochemical parameters and status of vitamins A and E.	The diet with buriti oil reduced the total cholesterol, LDL, triglycerides and aspartate transaminase enzyme levels	Aquino et al., 2015
To evaluate the cytotoxic activity of *M. flexuosa* fruits in tumor cells and healing activity *in vitro*	Nanoemulsions are potential adjuvants in breast cancer therapy and also show potential as a healing drug.	Sampaio, 2017
To investigate the phytochemical profile and antioxidant potential of *M. flexuosa* fruits and determine the bioaccessibility of phenolic compounds.	The epicarp (shell) revealed higher values of phenols, flavonoids, carotenoids, tannins and ascorbic acid when compared to pulp and endocarp. All samples showed the ability to sequester free radicals and phenolic compounds exhibited reduced bioaccessibility after simulated gastrointestinal digestion *in vitro*. Protective effect of rat blood cells against free radical-induced lysis.	Pereira-Freire et al., 2018
To study the effect of chitosan gel containing buriti oil as a healing agent	The chitosan gel and buriti oil promoted faster retraction and complete healing of the skin lesion.	Ferreira et al., 2020
To investigate the antidiabetic effect of buriti oil	Buriti oil at low concentrations (15 to 30 mg/mL) promoted the inhibition of α-amylase activity.	Oliveira et al., 2020
To investigate the effect of buriti oil on SARS-CoV-2 virus	Molecular docking and molecular dynamics interactions analyses revealed that compounds 13-cis-β-carotene, 9-cis-β-carotene and α-carotene exhibited interactions with 2GTB peptidase, suggesting promising candidates in the preparation of new drugs against SARS-CoV-2.	Costa et al., 2021
To study the photoprotective effect of nanoemulsions containing buriti oil.	Formulation with buriti oil was more stable, and exhibited greater photoprotection	Mansur et al., 2020
To evaluate the wound healing potential of a topical formulation containing buriti oil in a liquid crystalline phase consisting of murumuru butter.	The formulation containing 15% of *Mauritia flexuosa* oil exhibited a significant effect on the process of epithelial regeneration, causing early re-epithelialization of the injured tissue.	Silva et al., 2021
To investigate the oral anti-inflammatory activity of epicarp, mesocarp and endocarp obtained from *M. flexuosa* fruits using *in vivo* models to verify physiological benefits.	The mesocarp of the fruit has oral physiological benefits and the ability to modify biochemical and cellular steps of the inflammatory cascade, indicating that dietary supplements containing these fruits can be combined with pharmacological tools to improve or prevent inflammatory diseases.	Amorim et al., 2021
To present the physicochemical characteristics of a serum developed from buriti oil (*Mauritia flexuosa*) for older skin.	A density of 1 013g/cm^3 was observed at pH 6.1. It displayed standard viscosity, with average spreadability of 2.6088 cm. The serum presents physicochemical favorable properties, it is stable at room temperature.	Pereira Santos et al., 2021
To evaluate the effect of supplementation of crude Buriti pulp oil on metabolic disorders caused by a hyperlipidic diet.	The crude oil from the pulp aggravated the metabolic state caused by the hyperlipidic diet, worsening the risk factors associated with metabolic syndrome, such as abdominal circumference and retroperitoneal fat, serum levels of total cholesterol, uric acid, alanine transaminase, glucose and triglycerides, in addition to renal fat, and changes in markers of oxidative stress.	Aydos, 2021

References

Altman, A. and Mcclure, A. 2019. Edible product comprising plant oils orcreams and cooked legumes. The Aquafaba Company, LLC. Patent WO nº 2019164651- A1.

Alvarenga, C.A. 2004. Composição removedora de esmalte para unhas. Patente BR nº 0301845-8 A2.

Amorim, V.R. and Ferreira, P.M.P. 2020. Functional properties of *Mauritia flexuosa* fruit in experimental models of inflammation, oxidative stress and metabolic disorders. Doctorate degree. Thesis, Postgraduate Program in Biotechnology of the Northeast Biotechnology Network, Federal University of Piauí.

Amorim, V.R., Rodrigues, D.C.N., Nascimento Silva, J.N., Ramos, C.L.S., Lívia Almeida, L.M.N., Almeida, A.A.C., Pinheiro-Neto, F.P., Almeida, F.R.C., Rizzo, M.S., Pereira-Freire, J.A. and Ferreira, P.M.P. 2021. Anti-inflammatory mechanisms of fruits and by-products from *Mauritia flexuosa*, an exotic plant with functional benefits. J. of Tox Env Health. 84(11): 441–457.

Aquino, J.S., Pessoa, D.C.N.D. Oliveira, C.E.V. and Cavalheiro, J.M.O. 2012. Processamento de biscoitos adicionados de óleo de buriti (*Mauritia flexuosa* L.): Uma alternativa para o consumo de alimentos fontes de vitamina A na merenda escolar. Rev Nut. 25(6): 765–774.

Aquino, J., Soares, J., Magnani, M., Stamford, T., Mascarenhas, R., Tavares, R. and Stamford, T. 2015. Effects of dietary Brazilian palm oil (*Mauritia flexuosa* L.) on cholesterol profile and vitamin A and E status of rats. Mol. 20(5): 9054–9070.

Armando, G.C. 2006. Preservation of oleaginous fruit juice. Federal University of Piaui. PI 0500512-4 A2.

Aydos, L.R. 2021. Suplementação de óleo da polpa de buriti (*Mauritia flexuosa*) para prevenção das alterações metabólicas induzidas por dieta hiperlipídica em camundongos. Doutorado. Tese, Universidade Federal do Mato Grosso do Sul.

Baigent, C., Blackwell, L., Emberson, J., Holland, L.E., Reith, C., Bhala, N., Peto, R., Barnes, E.H., Keech, A., Simes, J. and Collins, R. 2010. Efficacy and safety of more intensive lowering of LDL cholesterol: a meta-analysis of data from 170,000 participants in 26 randomised trials. Lancet (London, England) 376(9753): 1670–1681.

Barichello, J.M., Ferreira, L.A. and Silva, L.C.G. 2019. Thermo-reversible micellar system of poloxamer 407, vegetable oil and insulin, production process, pharmaceutical composition and uses. Federal University of Ouro Preto.Patent BR nº10 2017 013407 5 A2.

Barrera-Arellano, D., Silva, C.R., Velazquez, M.D.C., Polezel, M.A. and Nogueira, C. 2005. Use of oil from the pulp of the fruits of palm trees of the genus Mauritia, as a potentiator of sun protection and as a source of carotenoids (pro-vitamin a) and vitamin e (tocopherols) natural antioxidants in cosmetic, hair care and pharmaceutical products. SUL AMÉRICA MARCAS E PATENTES S/C LTDA. PI 0303404-6 A2.

Bataglion, G.A., Silva, F.M.A., Eberlin, M.N., Koolen, H.H.F. 2014. Simultaneous quantification of phenolic compounds in buriti fruit (*Mauritia flexuosa f.*) by ultra-high performance liquid chromatography coupled to tandem mass spectrometry. Food Res. Int. 66: 396–400.

Bertoline Construção Naval a Amazônia LTDA. Bertolini, E. 2020. Balsa e processo para produção de polpa de frutas congeladas. Patente BR nº 10 2020 006660 9 A2.

Bertolini, E. 2020. Ferry and process for the production of frozen fruit pulp. Patente BR nº 10 2020 006660 9 A2. Bertoline Construção Naval a Amazônia LTDA.

Boeira, L.S. 2018. Balas de polpa de frutas e seu processo de produção. Instituto Federal de Educação Ciência e Tecnologia do Amazonas. Patente BR nº 10 2016 020796 7 A2.

Boeira, L.S. 2019. Balas de gelatina sem açúcar e com polpa de frutas e seu processo de produção. Instituto Federal de Educação Ciência e Tecnologia do Amazonas. Patente BR nº BR 10 2017 021942 9 A2.

Bonfim, M.R., Oliveira, A.S.B., Amaral, S.L. and Monteiro, H.L. 2015. Tratamento das dislipidemias com estatinas e exercícios físicos: evidências recentes das respostas musculares. Arq. Bras de Card 104: 324–33.

Caballero, N.E.D., Caversan, J. and Gaspari, P.D.M. 2015. Nanopartícula lipídica carreadora e seu uso. Universidade Estadual de Campinas. Patent BR nº 10 2013 033412 0 A2.

Camelo-Silva, C., Barroso, N.G., Barros, E.L.S., Sanches, M.A.R., Verruck, S. and Tussolini. 2020. Inovações e tendências na utilização do fruto de buriti (*Mauritia flexuosa* L) e tendências na utilização do fruto de buriti (*Mauritia flexuosa* L.) e seus subprodutos. In: Avanços em Ciência e Tecnologia de Alimentos - Volume 1. Edit Cien Dig 1: 253–262.

Cândido, T.L.N., Silva, M.R. and Agostini-Costa, T.S. 2015. Bioactive compounds and antioxidant capacity of buriti (*Mauritia flexuosa* L.f.) from the Cerrado and Amazon biomes. Food Chem. 177: 313–319. DOI: 10.1016/j.foodchem.2015.01.041.

Carmona-Hernandez, J.C., Le, M., Idárraga-Mejía, A.M. and González-Correa, C.H. 2021. Flavonoid/polyphenol ratio in *Mauritia flexuosa* and *Theobroma grandiflorum* as an indicator of effective antioxidant action. Mol. 26(21): 6431.

Carneiro, B.L.A. 2016. Estabilidade química e funcional dos compostos bioativos da polpa de buriti congelada, liofilizada e atomizada. Mestrado. Dissertação, Escola Superior de Agricultura "Luiz de Queiroz", Piracicaba.

Chaves, K.F., Barrera-Arellano, D. and Ribeiro, A.P.B. 2018. Potencial aplicação de organogéis lipídicos para a indústria de alimentos. Food Res. Int. 105: 863–872.

Cho, A.E., Jeon, S.M., Kim, M.J., Yeo, J.C.E.K., Choi, M.S. and Lee, M.K. 2010. Chlorogenic acid exhibits anti-obesity property and improves lipid metabolism in high-fat diet-induced-obese mice. Food and Chem. Tox. 48(3): 937–943.

Chu, A.J. 2022. Quarter-century explorations of bioactive polyphenols: diverse health benefits. Front. Biosci. (Landmark Ed.) 27(4): 134. https://doi.org/10.31083/j.fbl2704134.

Correia, A.G. 2006. Conservação de suco de frutos oleaginosos. Patente BR nº 0500512-4 A2.

Costa, A.N., de Sá, É.R.A., Bezerra, R.D.S., Souza, J.L. and Lima, F.D.C.A. 2021. Constituents of buriti oil (*Mauritia flexuosa* L.) like inhibitors of the SARS-Coronavirus main peptidase: an investigation by docking and molecular dynamics. J. Biomol. Struct. Dyn. Aug. 39(13): 4610–4617.

Cruz, M.B., Oliveira, W.S., Araújo, R.L., França, C.H. and Pertuzatti, P.B. 2020. Buriti (*Mauritia flexuosa* L.) pulp oil as an immunomodulator against enteropathogenic *Escherichia coli*. Ind. Crop and Prod. 149: 1–6.

Cymerys, M., Fernandes, N.M.P. and Rigamonte-Azevedo. 2005. Buriti: *Maurita flexuosa*. *In*: Shanley, P. and Medina, G. (eds.). Frutíferas e Plantas Úteis na Vida Amazônica. CIFOR e IMAZON, Belém. 1: 300.

Dahecha, I., Harrabia, B., Hambenb, K., Fekib, A., Mejdouba, H., Belghithc, H. and Belghitha, K.S. 2013. Antioxidant effect of nondigestible levan and its impact on cardiovascular disease and atherosclerosis. International J. of Biol. Macro. 58: 281–286.

Faludi, A.A., Izar, M.C.O., Saraiva, J.F.K., Chacra, A.P.M., Bianco, H.T., Afiune Neto, A., Bertolami, A., Pereira, A.C., Lottenberg, A.M., Sposito, A.C., Chagas, A.C.P. and Salgado, W. 2017. Atualização da Diretriz Brasileira de Dislipidemias e Prevenção da Aterosclerose. Arq Bras de Card 109(2): 70–76.

Ferreira, P.M.P., Freire, J.A.P., Oliveira, G.L.S., Medeiros, S.R.A., Joaquim Júnior, S.C. and Rolim, L.A. 2019. Antioxidant agent for application in drug, cosmetic and food formulations from the powdered epicarp of *Mauritia flexuosa*. Federal University of Piaui; Federal Institute of Piauí; Federal University of the São Francisco Valley. Patent BR nº 10 2018 005239 0 A2.

Ferreira, M.O.G., Lima, I.S., Ribeiro, A.B., Lobo, A.O., Rizzo, M.S., Osajima, J.A., Estevinho, L.M. and Silva-Filho, E.C. 2020. Biocompatible gels of chitosan-buriti oil for potential wound healing applications. Mater. 13: 1–17. DOI: 10.3390/MA13081977.

Franceschi, E., Anjos, H.A., Rodríguez, J.A.L., Macedo, M.L.H., Pagani, A.A.C., Barbosa, A.M. and Agenor Neto, G.S. 2021. Embalagem ativa a base de gelatina com óleo de buriti (*Mauritia flexuosa* l. F.) e processopara preparar a dita embalagem. Instituto de Tecnologia e Pesquisa, Universidade Tiradentes, Universidade Federal de Sergipe. Patente BR n°10 2020 003381 6 A2.

Gazel Filho, A.B. and Lima, J.A.S. 2001. Cultivation of Pupunheira (*Bactris gasipaes* H.B.K.) for Fruit Production in Amapa. EMBRAPA. Fol 2681: 1–11.

Gomes, R.C., Coimbra, K.L.F., Silva, A.S.S. and Silva Junior, C.S. 2020. Elaboração e caracterização do pão enriquecido com farinha à base de buriti (*Mauritia flexuosa* L.). Pubvet. 14(1): 1–5. DOI: https://doi.org/10.31533/pubvet.v14n1a497.1-5.

Igbal, M.P. 2014. Trans fatty acids—A risk factor for cardiovascular disease. Pak. J. Med. Sci. 30(1): 194–197.

Koolen, H.H.F., Silva, F.M.A., Gozzo, F.C., Souza, A.Q.L. and Souza, A.D.L. 2013. Antioxidant, antimicrobial activities and characterization of phenolic compounds from buriti (*Mauritia flexuosa* L.). Food Res Int. 51(2): 467–473.

Leitner, B., Walter, Andrea, Kalensky, Karin, Kohler, Astrid and Granzow, Silke. 2010. A composição cosmética, útil, por exemplo, como brilho labial, batom, sombra para os olhos, rouge, pó/creme para a pele e rímel, compreende uma combinação de dois extratos de plantas diferentes, por exemplo, *Centella asiática* e *Mauritia flexuosa*, e um excipiente. DE102008044701, mar, 04, 2010.

Longo, M.A. 2020. Processo de obtenção de protetor solar comestível e produtos obtidos. Vilage Marcas e Patentes LTDA. Patente BR 10 2018 074324 4-A2.

Macedo, G.A., Speranza, P., Macedo, J.A., Falcão, A.O. and Ribeiro, A.P.B. 2016. Enzymatic interesterification process, interesterified lipid composition, its uses, pharmaceutical composition, Biotechnological Applications of Flours and By-products of *Mauritia flexuosa* L. 55 food product and cosmetic composition. Campinas State University. Patent BR nº 10 2014 020266 8 A2.

Manhães, L.R.T. and Sabba-Srur, A.U.O. 2011. Centesimal composition and bioactive compounds in fruits of buriti collected in Pará. Ciência e Tecnologia de Alimentos, Campinas 31(4): 856–863.

Mansur, M.C.P.P.R. et al. 2020. Photoprotective nanoemulsions containing microbial carotenoids and buriti oil: Efficacy and safety study. Arabian J. of Chem. 13: 6741–6752. DOI: 10.1016/j.arabjc.2020.06.028.

Martins, R.C., Filgueiras, T.S. and Ulysses, P. 2012. Ethnobotany of *Mauritia flexuosa* (*Arecaceae*) in a maroon community in central Brazil. Eco Bot. 66(1): 91–98.

Melo-Cavalcante, A.A.C., Sousa, L.R., Alencar, M.V.O.B., Santos, J.V.O., Mata, A.M.O.F., Paz, M.F.C.J., Ricardo, M.C., Nunes, N.M.F., Ahmed, M.L., Mendes, A.N., Goncalves, J.C.R., Silva, F.C.C., Ferreira, P.M.P. and Sousa, J.M.C. 2019. Retinol palmitate and ascorbic acid: Role in onc prev and ther. Biomed. Pharmacother. 109: 1394–1405.

Milanez, J.T., Neves L.C., Da Silva, P.M.C., Bastos, V.J., Shahab, M., Colombo, R.C. and Roberto, S.R. 2016. Pre-harvest studies of buriti (*Mauritia flexuosa* L. f.), a Brazilian native fruit, for the characterization of ideal harvest point and ripening stages. Sci. Hortic. 202: 77–82.

Milanez, J.T., Neves, L.C., Colombo, R.C., Shahab, M. and Roberto, S.R. 2018. Bioactive compounds and antioxidant activity of buriti fruits, during the postharvest, harvested at different ripening stages. Scit Hortic. 227: 10–21. DOI: 10.1016/j.scienta.2017.08.045.

Morais, R.A., Melo, K.K.S., Oliveira, T.T.B., Teles, J.S., Peluzio, J.M.G. and Martins, A.S. 2019. Caracterização Química, física e tecnológia da farinha obtida a partir da casca de Buriti (*Mauritia flexuosa* L. f.). Braz. Journ. Develop 5(11): 23307–23322.

Mudgil, D. and Barak, S. 2013. Composition, properties and health benefits of indigestible carbohydrate polymers as dietary fiber: A review. Int. J. Biol. Mac. 61: 1–6.

Murta, M.M., Drummond, A.L., Durães, J.A., Sales, M.J.A., Moreira, S.G.C. and Pimentel, T.A.P.F. 2006. Process for the preparation of photoprotective and photoluminescent composites from the doping of polystyrene and poly (methyl methacrylate) with buriti oil (*Mauritia flexuosa* L.). Patente BR nº 0403407-4 B1.

Noce, A., Di Lauro, M., Di Daniele, F., Zaitseva, A.P., Marrone, G., Borboni, P. and Di Daniele, N. 2021. Compostos bioativos naturais úteis no manejo clínico da síndrome metabólica. Nutrientes 13(2): 630. https://doi.org/10.3390/nu13020630.

Nogueira, J.B.C. and Quintans Júnior, L.J. 2019. Efeito da administração intravenosa repetida de lidocaína associada à pregabalina em pacientes fibromiálgicos: evidências sobre o estresse oxidativo. 57 f. Doutorado.Tese,Programa de Pós-graduação em Ciências da Saúde – Universidade Federal de Sergipe, Aracaju.

Nonato, C.D.F.A., Camilo, C.J., Leite, D.O.D., da Silva Neto, J.F., Costa, L.R., Rodrigues, F.F.G. and da Costa, J.G.M. 2021. Composição centesimal e avaliação antioxidante da polpa dos frutos de *Mauritia flexuosa* L. f. do Cariri cearense. Amb: Gest. Desenvolv.

Nonato, I.A. Castro, A.S.B. Oliveira, T.T. and Carvalho, C.A. 2015. Atividade farmacológica de extratos de abacate (*Percea americana miller*), acerola (Malpighia emarginata dc) e do flavonoide naringina no tratamento da dislipidemia. Ciênc. Nat. 37(3): 747.

Oliveira, N.M.C.A. 2019. Manufacturing process for formulating mineral, animal and vegetable oils various ozonized products and obtained product. Patent BR nº 13 2018 003837 0 E2.

Oliveira, P.C., Pierluigi, S. and Barbosa, S.N. 2001. Natural nutritional supplement based on alternative foods and their preparation process. Federal University of Piaui, PI904761-6 A2, 27 out. 1999, 05 jun.

Oliveira, R.M.M., Pereira, F.T., Pereira, E.C. and Mendonça, C.J.S. 2020. Buriti oil: Nutritional quality index and antioxidant and antidiabetic effect. Rev Vir de Quí. 12: 2–12. DOI: 10.21577/1984-6835.20200002.

Pereira-Freire, J.A., Barros, K.B.N.T., Lima, L.K.F., Martins, J.M., Araújo, Y.C., Da Silva Oliveira, G.L., De Souza Aquino, J. and Ferreira, P.M.P.F. 2016. Phytochemistry profile, nutritional properties and pharmacological activities of *Mauritia flexuosa*. J. of Food Scie. 81: 2611–2622.

Pereira-Freire, J.A. 2017. Caracterização nutricional, potencial quimiopreventivo e toxicidade de *Mauritia flexuosa* (buriti): Incentivo à biotecnologia sustentável e bioprospecção de frutos regionais. Doutorado. Tese, Programa de Pós-Graduação em Biotecnologia da Rede Nordeste de Biotecnologia - RENORBIO, Universidade Federal do Piauí, Teresina-PI.

Pereira-Freire, J.A., Oliveira, G.L.S., Lima, L.K.F., Ramos, C.L.S., Medeiros, S.R.A., Lima, A.C.S., Teixeira, S.A., Oliveira, G.A.L., Nunes, N.M.F., Amorim, V.R., Lopes, L.S., Rolim, L.A., Costa-Junior, J.S. and Ferreira, P.M.P. 2018. *In vitro* and *ex vivo* chemopreventive action of *Mauritia flexuosa* products. Evi-Based Comp and Alt. Med. 2018: 1–12.

Pereira Santos, F.D., Guelma Belfort Santos, M., Brasil Lucena, V. and Alves Chaves, J. 2021. Características físico-químicas de um sérum desenvolvido à base do óleo de buriti (*Mauritia flexuosa*) para pele idosa. Rev. Enf. Atual. In Der. 95(33): e-021002.

Pereira-Freire, J.A.P., Aquino, J.S., Campos, A.R.N., Viana, V.G.F., Costa Júnior, J.S., Silva, J.N., Moura, A.K.S., Cito, A.M.G.L., Moreira-Araújo, R.S.R., Frota, K.M.G., Medeiros, S.R.A. and Ferreira, P.M.P. 2022. Nutritional, physicochemical and structural parameters of *Mauritia flexuosa* fruits and by-products for biotechnological exploration of sustainable goods. Food Tec and Biot. 60: 100–112.

Pezoti, O., Cazetta, A.L., Souza, I.P.A.F., Bedin, K.C., Martins, A.C., Silva, T.L. and Almeida, V.C. 2014. Adsorption studies of methylene blue onto ZnCl2-activated carbon produced from buriti shells (*Mauritia flexuosa* L.). J. of Ind. and Eng. Chem. 20: 4401–4407. DOI: 10.1016/j.jiec.2014.02.007.

Queiroz, M.L.M., Gomes, T.F., Cabral, F.F.C., Lopes, I.S., Silva, I.C.C. and Silva, S.L.P. 2017. Cereal bar and method to manufacture a cereal bar based on buriti mesocarp and pumpkin seeds. State University of Maranhao. Patent BR nº 10 2016 003485 0 A2.

Ramírez-Moreno, E., Arias-Rico, J., Jiménez-Sánchez, R.C., Estrada-Luna, D., Jiménez-Osorio, A.S., Zafra-Rojas, Q.Y., Ariza-Ortega, J.A., Flores-Chávez, O.R., Morales-Castillejos, L. and Sandoval-Gallegos, E.M. 2022. Role of bioactive compounds in obesity: metabolic mechanism focused on inflammation. Foods 11: 1232. https://doi.org/10.3390/ foods11091232.

Resende, L.M., Franca, A.S. and Oliveira, L.S. 2019. Buriti (*Mauritia flexuosa* L. f.) fruit by-products flours: Evaluation as source of dietary fibers and natural antioxidants. Food Chem. 270: 53–60. DOI: 10.1016/j.foodchem.2018.07.079.

Rocha, C.Q., Figueirêdo, P.M.S., Ferreira, L.S., Silva, A.S., Mubárack, T.C., Sodré, A.J.S. and Oliveira, A.C.C. 2020. Antimicrobial herbal medicine obtained from the extract of the leaves of *Mauritia flexuosa* (buriti). Federal University of Maranhao. Patent BR nº 10 2018 016091 5-A2.

Rodrigues, T.T., Silva, R.P.M. and Rusleyd, M.M.A. 2021. Buriti (*Mauritia flexuosa*) as a medicinal plant for better aging. Rev Cient Saúde 3(1): 23–46.

Romero, A.B.R., Carvalho e Martins, M.C., Nunes, P.H.M., Ferreira, N.R.T., Brito, A.K.S., Cunha, P.F.M., Lima, A., Assis, R.C. and Araújo, E.M. 2015. *In vitro* and *in vivo* antioxidant activity of Buriti fruit (*Mauritia flexuosa* L. f.). Nut Hosp. 32(5): 2153–2161.

Rosso, V.V. and Mercadante, A.Z. 2007. Identification and quantification of carotenoids, by HPLC-PDA-MS/MS, from Amazonian fruits. Jour. of Agr. and Food Chem. 55(13): 5062–5072. DOI: https://doi.org/10.1021/jf0705421.

Rudke, A.R., Mazzutti, S., Andrade. K.S., Vitali, L. and Ferreira, S.R.S. 2019. Optimization of green PLE method applied for the recovery of antioxidant compounds from buriti (*Mauritia flexuosa* L.) shell. Food Chem. 298: 125061. DOI: 10.1016/j.foodchem.2019.125061.

Saad, S.M.I. and Vasconcelos, B.G. 2011. Processo de fabricação de gelados comestíveis simbióticos, à base de frutas, não-lácteo, produtos gelados comestíveis e seu uso. Universidade de São Paulo; Fundação FAPESP. Patent BR nº PI0905343A2.

Sampaio, M.B. 2011. Boas práticas de manejo para o extrativismo sustentável do buriti. Brasília – DF. Ins Soc, Pop e Nat (ISPN) 1–80.

Sampaio, M.B. 2012. Manual Tecnológico de Aproveitamento Integral do Fruto e da Folha do Buriti (*Mauritia flexuosa*). Brasília – DF. Ins Soc, Pop e Nat (ISPN) 1–76.

Sampaio, M.C. 2017. Desenvolvimento e caracterização de nanoemulsões à base de óleo de buriti (*Mauritia flexuosa*) para avaliação de efeitos biológicos em células de câncer de mama *in vitro*. Mestrado. Dissertação, Programa de Pós-graduação em em Nanociência e Nanobiotecnologia)—Universidade de Brasília, DF. Disponível em: <https://repositorio.unb.br/handle/10482/23249>. Acesso em: 31 de ago. de 2021.

Santos, D.B., Almeida, A.A.C. and Ferreira, P.M.P. 2019. Fisiopatologia da Doença de Alzheimer. *In*: Feitosa, C.M., Bonfim, K.L.F. (Org.). Acompanhamento Farmacoterapêutico e suas Implicações na Terapia da Doença de Alzheimer. 1° ed. Ponta Grossa: Atena Editora. 1: 15–26.

Santos, R.D., Gagliardi, A.C.M., Xavier, H.T., Magnoni, C.D., Cassani, R., Lottenberg, A.M., Casella Filho, A., Araújo, D.B., Cesena, F.Y., Alves, R.J., Fenelon, G., Nishioka, S.A.D., Faludi, A.A., Geloneze, B., Scherr, C., Kovacs, C., Tomazzela, C., Carla, C., Barrera-Arellano, D., Cintra, D., Quintão, E., Nakandakare, E.R., Fonseca, F.A.H., Pimentel, I., Santos, J.E., Bertolami, M.C., Rogero, M., Izar, M.C.O., Nakasato, M., Damasceno, N.R.T., Maranhão, R., Cassani, R.S.L., Perim, R. and Ramos, S.L. 2013. I Diretriz sobre o consumo de Gorduras e Saúde Cardiovascular. Arq. Bras de Card. 100(1): 1–40.

Silva, Y.O., Lindoso, D.P., Nunes, G.S. and Ivone Rosa, I.G. 2019. Cake enriched with buriti. Federal University of Maranhao. Patent BR n° 10 2017 015017 8 A2.

Silva, C.S.M., Araújo, J.A., Silveira, T.S., Castro, K.C.F., Baratto, L.C., Kaminski, R.C.K., Santos, G.B. and Nunes, K.M. 2021. Atividade de cicatrização de feridas de formulações tópicas contendo óleo de *Mauritia flexuosa*. Rev. Bras. Farmacogn. 31: 225–231. DOI: https://doi.org/10.1007/s43450-021-00149-2.

Silverman, M.G., Ference, B.A., IM, K., Wiviott, S.D., Giugliano, R.P., Grundy, S.M., Braunwald, E. and Sabatine, M.S. 2016. Association between lowering LDL-C and cardiovascular risk reduction among different therapeutic interventions: a systematic review and meta-analysis. JAMA 316(12): 1289–1297.

Siqueira, E.P., Andrade, A.A., Souza-Fagundes, E.M., Ramos, J.P., Kohlhoff, M., Nunes, Y.R. and Cota, B.B. 2014. *In vitro* antibacterial action on methicillin-susceptible (MSSA) and methicillin-resistant (MRSA) *Staphylococcus aureus* and antitumor potential of *Mauritia flexuosa* L. f. J. of Med. Plants Res. 8(48): 1408–1417.

Soares, N.R., Nicolau, E.S., Damiani, C. and Souza, A.R.M. 2018. Antimicrobial liquid soap based on buriti oil. Goias Federal University. Patent BR n°10 2016 015728 5 A2.

Sousa, P.B., Silva, E.F., Monção. E.C., Silva, J.N., Silva, M.J.M. and Sousa, M.M. 2010. Fenólicos totais, carotenóides e capacidade antioxidante de raspas de buriti (*Mauritia flexuosa* L.) *in natura* comercializadas em Teresina-Piauí. Instituto Federal de Educação, Ciência e Tecnologia do Piauí. Teresina. Disponível em: <http:// congressos.ifal.edu.br/index.php/connepi/ CONNEPI2010/paper/viewFile/1537/41>. Acesso em: 31 de ago. 2021.

Souza, D. 2008. Estudo das propriedades físicas de polpas e néctares de pequenos frutos.2008. Mestrado. Dissertação, Programa de Pós-graduação em Engenharia Química – Universidade Federal do Rio Grande do Sul, Porto Alegre.

Srur, A.U.O.S. and Manhães, L.R.T. 2009. S. Barra de cereais com banana e polpa de buriti desidratada. Patente BR n° 0800470-6 A2.

Vasconcelos, P.R.L. 2010a. Preparation based on vegetable oils used as a nutritional supplement. BRPI1000759A2.

Vasconcelos, P.R.L. 2010b. Fish and plant-oil-based preparations used as a nutritional supplement. Wettor Bureau de Apoio Empresaria/S/S Ltda. WO2011113128.

Vaz, S.M.S. 2016. Fitoterapia no controle das dislipidémias. Mestrado. Dissertação, Programa de Mestrado Integrado em Ciências Farmacêuticas - Universidade de Coimbra, Portugal.

Wong, B.W., Meredith, A., Lin, D. and Mcmanus, B.M. 2012. The biological role of inflammation in atherosclerosis. Can. J. Cardiol. 6(28): 631–641.

Chapter 5

Sustainable Use of Food Waste in the Preparation of Biodegradable Packaging for the Food Industry

Alyandra de Sousa Nascimento,[2] *Maria Crisnanda Almeida Marques,*[1] *Patrícia e Silva Alves,*[2] *Maria das Dores Alves de Oliveira*[2] and *Chistiane Mendes Feitosa*[2]

Introduction

Currently, the main material used in the manufacture of packaging is plastic, requiring 39.6% of the overall production of this material, its main purpose is to protect its contents from any damage, is indispensable to most products such as food (Ma et al., 2020). Among the problems that come from this, the main one focuses on waste management, because without a proper purpose, it can be responsible for the pollution of oceans, thus causing great damage to the environment.

In relation to plastic recycling, it is observed that the production of virgin plastic is higher than the reuse. This fact is due to a high cost of production, as well as difficult implementation. In this sense, as an alternative, there is a variety of plastic polymers from waste, in which the production costs of its virgin material are lower, making the reuse economically unfeasible (Bassi et al., 2020).

According to the Ministry of Environment (2012), waste synthetic plastics take more than 400 years to decompose in the environment. As an alternative, the need arises for the development of biodegradable materials from natural sources,

[1] Department of Pharmacy, Federal University of Piauí, 64049-550 Teresina-PI, Brazil
[2] Department of Chemistry, Federal University of Piauí, 64049-550 Teresina-PI, Brazil.
* Corresponding author: chistiane@ufpi.edu.br

such as biopolymers extracted from by-products of the agricultural industry or food waste, because they are of natural origin, these materials in the environment have an accelerated degradation process that through enzymatic catalysis induced by microorganisms or chemical processes generate natural products during degradation, causing a positive impact on the environment (Zhong et al., 2020). Another alternative to reduce these solid particles is the use of edible coatings in food preservation. Cecatto and Lenz (2022), when evaluating the conservation of strawberry in cassava starch as an edible, viable and efficient covering, concluded that it was possible to keep the fruit preserved for longer, thus maintaining its quality and the characteristics of the fruits stable during storage.

Biodegradable films have a potential application in packaging, especially for food, because most of them are derived from food industry waste, having advantages such as the possibility of bioactive properties, antimicrobial and antioxidant activity, besides actively participating in the conservation of products providing benefits in the marketing and consumption of food (Jafarzadeh et al., 2020).

Given the above, this chapter aims to review the options of food waste that can be used in the development of biodegradable films for the production of food packaging, generating a positive impact on the environment and in terms of food preservation.

Food packaging and preservation

The primary function of food packaging is conservation, which is responsible for protecting the food from physical, chemical and microbiological damage during all stages of processing until consumption. Currently packaging carries other functions besides conservation, such as causing a positive impact on the environment and being economically viable (Topuz and Uyar, 2020). The shelf life of the product is closely linked to the packaging, which must be planned and adapted to the needs of each food category since its poor planning and adequacy can lead to waste of food (Williams et al., 2020).

According to the FAO (Food and Agriculture Organization of the United Nations), around one-quarter to one-third of food produced globally is wasted, impacting the amount of food available for food, the environment and producers/consumers. These food wastes are considered plant parts such as stems, roots, barks, seeds and the by-products of processing, as well the animal wastes (Arun et al., 2020).

The complexity of the food production and consumption chain makes it difficult to control waste, highlighting many responsible parties throughout the process and the production of food waste is observed from the post-harvest stage to the consumption stage, however, when evaluating waste from the perspective of household waste production, the development of efficient packaging with properties and from food waste itself is an alternative to reduce waste and direct food industry waste (Poyatos-Racionero et al., 2018).

In Fig. 5.1, it is possible to observe all food waste generation and the direction for packaging production, starting from harvest, processing, distribution and consumption, and ending up in the sustainable production of packaging.

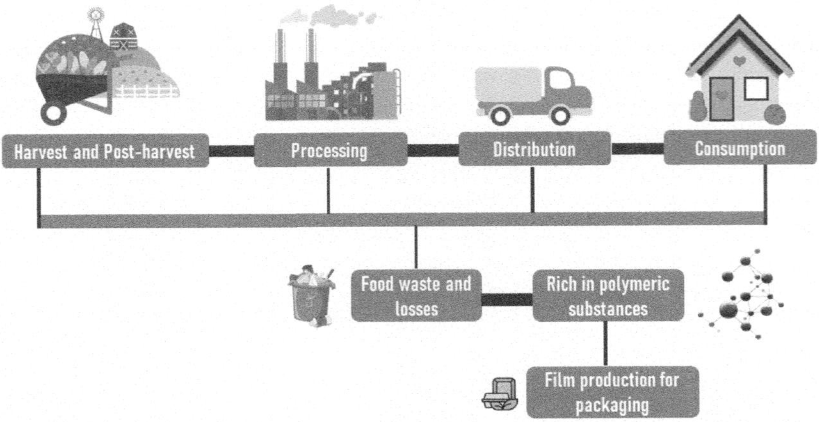

Figure 5.1. Generation of food waste and direction for packaging production. Source: Own Authorship, 2021.

Food waste for packaging application

The unavoidable food waste, or by definition parts that would not be used for consumption, as well as avoidable waste generate waste management problems and lately has drawn attention, because most of it consists of organic fractions rich in substances such as carbohydrates and proteins, presenting the possibility of reuse and application (Mak et al., 2020). Combined with the need for biodegradable packaging of biological origin, agricultural and food waste has shown great potential from which polymers can be extracted for the production of naturally biodegradable packaging with added value, and can present or be incorporated with conservation properties (Guillard et al., 2018).

Polymers of natural origin are called as biopolymers, which are extracted from biological materials. Due to their biodegradable property, they have become an alternative to plastics that are produced from non-renewable sources, which take years to decompose, thus causing damage to the environment (George et al., 2020). They can be of plant origin (cellulose, starch, pectin), animal origin (gelatin, chitosan, casein, whey), from fermentation by-products (polyhydroxybutyrate (PHB) and polyhydroxyalkanoates (PHA)) or obtained through chemical synthesis (Kabir et al., 2020).

In Fig. 5.2, a schematic of the preparation of chitin and chitosan from the exoskeleton (shells) of crustaceans and fungi and their applications namely: pharmaceutical uses, therapeutic uses and films for coating has been shown. The chitin is separated from other components of the shell by a chemical process that involves the steps of demineralization and deproteinization of the shells with diluted solutions of Hydrochloric Acid (HCl) and Sodium Hydroxide (NaOH), followed by decolorization with Potassium Permanganate ($KMnO_4$) and oxalic acid, for example. The chitin obtained, the biopolymer containing acetyl groups ($NHCOCH_3$), is deacetylated with concentrated NaOH solution, producing chitosan (Azevedo et al., 2007). Chitosan, in turn, is a natural product obtained from the chitin of crustacean shells. A scheme for obtaining these materials is shown in Fig. 5.2. Furthermore, chitosan

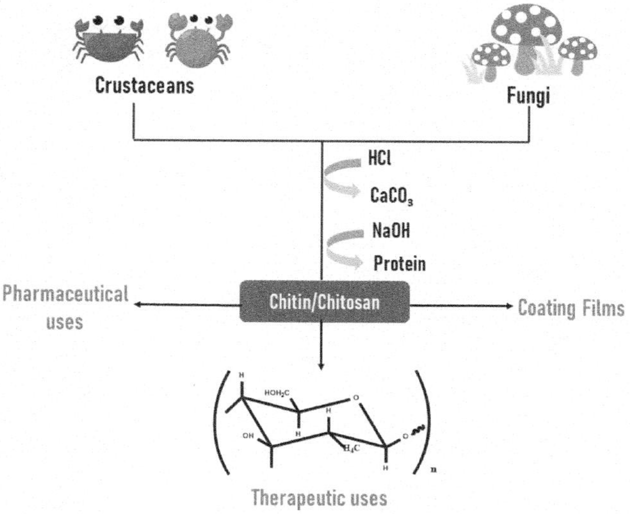

Figure 5.2. Scheme of chitin and chitosan preparation from crustacean and fungal shells and their applications. Source: Own Authorship, 2021.

is considered a natural product of low cost, renewable and biodegradable, being of great economic and environmental importance.

Agricultural wastes that comprise from post-harvest to processing, and post-consumer wastes are a primary source for extraction of polymers such as starch, cellulose, pectin and fibers, which can be employed in the production of biodegradable packaging and can be applied in the food industry itself due to the natural origin (Maraveas, 2020). Table 5.1 shows some food waste that has been used for the production of biodegradable packaging films.

In Fig. 5.3 the chemical structure of cellulose, chitin and chitosan, used in most of biofilm packaging is shown. Starch is composed of anhydroglucose units—amylase and amylopectin—has been used for various applications including wound dressing, tissue engineering scaffolds, drug delivery systems, bone replacement implants and substrate for cell seeding due to its biodegradability, availability, biocompatibility and economic feasibility (Hassan et al., 2018; Torres et al., 2011).

Food waste has great potential for extraction of substances, mostly rich in polymers such as starch, cellulose, fibers and other substances that enable the formation of films for the production of biodegradable packaging, and may also possess or be added active properties (De Moraes Crizel et al., 2016). Table 5.1 describes the above presented examples of substances extracted from waste for the production of biofilms, which can be classified into two groups as shown in (Fig. 5.4).

The first group is represented by polysaccharides, which are polymeric carbohydrates. This biopolymer presents high compatibility, low toxicity and biodegradability, important characteristics for its application in packaging. Plant polysaccharides are more abundant and are present in the cell wall of plants (Shao et al., 2020; Yang et al., 2020). The most abundant plant polysaccharide is starch, which can be extracted from waste products such as jackfruit seeds and babassu

Table 5.1. Food waste used in the production of packaging. Source: Own Authorship, 2021.

Food residue	Extracted substance	Produced packaging	Reference
Raw shrimp shell and rice straw	Chitosan and Cellulose	Cast thin film	(Elhussieny et al., 2020)
Fish skin *Centrolophus niger* (Jellyfish)	Collagen	Collagen-chitosan mixed film	(Bhuimbar and Bhagwat, 2019)
Almond shell and rice hulls	Fibers	PHB and fiber composite	(Sánchez-Safont et al., 2018)
Lemon peel	Pectin	Starch and pectin film	(Dash et al., 2019)
Potato peel	Flour	Film	(Xie et al., 2020)
Jackfruit Seeds (*Artocarpus heterophyllus*)	Starch	Film	(Santana et al., 2018)
Pomegranate Bark (*Punica granatum*)	Flour	Film with fish gelatin	(Hanani et al., 2019)
Babassu Mesocarp Flour (*Attalea speciosa*)	Starch	Flour and Starch Film	(Maniglia et al., 2017)

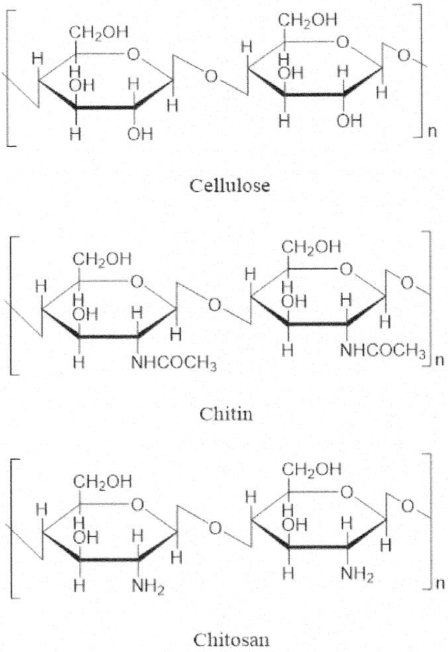

Figure 5.3. Chemical structure of cellulose, chitin, and chitosan. Source: Own Author, 2021.

mesocarp. Although starch is a low-cost and renewable source, this material produces a brittle film, to improve this characteristic a plasticizer is added to the production of the film making it more pliable for packaging production (Zhong et al., 2020). Jackfruit seed is considered a waste product and represents about 15% of the total weight of the fruit, the extracted starch is rich in amylose, a basic substance for film

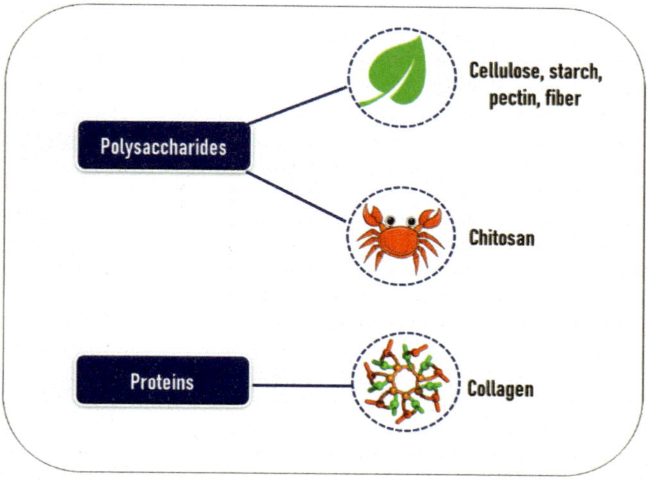

Figure 5.4. Classification of substances extracted from food waste. Source: Own Authorship, 2021.

formation. A starch-based biofilm with clove essential oil showed good efficiency in fruit preservation, preventing the loss of water from the organic vegetable and the proliferation of microorganisms (Costa et al., 2022). In another study, the film produced from jackfruit seed starch using glycerol as a plasticizer showed relatively high mechanical stability with moderate water vapor permeability and low opacity (Santana et al., 2018).

Maniglia et al. (2017), produced films from babassu mesocarp flour and the starch extracted from it, obtaining films with different characteristics, where the film prepared with starch from water maceration showed the best properties, such as lower water vapor permeability and low hydrophilicity, babassu films still present antioxidant activity, which is another advantage of its application as food packaging.

Food waste rich in carbohydrates can be transformed into flour for further application, such as potato peel flour used to produce film intended for food packaging, producing films reinforced with bacterial cellulose, with the addition of antioxidant activity through curcumin, showing the potential of this film for packaging, and disposing of a large amount of this waste (Xie et al., 2020). Other polymeric substances such as pectin and fibers can also be found in waste and intended for the production of biodegradable films.

Cellulose is another polysaccharide of plant origin, abundant, of renewable origin and with low cost, its characteristics provide good film formation. Chitosan, after cellulose is the most abundant polymer, is produced by deacetylation of chitin, present in the exoskeleton of crustaceans and in the cell wall of some fungi (Fig. 5.2), presenting great potential for application in film production (Negm et al., 2020; Satani et al., 2020). Rice straw and shrimp shells, both waste materials, enabled the extraction of cellulose and chitosan respectively, employed in the production of a thin film by casting, producing a film favorable to the production of packaging because combined these two materials the film provided the mechanical performance suitable for this purpose (Elhussieny et al., 2020).

Proteins can also be targeted for biofilm production, they are present in animal waste such as from fish processing. Fish skin is rich in collagen, this substance presents emulsifying properties and combines the filmogenic properties of chitosan, forming films with good properties intended for packaging (Bhuimbar et al., 2019). Kaczmarek-Szczepańska et al., 2022 carried out a study with chitosan modified with phenolic acids in the elaboration of packaging and the results were quite promising in terms of physical and thermal properties, where the application of chitosan with ferulic acid was the most relevant.

The schematic of the production of a biofilm is shown below (Fig. 5.5), prepared by Hanani and colleagues (2019). In this research, they used an extract from the seeds of the pomegranate (*Punica granatum* L.) species (EPG) in the antioxidant and antimicrobial properties of fish gelatin films as active packaging. Biofilm production was carried out from the aqueous extract of pomegranate seeds. Three formulations were processed by adding EPG to fish gelatin (0, 10, 30% by weight) by the casting technique, exhibiting films with 10 wt.% PSP as the best performance. The addition of EPG decreased elongation at break and increased stiffness in the fish gelatin films, particularly for 30% by weight. A good match between fish gelatin and EPG was observed by SEM. No significant difference ($p < 0.05$) was obtained for oxygen barrier properties and water vapor permeability compared to the control with PSP incorporation, while water resistance increased considerably and transparency values decreased ($p < 0.05$). High thermal stability of the films and inhibition against *S. aureus* was observed. The addition of PSP at 10% by weight % in FG was shown as a potential strategy to maintain the integrity of the material and protect the food against lipid oxidation, reducing large amounts of pomegranate and fish residues.

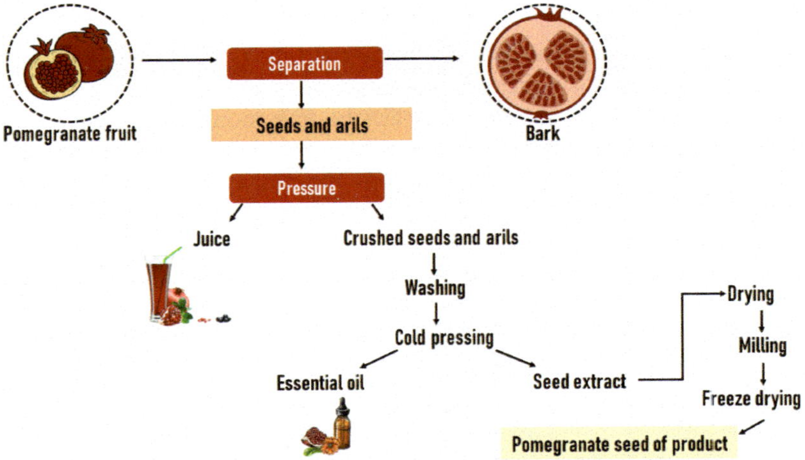

Figure 5.5. Production of a biofilm from the extract of the seeds of the pomegranate species. Source: Own Authorship, 2021.

Active packaging

Based on the principle that packaging is responsible for the conservation, active packaging has a greater interaction with the product, expanding the shelf life and preserving the sensory characteristics of the food. One way to provide these benefits would be through the addition of activities such as antimicrobial, antioxidant and atmosphere control with the ability to interact and influence (Aragüez et al., 2020; De Carvalho and Junior, 2020).

For the development of active packaging, a preference for polymers' natural origin is observed, as besides being biodegradable, it is possible to develop packaging adapted to the conservation needs of each food category, the added active substances are mostly extracted from natural products and the polymers themselves can present these bioactive properties (Alizadeh-Sani et al., 2020).

Food residues, besides being sources of filmogenic substances, also present active substances that can be extracted and incorporated into biodegradable films, such as pomegranate peel flour, which when incorporated into fish gelatin film, added higher antioxidant activity compared to the film without its addition, as well as antimicrobial activity that showed larger inhibition zones as the concentration of the flour increased in the film, with higher activity for *S. aureus* (Hanani et al., 2019). A film prepared with starch and chitosan plus grapefruit seed extract retarded fungal growth by 20 d on the bread sample, due to the phenolic compounds present in the extract, presenting biodegradable packaging with fungicidal properties (Jha, 2020).

The flour and starch of babassu mesocarp are rich in phenolic compounds, during the preparation of the film the content of these substances decreases, but did not affect their antioxidant activity, making the film a good material for packaging food sensitive to oxidation, extending the shelf life (Maniglia et al., 2017). Curcumin (Fig. 5.6) has known antioxidant activity and was incorporated into potato peel starch and bacterial cellulose film for this purpose, obtaining films with antioxidant properties and able to prevent lipid oxidation in fresh pork meat (Xie et al., 2020).

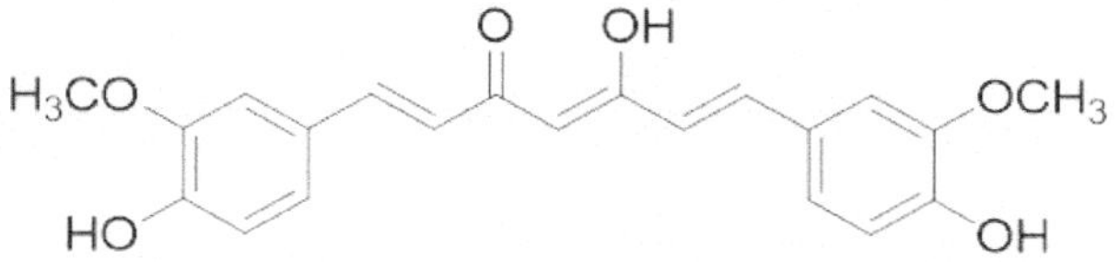

Figure 5.6. Chemical structure of curcumin. Source: Own Authorship, 2021.

Life cycle assessment

Life cycle assessment is an approach for evaluating the environmental performance of a product, process or service by showing the impact they may have throughout its life cycle. This is done by assessing the impact of raw material extraction to the final disposal (Yossef and Hot, 2021).

The packaging size of certain products is considered one of the main reported reasons for food waste mentioned by consumers, often these are too large and not suitable for people living alone or as couples, while the prices of non-prepackaged food or smaller packaging are comparatively high (Evans, 2011; Graham-Rowe et al., 2014). According to Williams et al. (2012), up to 20–25% of food waste can be related to packaging sizes that are too large and packages that are difficult to empty. In addition, packaging affects the environment because of the time it requires to be completely degraded, when disposed of and also because of the processes used to recycle it (Suput et al., 2013).

Conditions based on the environmental impact generated by packaging and the potential for appropriate innovations to reduce food waste direct future strategies for improving sustainability by making the use of sustainable packaging. Packaging reduction and a shift to alternative materials and/or technologies should be specially addressed for products characterized by a high relative packaging impact (Licciardello, 2017). Therefore, for a reduction of the environmental impact caused by the life cycle of food packaging, it is indispensable that it is able to reduce food waste, in addition to being sustainable.

Sustainability

Food packaging is an area of constant evolution due to the introduction of new food technologies, and a continuous change in consumer habits, lifestyle and environmental influences (Del Borghi et al., 2021; Reinisch, 2022). Such constant change occurs to meet industry requirements and consumer needs, maintain food safety and reduce environmental impacts (Marsh and Bugusu, 2007).

For this reason of constant change, over the years there has been a growing interest for such environmental protection around the world, and the need for the use of green packaging arose, it has great importance to reduce the impact of waste and pollution and promote sustainable development (Wong et al., 2012).

Green packaging, also known as 'sustainable packaging', or 'recyclable packaging', makes use of eco-friendly materials for packaging purposes, with the motive that its products are effective and safe for human health and the environment (Pauer et al., 2019).

Companies in this way, are encouraged to promote sustainable packaging not only by the increasing importance that consumers place on the environment, but also by new laws, regulations, taxation and other actions promoted by governments to make packaging sustainable and eco-friendly (Nguyen et al., 2020; Qing and Guirong, 2012). In addition, there are many campaigns promoting environmentally sustainable food packaging. Therefore, green packaging aims to pollute as little as possible, not only the food but also the environment, after use (Del Borghi et al., 2021).

Recycling

Due to the large increase in solid waste the need to reuse these materials and make them marketable products arose, reducing the amount in dumps, preserving it and saving raw materials and energy (Spinacé and De Paoli, 2005). Recycling is understood as the physical, physical-chemical and biological transformation of solid waste, and the recycling process involves three main steps: separation, identification and cleaning (De Oliveira et al., 2015; Franchetti and Marconato, 2006; Lin et al., 2022).

Among the various types of recycling, plastic recycling can be highlighted, since it is a material commonly found in food packaging. Plastic recycling can be divided into three types: mechanical recycling, which transforms the plastic waste into granules that can be reused in the manufacture of other materials; chemical recycling, which involves recovering and purifying the starting monomers to form new polymers; and energetic recycling, which consists of recycling the plastic waste through combustion or pyrolysis, recovering the energy that was released (Forlin and Faria, 2002; Caraschi and Leão, 2002).

Mechanical recycling is used the most by industries due to cheaper labor, lower investment and a large volume of the post-consumer polymer. However, it is worth pointing out that the use of recycled packages in the food industry is restricted, due to possible contamination of the package with food from misuse, contact with cleaning products or contact with toxic products or solvents. However, there are technologies capable of decontaminating these packages, called superclean, where a polymer (usually PET) goes through a washing process with non-caustic substances at high temperatures and under pressure. The material is then subjected to a vacuum for a period of time and then undergoes filtration (Cruz et al., 2011). Given this, it is worth noting that it is possible to search for viable alternatives for the preservation of the environment through the recycling of plastic waste.

Green and biodegradable polymers

Polymers are macromolecules formed from monomers (small molecule units) and the term comes from the Greek poly- and meros-parts/units (Franchetti and Marconato, 2006). There are two types of polymers, as regards to their origin, the natural ones that come from nature, such as silk, cellulose and cotton fibers, and the synthetic ones, which are obtained through industry, such as polypropylene (PP), polyethylene terephthalate (PET), polyethylene (PE) and polyvinyl chloride (PVC) (Spinacé and De Paoli, 2005).

For their physical behavior, they can be classified into thermoplastics (plastics), thermosets, rubbers and fibers, and are quite resistant to natural degradation, which can take years to decompose and increase the accumulation of waste in landfills or municipal dumps (Spinacé and De Paoli, 2005). Due to these factors the need arose to minimize the impacts caused to the environment, adopting more sustainable measures, such as the manufacture of green polymers and biodegradable polymers. Biodegradable plastics are among the most promising materials to replace conventional petroleum-based plastics that have caused many adverse impacts on the environment, such as pollution (land, water, etc.), and global warming. Among a range of biodegradable plastics, Poly Lactic Acid (PLA) is not only widely available but also safe to be decomposed after its usage without polluting the environment (Taib et al., 2022).

A green polymer is one that in its production or degradation process does not generate environmental impact and is less dependent on raw materials derived from fossil fuels. These materials can be replaced by polysaccharides, starch or cellulose, as they are renewable sources of polymers (Brito et al., 2011; Kobayashi, 2017).

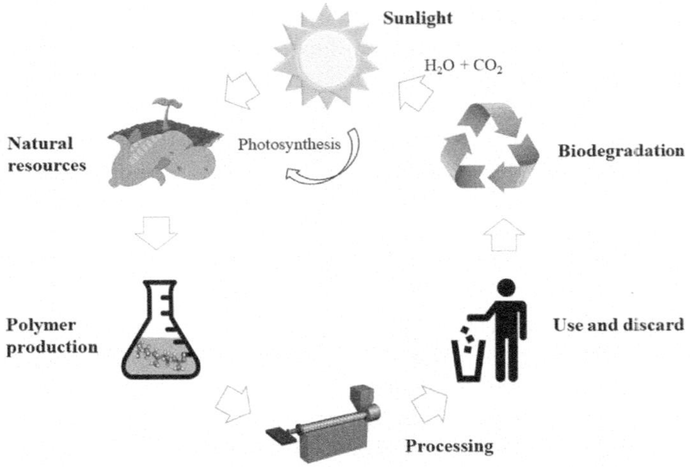

Figure 5.7. Life cycle of biodegradable polymers obtained from renewable sources. Adapted from Brito et al. (2011).

Biodegradable polymers, on the other hand, are those that are present and are degraded by bacteria, fungi and algae and the degradation occurs through the action of enzymes (Fechine, 2013). They can be of natural origins such as corn, cellulose, potato or sugar cane, can be of animal origin, from chitin, chitosan or proteins and synthetic such as poly (lactic acid), poly (glycolic acid), poly (polyglycolic-lactic acid) and poly (-caprolactone). The advantages of biodegradable polymers from renewable sources are because they reduce the impact on the environment due to the release of CO_2 into the atmosphere, where it is absorbed by plants, in addition to forming a closed life cycle, as shown in (Fig. 5.7) (Franchetti and Marconato, 2006; Brito et al., 2011).

Conclusion

Plastic packaging is gradually being replaced by packaging with less environmental impact, that is more compatible with food and has superior conservation capacity. One source for the development of new materials is food waste, most of which is not reused, and represents a source of polymeric substances with active properties. These biodegradable polymers solve the problem of plastic pollution, dispose of food waste and compose active packaging that reduces food waste, generating a positive impact throughout its development chain.

References

Alizadeh-Sani, M., Mohammadian, E. and McClements, D.J. 2020. Eco-friendly active packaging consisting of nanostructured biopolymer matrix reinforced with TiO_2 and essential oil: Application for preservation of refrigerated meat. Food Chem. 322: 126782.

Aragüez, L., Colombo, A., Borneo, R. and Aguirre, A. 2020. Active packaging from triticale flour films for prolonging storage life of cherry tomato. Food Packag. Shelf Life. 25: 100520.

Arun, K.B., Madhavan, A., Sindhu, R., Binod, P., Pandey, A.R.R. and Sirohi, R. 2020. Remodeling agro-industrial and food wastes into value-added bioactives and biopolymers. Ind. Crops Prod. 154: 112621.

Azevedo, V., Chaves, S., Bezerra, D., Lia Fook, M. and Costa, A. 2007. Quitina e Quitosana: aplicações como biomateriais. REletrônicaMP 2: 27–34.

Bassi, S.A., Boldrin, A., Faraca, G. and Astrup, T.F. 2020. Extended producer responsibility: How to unlock the environmental and economic potential of plastic packaging waste? Resour. Conserv. Recycl. 162: 105030.

Benítez, R.O. 2020. Perdas e desperdícios de alimentos na América Latina e no Caribe. FAO. Available in: <http://www.fao.org/americas/noticias/ver/pt/c/239394/>. Access on: 20/07/2020.

Bhuimbar, M.V., Bhagwat, P.K. and Dandge, P.B. 2019. Extraction and characterization of acid soluble collagen from fish waste: Development of collagen-chitosan blend as food packaging film. J. Environ. Chem. Eng. 7: 102983.

Brito, G., Agrawal, P., Araújo, E. and Mélo, T. 2011. Biopolímeros, polímeros biodegradáveis e polímeros verdes. REMP 6: 127–139.

Caraschi, J.C. and Leão, A.L. 2002. Avaliação das propriedades mecânicas dos plásticos reciclados provenientes de resíduos sólidos urbanos. Acta Sci. Technol. 24: 1599–1602.

Cecatto, A.P. and Lenz, R.G. 2022. Comportamento pós-colheita de morangos frescos com o uso de cobertura comestível Postharvest behavior of fresh strawberries with the use of edible coating. Braz. J. Dev. 8: 10623–10636.

Costa, M.L.X., de Aquino, A.A., Rocha, V.C.F., de Castro Ferreira, T.A.P., de Souza Moreira, E., de Barros Filho, C.J. and Moura, A.H. 2022. Conservação pós-colheita de tomate-cereja orgânico embalados com filme ativo biodegradável à base de amido e óleo essencial de cravo-da-Índia. Concilium 22: 387–400.

Cruz, S.A., Oliveira, É.C., Oliveira, F., Garcia, P.S. and Kaneko, M.L. 2011. Polímeros reciclados para contato com alimentos. Polimeros 21: 340–345.

Dash, K.K., Ali, N.A., Das, D. and Mohanta, D. 2019. Thorough evaluation of sweet potato starch and lemon-waste pectin based-edible films with nano-titania inclusions for food packaging applications. Int. J. Biol. Macromol. 139: 449–458.

De Carvalho, A.P.A. and Junior, C.A.C. 2020. Green strategies for active food packagings: A systematic review on active properties of graphene-based nanomaterials and biodegradable polymers. Trends Food Sci. Technol. 103: 130–143.

Del Borghi, A., Parodi, S., Moreschi, L. and Gallo, M. 2021. Sustainable packaging: an evaluation of crates for food through a life cycle approach. Int. J. Life Cycle Assess. 26: 753–766.

De Moraes Crizel, T., Costa, T.M.H., de Oliveira Rios, A. and Flôres, S.H. 2016. Valorization of food-grade industrial waste in the obtaining active biodegradable films for packaging. Ind. Crops Prod. 87: 218–228.

De Oliveira Neto, G.C., Shibao, F.Y., Godinho Filho, M. and Chaves, L.E.C. 2015. Produção mais limpa: estudo da vantagem ambiental e econômica na reciclagem de polímeros. Interciencia 40: 364–373.

Elhussieny, A., Faisal, M., D'Angelo, G., Aboulkhair, N.T., Everitt, N.M. and Fahim, I.S. 2020. Valorisation of shrimp and rice straw waste into food packaging applications. Ain Shams Eng. J. 11: 1219–1226.

Evans, D. 2011. Beyond the throwaway society: ordinary domestic practice and a sociological approach to household food waste. Sociology 46: 41–56.

Fechine, G.J.M. 2013. Polímeros biodegradáveis: tipos, mecanismos, normas e mercado mundial. SciELO-Editora Mackenzie.

Forlin, F.J. and Faria, J.D.A.F. 2002. Considerações sobre a reciclagem de embalagens plásticas. Polimeros 12: 1–10.

Franchetti, S.M.M. and Marconato, J.C. 2006. Polímeros biodegradáveis-uma solução parcial para diminuir a quantidade dos resíduos plásticos. Quím. Nova 29: 811–816.

George, A.M., Peddireddy, S.P.R., Thakur, G. and Rodrigues, F.C. 2020. Biopolymer-based scaffolds: Development and biomedical applications. In Biopolymer-Based Formulations. 717–749.

Graham-Rowe, E., Jessop, D.C. and Sparks, P. 2014. Identifying motivations and barriers to minimising household food waste. Resour. Conserv. Recycl. 84: 15–23.

Guillard, V., Gaucel, S., Fornaciari, C., Angellier-Coussy, H., Buche, P. and Gontard, N. 2018. The next generation of sustainable food packaging to preserve our environment in a circular economy context. Front. Nutr. 5: 121.

Hanani, Z.N., Yee, F.C. and Nor-Khaizura, M. 2019. Effect of pomegranate (*Punica granatum* L.) peel powder on the antioxidant and antimicrobial properties of fish gelatin films as active packaging. Food Hydrocoll. 89: 253–259.

Hassan, A., Niazi, M.B.K., Hussain, A., Farrukh, S. and Ahmad, T. 2018. Development of anti-bacterial PVA/starch based hydrogel membrane for wound dressing. J. Polym. Environ. 26: 235–243.

Jafarzadeh, S., Jafari, S.M., Salehabadi, A., Nafchi, A.M., Kumar, U.S.U. and Khalil, H.A. 2020. Biodegradable green packaging with antimicrobial functions based on the bioactive compounds from tropical plants and their by-products. Trends Food Sci. Technol. 100: 262–277.

Jha, P. 2020. Effect of plasticizer and antimicrobial agents on functional properties of bionanocomposite films based on corn starch-chitosan for food packaging applications. International Journal of Biological Macromolecules 160: 571–582.

Kabir, E., Kaur, R., Lee, J., Kim, K.H. and Kwon, E.E. 2020. Prospects of biopolymer technology as an alternative option for non-degradable plastics and sustainable management of plastic wastes. J. Clean. Prod. 258: 120536.

Kaczmarek-Szczepańska, B., Zasada, L. and Grabska-Zielińska, S. 2022. The physicochemical, antioxidant, and color properties of thin films based on chitosan modified by different phenolic acids. Coatings 12: 126.

Kobayashi, S. 2017. Green polymer chemistry: new methods of polymer synthesis using renewable starting materials. Struct. Chem. 28: 461–474.

Licciardello, F. 2017. Packaging, blessing in disguise. Review on its diverse contribution to food sustainability. Trends Food Sci. Technol. 65: 32–39.

Lin, H., Luo, S., Zhang, H. and Ye, J. 2022. Em direção à reciclagem de carbono movida a energia solar. Joule.

Ma, X., Park, C. and Moultrie, J. 2020. Factors for eliminating plastic in packaging: The European FMCG experts' view. J. Clean. Prod. 256: 120492.

Mak, T.M., Xiong, X., Tsang, D.C., Iris, K. and Poon, C.S. 2020. Sustainable food waste management towards circular bioeconomy: Policy review, limitations and opportunities. Bioresour. Technol. 297: 122497.

Maniglia, B.C., Tessaro, L., Lucas, A.A. and Tapia-Blácido, D.R. 2017. Bioactive films based on babassu mesocarp flour and starch. Food Hydrocoll. 70: 383–391.

Maraveas, C. 2020. Production of sustainable and biodegradable polymers from agricultural waste. Polymers 12: 1127.

Marsh, K. and Bugusu, B. 2007. Food packaging—roles, materials, and environmental issues. J. Food Sci. 72: R39–R55.

Ministério do Meio Ambiente. Available in: <http://www.mma.gov.br/responsabilidade-socioambiental/producao-e-consumo-sustentavel/consumo-consciente-de-embalagem/impacto-das-embalagens-no-meio-ambiente>. Access: 22 oct. 2021.

Negm, N.A., Hefni, H.H., Abd-Elaal, A.A., Badr, E.A. and Abou Kana, M.T. 2020. Advancement on modification of chitosan biopolymer and its potential applications. Int. J. Biol. Macromol. 152: 681–702.

Nguyen, A.T., Parker, L., Brennan, L. and Lockrey, S. 2020. A consumer definition of eco-friendly packaging. J. Clean. Prod. 252: 119792.

Pauer, E., Wohner, B., Heinrich, V. and Tacker, M. 2019. Assessing the environmental sustainability of food packaging: An extended life cycle assessment including packaging-related food losses and waste and circularity assessment. Sustainability 11: 925.

Poyatos-Racionero, E., Ros-Lis, J.V., Vivancos, J.L. and Martínez-Máñez, R. 2018. Recent advances on intelligent packaging as tools to reduce food waste. J. Clean. Prod. 172: 3398–3409.

Qing, G. and Guirong, Z. 2012. The green packaging management for the logistics enterprises. Paper presented at the 2012 International Conference on Information Management, Innovation Management and Industrial Engineering 1: 134–137.

Reinisch, M.A. 2022. Principais embalagens de alimentos por segmentação da indústria alimentícia: uma abordagem atual e sustentável.

Sánchez-Safont, E.L., Aldureid, A., Lagarón, J.M., Gámez-Pérez, J. and Cabedo, L. 2018. Biocomposites of different lignocellulosic wastes for sustainable food packaging applications. Compos B Eng. 145: 215–225.

Santana, R.F., Bonomo, R.C.F., Gandolfi, O.R.R., Rodrigues, L.B., Santos, L.S., dos Santos Pires, A.C., Pires, A.C.S., Oliveira, C.P., Fontan, R.C.I. and Veloso, C.M. 2018. Characterization of starch-based bioplastics from jackfruit seed plasticized with glycerol. J. Food Sci. Technol. 55: 278–286.

Satani, H., Kuwata, M. and Shimizu, A. 2020. Simple and environmentally friendly preparation of cellulose hydrogels using an ionic liquid. Carbohydr. Res. 494: 108054.

Sganzerla, W.G., Rosa, G.B., Ferreira, A.L.A., da Rosa, C.G., Beling, P.C., Xavier, L.O., Hansen, C.M., Ferrareze, J.P., Nunes, M.R., Barreto, P.L.M. and Veeck, A.P.L. 2020. Bioactive food packaging based on starch, citric pectin and functionalized with Acca sellowiana waste by-product: Characterization and application in the postharvest conservation of apple. Int. J. Biol. Macromol. 147: 295–303.

Shao, P., Feng, J., Sun, P., Xiang, N., Lu, B. and Qiu, D. 2020. Recent advances in improving stability of food emulsion by plant polysaccharides. Food Res. Int. 137: 109376.

Spinacé, M. and De Paoli, M.A. 2005. A tecnologia da reciclagem de polímeros. Quím. Nova 28: 65–72.

Šuput, D.Z., Lazić, V.L., Lević, L.B., Krkić, N.M., Tomović, V.M. and Pezo, L.L. 2013. Characteristics of meat packaging materials and their environmental suitability assessment. Hem. Ind. 67: 615–620.

Taib, N.A.A.B., Rahman, M.R., Huda, D., Kuok, K.K., Hamdan, S., Bakri, M.K.B. and Khan, A. 2022. A review on poly lactic acid (PLA) as a biodegradable polymer. Polym. Bull. 1–35.

Topuz, F. and Uyar, T. 2020. Antioxidant, antibacterial and antifungal electrospun nanofibers for food packaging applications. Food Res. Int. 130: 108927.

Torres, F., Troncoso, O., Torres, C., Díaz, D. and Amaya, E. 2011. Biodegradability and mechanical properties of starch films from Andean crops. Int. J. Biol. Macromol. 48: 603–606.

Williams, H., Wikström, F., Otterbring, T., Löfgren, M. and Gustafsson, A. 2012. Reasons for household food waste with special attention to packaging. J. Clean. Prod. 24: 141–148.

Williams, H., Lindström, A., Trischler, J., Wikström, F. and Rowe, Z. 2020. Avoiding food becoming waste in households—The role of packaging in consumers' practices across different food categories. J. Clean. Prod. 265: 121775.

Wong, C.W., Lai, K.H., Shang, K.C., Lu, C.S. and Leung, T. 2012. Green operations and the moderating role of environmental management capability of suppliers on manufacturing firm performance. Int. J. Prod. Econ. 140: 283–294.

Xie, Y., Niu, X., Yang, J., Fan, R., Shi, J., Ullah, N., Feng, X. and Chen, L. 2020. Active biodegradable films based on the whole potato peel incorporated with bacterial cellulose and curcumin. Int. J. Biol. Macromol. 150: 480–491.

Yang, X., Li, A., Li, X., Sun, L. and Guo, Y. 2020. An overview of classifications, properties of food polysaccharides and their links to applications in improving food textures. Trends Food Sci. Technol. 102: 1–15.

Yossef, D. and Hot, D. 2021. Comparative life cycle assessment of organic building materials. Energy Effic. 27: 1–48.

Zhong, Y., Godwin, P., Jin, Y. and Xiao, H. 2020. Biodegradable polymers and green-based antimicrobial packaging materials: A mini-review. Advanced Industrial and Engineering Polymer Research 3: 27–35.

Chapter 6

Edible Films and Coatings

Raw Material, Properties, and its Application in Food

Gabriela Almeida de Paula,[1,*] *Sandra Helena de Mesquita Pinheiro,*[2] *Lindalva de Moura Rocha,*[3] *Ronnyely Suerda Cunha Silva,*[4] *Whellyda Katrynne Silva Oliveira,*[4] *Francisco Eduardo Pareira Rocha,*[5] *José Ribeiro dos Santos Júnior*[6] and *Erivan Santos Lima*[7]

Introduction

For a long time, people have been worrying about finding methods to conserve and protect food. With that in mind, they had the idea of using some kind of coating that could protect the food from deterioration and mechanical damage. Many materials, some descending from renewable sources, have been used like: plastics, metals, wood, glass, paper, etc.

As the years passed by, the world population has been increasing and consequently the consumption of food. The growth in food consumption led to an increase in the use of packaging for storage and transportation. As the vast majority of these packaging

[1] Assistant professor at Federal University of Piauí. Doctor in Biotechnology trom the Northeast Network of Biotechnology/UFC, Fortaleza/CE.

[2] Associate Professor at Federal University of Delta do Parnaíba. Doctor in Food Sciences and Technology from the Federal University of Viçosa-MG.

[3] Doctoral student in Biotecnology from the Northeast Network of Biotechnology/UFPI, Teresina-PI, Brazil.

[4] Bachelors of Nutrition from the Federal University of Piauí, Teresina-PI.

[5] Bachelors of Fishing Engineering from the Federal University of Delta do Parnaíba, Parnaíba-PI.

[6] Full Professor at Federal University of Piauí. Doctor in Chemistry (Physicochemical) from the University of São Paulo.

[7] Graduating in Fishing Engineering from Delta do Parnaíba Federal University, Parnaíba-PI.

* Corresponding author: gabrieladepaula@ufpi.edu.br

is not recyclable or biodegradable, the tendency is for them to keep accumulating in big landfills or simply being thrown into the environment, causing the pollution of cities, rivers, lakes, oceans, forests, etc. This enormous pollution of the environment may cause a variety of dangers for the population like the emergence of several diseases, the contamination of potable water sources, the extinction of animals and plant species, increased pest population and other dangers.

The quality of a food product depends on its sensorial, nutritional and hygienic characteristics that change during storage and commercialization. Many chemical and physical processes have been developed to preserve the quality of food. However, it is necessary that appropriate packing for the conservation and commercialization of the product, for it has a preponderant role in the maintenance of the food's quality (Debeaufort et al., 1998).

The use of edible films and coating has become a topic of great interest due to the potential of these elements to avoid the deterioration of the food and due to their biodegradable characteristic (McHugh et al., 1996). This way, research on packaging has been focused on edible coating and films based on biopolymers like proteins, polysaccharides and lipids that are completely biodegradable within a considerable short period of time, greatly contributing to the decrease of environmental pollution (Han, 2000). Edible films and coatings have been shown to be an effective preservation technique for fruits and vegetables to keep their fresh look, firmness and luster, thus increasing their commercial value (Sotero, 2000).

The concept of applying edible films and coating on food is not new. Edible films have been used for a long time, for example the coating of sausage was initially made of animal intestines (Lawton, 1996). Since the XII century, the Chinese applied wax to oranges and lemons to increase the shelf life of these fruits (Bourtoom, 2008).

The first document about edible films and coatings was around 1800 (Debeaufort et al., 1998). From 1930, beeswax, paraffin, carnauba, mineral and vegetable oil were used in fruit conservation. Wax emulsion and oil have been used on dry fruits to increase their appearance, color and shine, control their ripening, and delay water loss (Klahorst, 1999). Around 100 different types of edible coatings and films were patented since 1950. Nevertheless, the use of edible packaging on processed food especially those with high humidity content has received little attention (Han, 2000).

Since the beginning of the mid-20th century, coatings have been used to avoid water loss and to shine fruits and vegetables. Edible films and coatings form a barrier to chemical, bodily and biological changes. When buying fruits and vegetables, the consumer decides what is healthy based on its freshness and appearance (Kader, 2002). The most difficult and common problem is to maintain and better manage the deterioration and the growth of pathogenic microorganisms at the fruit market. A more efficient way to solve this problem is the use of food using edible films and coatings (Kader, 2002). A healthy edible coating offers extra protection to the product and also offers the same influence during the storage in the modified atmosphere without altering the internal composition of the gases. These days, several edible coatings are applied in an efficient way for the preservation of fruit and vegetables, including, orange, apple, grapefruit, cherry, cucumber, strawberry, tomato and bell pepper being stored with success.

The edible films and coating are defined by two principles. First, the edible term in the compounds used in making the packing are GRAS (Generally Recognized As Safe) by the FDA (Food and Drug Administration), and processed within the Good Manufacturing Practices (BPF), established for food. Second, these films and coating must be made of a polymer, typically a biopolymer, since the long chain is necessary to give some insolubility and stability to the matrix of the packaging in an aqueous media (Krochta et al., 1994; Krochta and DeMulder-Johnston, 1997). The edible films and coatings are presented in different ways. As a film, it is a thin skin formed separately from the food and then applied to it. As a coating or a cover, it is a suspension or emulsion applied directly over the surface of the food occurring after drying, the formation of thin skin on the product. The thin covering on the food product acts as a barrier to the external elements like humidity, oils and organic vapor, protecting the product and extending its shelf life (Salleh, 2013).

The objective of this chapter is to review various types of natural biopolymers such as polysaccharides, proteins and lipids as raw material sources for edible films and coatings, including their properties and application in foods.

The importance of edible films and coatings

The films are used to improve the sensorial characteristics of packaged food. Films/coatings play a function as conveyors of different compounds like colorings, flavorings, sweeteners, antimicrobials and antioxidants (Han and Gennadios, 2005; Vasconez et al., 2009). The films can be implemented like heterogeneous internal ingredients at the interfaces between distinct layers of food compounds. These films are the correct way to stop deteriorating moisture of the internal compounds and the movement of the solute in several food products including sweets, pies and pizzas (Bourtoom, 2008). Another great use for the edible films and coating is the utilization of ingredients of multilayered food packaging together with inedible films and coatings. In this type of film, the edible films and coats create the internal layers with direct interaction with food ingredients. The functional activity and the permeability of the edible films and coatings are usually lesser than the artificial films and can produce lead contamination (Kester and Fennema, 1986). Beyond its efficiencies like selective barriers for the migration of the gases, humidity and solute, the edible coating can effectively reduce the microbial growth in solids and semi-solid food products, decreasing the diffusion rate of the antimicrobial agents of the coating material for food products (Aloui and Khwaldia, 2016).

Edible films and coatings act as a blocker of oxygen, moisture and solute disposition for food without altering the original ingredients (Debeaufort et al., 1998; Krochta, 2002). The coating material can be a whole meal or can be synthesized as a continuous coating between food additives (Guilbert et al., 1996). Edible films and coating have acquired a great interest in the past few years due to their benefits over artificial films in promising results in food preservation (Galus and Kadzinska 2015). The main purpose to form edible films instead of conventional packaging is that these can be used with the packaged product (Kuorwel et al., 2015). The motivation for the increase of the interest and the research activity in the edible coating is due to the crescent need of the consumer for safe products, healthy and stable foods and also the awareness of harmful environmental effects of non-biodegradable waste resulting from packaging (Debeaufort et al., 1998; De Azeredo, 2012). The edible films are made entirely from renewable and safe materials in the preparation of their components. The edible coatings are transported in liquid form, however, the edible films are acquired as strong bases and then implemented in food products (Falguera et al., 2011). Edible coatings are being applied to food by spraying, immersion and more recently electrospraying which produces fine products and uniform coating (Khan et al., 2012; Khan et al., 2013).

The increase in consumer demand for natural food preservation has led to the improvement of alternative methods of protection, for example, the use of biopolymers that one can use from renewable resources or industrial sub-products (Campos et al., 2011). The use of edible films and coatings is a topic of great interest due to their characteristics to prolong the service life of industrialized food. The service life of the product can be prolonged by decreasing the respiration, with the help of edible coatings (Baldwin, 1995). Edible coatings are biodegradable helping the environment to reduce the use of plastic packaging. The increase in the service life of food products is very important because an extension of some days of service life can represent a significant economic advantage for food companies.

The raw material used in the formulation of edible films and coatings

According to their composition, edible films and coatings can be classified into three categories: Hidrocolloidal–films based on polysaccharides or proteins. They feature low permeability to oxygen, carbon dioxide and lipids. However, due to its hydrophilic nature, have a low barrier to water vapor (Labuza and Breene, 1989). Lipids, due to their hydrophilic nature show low permeability to water (Han, 2000). There are many lipids in crystalline form with low permeability to gases and water vapor. Compound–based on proteins with lipids or polysaccharides with lipids. They can exist as separate layers or be associated, where both the components are added to the film. The researchers have focused on compound packaging, because they match the advantages of every component, thus reducing their disadvantages (Labuza and Breene, 1989).

The base of the edible films and coating are biopolymers like polysaccharides, proteins and lipids derived from various natural sources.

Polysaccharides-based coatings

Polysaccharides are polymers of natural occurrence, widely used to prepare edible films and coatings including starch, cellulose, pectin and their derivatives, pullulan, alginates and chitosan (Krochta and Mulder-Johnston, 1997). Polysaccharides are coatings based mainly to be efficient oxygen blockers due to their well-ordered hydrogen bond in a chain format (Table 6.1). But they do not act well as a humidity barrier because of their hydrophilic nature (Mchugh and Krochta, 1994; Yang and Paulson, 2000). Despite the fact that the coatings that use polysaccharides have poor barrier properties to water vapor, these coatings are a reason that merchants

Table 6.1. Some coatings based on polysaccharide for different products with significant function.

Coatings with the suitable additives		
Coatings of k-carrageenan or tapioca starch with potassium sorbate, ascorbic acid, citric acid, glycerol.	Pumpkin	Fortifies pumpkin color and antimicrobial activity
Alginate, pectin and gelatin coatings with deacytaled gellan gum, methoxy pectin, esterified potassium salt, glycerol, calcium chloride additives.	Minimally processed melon	The effects of edible coatings based of alginate, pectin and gellan in the shelf life of the minimally processed melon.
Starch-based coatings with D-glucose, silver nitrate, EDTA, trichloroacetic acid additives.	Chicken sausage	Increase the antimicrobial activity.
Polysaccharide from *A. occidentale* L. tree gum with Sorbitol, Tween 80 additives.	Apples	Increase the permeability to water vapor, opacity and mechanical properties.
Hydroxypropylmethylcellulose (HPMC) and beeswax coatings with oleic acid, glycerol additives.	Cherry tomato	Weight loss, skin color, fruit firmness, respiration rate and improved sensorial attributes.
Coatings of Pectin and Sodium Alginate with Essential Oils (citral and eugenol) as additives.	Raspberry	They control color, the concentration of soluble solids, weight loss, the equivalent antioxidant capacity of the trolox, microbial growth and flavor.
Starch-based coatings with sodium hydroxide solution, glycerol, sorbitol, sunflower oil additives.	Brussels sprouts	Increasing the shelf life, optimizing the weight loss, color of the surface and texture.
Chitosan-based coatings with additives of glycerol, gallic acid and powdered green tea	Walnut core	The effects on lipid oxidation, fungal growth and sensorial properties.
Chitosan gelatin-based coatings with glycerol additives.	Meat	Effects of meat on color preservation and lipid oxidation during retail display.
Coatings of cellulose derivatives (hydroxypropylmethyl cellulose, methylcellulose) with sorbitol additives.	Potato	Potato coating is used to reduce oil absorption in fried products (water and lipid content).
Pectin-based coatings with green tea powder, polyethylene glycol additives.	Pork patty	Maintains the physical-chemical, microbiological and sensorial qualities of cooked pork burger.
Pullulan-based coatings with glutathione, a-amylase and ethanol additives.	Apples	Color, appearance and sensorial attributes during storage in hypothermia.
Chitosan and pectin coatings with trans-cinnamaldehyde and beta cyclodextrin hydrate additives.	Minimally processed melon	Extend the shelf life of melons stored at 4°C
Xanthan gum coatings with cinnamic acid additives	Minimally processed pear	Quality attributes (color, firmness, weight loss, water loss) of the pear stored at 4°C
Chitosan coatings with glacial acetic acid, NaOH.	Mango	The sensorial qualities of mango flavor, color and water loss were examined.
Coatings of sodium alginate with glycerol, calcium chloride, N-acetylcysteine.	Apples	Shelf life during storage color, firmness, sensorial and antimicrobial quality.

are concerned about food suffering from humidity loss (Kester and Fennema, 1986). The polysaccharide coatings are colorless, have an aspect free of oil and a lesser conductivity to heat, and can be applied to prolong the service life of fruits, vegetables, seafood or products based on meat, reducing significantly the dehydration, darkening of the surface and oxidative rancidity (Table 6.1).

Starch is one of the most important and abundant polysaccharides in nature and has been a subject of numerous investigations that report, among others, its capacity to form coatings for application in the food industry (Arvanitoyannis et al., 1998). The use of starch can be an interesting solution for edible films and coatings due to its low cost, high production, biodegradability, edibility and easy manipulation (McHugh and Krochta, 1994). Starch is obtained in a granular shape from renewable resources like corn, yam, manioc, potato, cereals and vegetables. It is the correct material for a variety of uses in the industry (Petersen et al., 1999). According to Lawton, 1996, the films, depending on the source of the starch, present different properties assigned to the content of amylose in the starch that carries between 18 and 30%. The higher the content, the better will be the film formation (Petersen et al., 1999). Yet, due to the hydrophilic aspect the films and coatings based on starch present low barrier properties to water steam, but strategies are being developed to increase the humidity stability of these films and coatings (Petersen et al., 1999). The addition of plasticizers, like sorbitol and glycerol to films and coatings based on corn and potato starch with different content of amylose improved their barrier properties to water vapor. So, the higher the concentration of plasticizers in the coatings, the lesser the permeability of water vapor allowing the coatings with sorbitol with the lowest values of permeability. The coatings containing a higher content of amylose present lesser permeability to water vapor (García et al., 1999). The addition of sunflower oil to the coatings based on corn starch with different content of amylose, significantly decreased the permeability to water vapor; when lipids and plasticizers were added at the same time the permeability was even smaller (García et al., 2000).

In Brazil, the main sources of starch for the food industry are corn (Zea mays) and manioc (*Manihot utilíssima*). Yam (*Dioscorea alata*) is considered as an alternative source because of many desirable properties of its starch like high-temperature stability and low pH (Alves et al., 1999). It is a good source of starch for the production of edible films and coatings because its starch on a dry base contains 30% of amylose, and this polysaccharide on the starch is responsible for the formation of the films. These based on yam starch (3,3% w/w of starch and 2% w/w of glycerol) present a homogeneous matrix with a stable structure under ambient conditions, permeability to water vapor of $1{,}55 \times 10^{-10}$ g^{-1} $^{-1}$ s Pa^{-1} at 25°C and 75% of relative humidity, much smaller than that presented by other edible films based on gluten of wheat/glycerol ($7{,}0 \times 10^{-10}$ g^{-1} m^{-1} s^{-1} Pa^{-1}) and hydroxypropyl methylcellulose/glycerol ($1{,}9 \times 10^{-10}$ g^{-1} m^{-1} s^{-1} Pa^{-1}) at the same temperature and relative humidity (McHugh and Krochta, 1994). This way, these films have great potential for application in the food industry, beyond being biodegradable. Other polysaccharides used in the production of edible films and coatings are cellulose ethers, polymers obtained by the partial replacing of hydroxyl groups at the cellulose that is water-soluble and possess good film-forming properties. Among them are methylcellulose (MC), hydroxypropyl cellulose (HPC), hydroxypropyl methylcellulose (HPMC) and carboxymethylcellulose (CMC) (Gennadios et al., 1997; Petersen et al., 1999).

Coatings based on proteins

Proteins usually occur like fibrous or globular proteins. The fibrous proteins are water-soluble and play a function as the primary structural material in animal tissue, while globular proteins are water-soluble, as also are soluble in acid, base or salt solution and explain many functions on living systems. Fibrous proteins are closely associated with each other by the H bond to the fibers made. In globular proteins the ionic, covalent and hydrogen bonds are present in folded and complex structures (Scope, 1994). The physicochemical characteristics of the proteins are totally dependent on the arrangement of substituting amino acids of acid and its relative amount next to the current polymer. Different types of globular proteins like soy protein, wheat gluten, whey protein and corn zein were studied for their film/coating forming properties (Table 6.2). The protein solution or dispersion is used to synthetize films, coatings and solvents that are used for this purpose are commonly limited to ethanol, water or ethanol-water combination (Kester and Fennema, 1986).

Normally, the denaturation of the protein is made by the use of acid, base, solvent and heat to mold the prolonged extra systems that can be necessary for film formation. Once stretched, the protein chains remain connected by the hydrogen, ionic and covalent bonds. The arrangement of polar groups next to the polymer

Table 6.2. Some protein-based coatings for different products with significant functions.

Coatings with suitable additives	Food product	Significant functions
Candelilla wax coating with guar gum and glycerol additives	Strawberry	Antifungal characteristics to increase the post-harvest life.
Covering of lipid base (sunflower oil and chocolate) with stearic acid, polyglycerol polyricinoleate (PGPR) and butter additives.	Hygroscopic candy pills	Properties of humidity barrier
Coating candelilla wax with jojoba oils, ellagic acid additives.	Delicious golden apple	Sensorial qualities, weight loss, fungal strain and microbiological activities
Beeswax coatings with glycerol additives	Dry sausage	Weight loss reduction
Candelia wax coating	Brussels sprout	Reduces weight loss, retention of vitamin C and polyphenols, limited softening during storage and improves the overall appearance.
Candelilla wax coatings with mineral oil	Guava	Ethylene emission of weight loss, brightness, retention of reduction in weight loss, color retention
Candelilla wax coatings with mineral oil	Persian limes	
Carnauba wax	Apple	Increased resistance to water vapor and reduced weight loss.
Carnauba wax	Nuts and pine nuts	Oxidative and hydrolytic rancidity, improves softness, flavor and overall appearance.

chain increases the chances of the necessary interactions. These chain interactions determine the resistance of the edible film, higher yields make the films stronger, but less permeable to vapor, liquids and gas. (Kester and Fennema, 1996). Therefore, protein-based films or coatings are considered highly efficient oxygen blockers, even at low relative humidity (RU). Different types of protein have been used to produce edible films/coatings that comprise whey protein, wheat gluten, gelatin, corn zein, casein and soy protein (Bourtoom, 2008; Gennadios et al., 1993).

Among the proteins, the ones with animal origins like collagen, gelatin, casein, cheese whey protein and the ones with plant origins like zein, derived from corn, the soy isolated protein and the wheat gluten protein are used (Gennadios and Weller, 1990; Lawton, 1996). The latter consists of two protein fractions, the gliadin and the glutenin. The gliadin, soluble in 70% ethanol, corresponds to the viscous component of the gluten, while the glutenin insoluble in 70% ethanol contributes to its elastic properties (Morel et al., 2000). The zein is a protein of hydrophobic character, due to the high concentration of non-polar amino acids like leucine, alanine and proline. It produces films with excellent humidity barrier properties. The zein-based coatings form a tasteless, hard and transparent cover and present stability in high humidity and heat is normally used for covering sweets, dried fruits, nuts, rice, seeds and pharmaceutical tablets. An additional benefit of the zein is one's natural resistance to bacterial and some insect attacks. The soy soluble proteins are also used in bakery products. The edible coatings in these products are frequently used to shine or as waterproofing in seeds nuts and whole grains (Krochta and DeMulder-Johnston, 1997). Researchers have been using milk proteins like casein and whey protein in the elaboration of edible and biodegradable films. This is a new application of the milk-derived ingredients, supplying an alternative to the dairy products industry, that is, adding value to dairy waste (Chen, 1995; Krochta and DeMulder-Johnston, 1997). The milk protein is made up of 80% casein and 20% whey protein. These are obtained from the cheese serum and separated from the lactose by ultrafiltration then they are evaporated and lyophilized. Whey proteins are made up of five fractions: α-Lactoalbumin (αLa), β-Lactoglobulin (β-14), Bovine Serum Albumin (BSA), immunoglobulin (Ig), peptone-protease, the β-Lg being the component with higher concentration, representing 50 to 60% of the proteins. Commercially, it is produced Whey Protein Concentrate (WPC) with 80% of proteins and Whey Protein Isolate (WPI) with a protein content higher than 90% (Chen, 1995). The isolated whey protein has the advantage of being highly soluble in water on an extensive range of pH and producing shiny, flexible films with excellent properties of oxygen, aroma and oils barrier in low and intermediate relative humidity (Krochta and DeMulder-Johnston, 1977; Miller and Krochta, 1997; Sothornvit and Krochta, 2001). Edible films using individual fractions of whey protein have been poorly studied due to the high cost of purification and separation. Beyond this, research made by Mate and Krochta (1994), comparing the properties of films made of whey protein isolated and films made of βLactoglobulin, shows that there are not significant differences between them ($P < 0{,}01$) in permeability to water vapor and oxygen. Other proteins of animal origin, like gelatin and collagen, have been used to produce edible films. The egg albumin is efficient to shine bakery products (Krochta and DeMulder-Johnston, 1997).

Lipid-based films and coatings

Lipids are excellent barriers to humidity (Debeaufort and Voilley, 2009). Lipids, when mixed with the proteins and polysaccharides produce coatings with superior mechanical and barrier resistance. Lipid-based compounds are applied like protective film/coating composed of herbal wax, acetylate monoglycerides and surfactants. The simplest compound lipids are paraffin and beeswax; Lipid-based films and coatings are considered highly efficient to block the entrance of humidity due to its low polarity. Normally the films or the coatings made of lipids are extra brittle and thicker due to its hydrophobicity (Perez-Gago et al., 2002). With the increase of concentration of hydrophobicity phase, there is a reduction of the permeability to water vapor (Valencia-Chamorro et al., 2010). However, it was reported that lipids containing coatings and films can damage the appearance and the brightness of coated food products (Perez-Gago and Roja, 2002; Valencia-Chamorro, 2010).

Among lipids are beeswax, paraffin, carnauba, mineral and plant oils. Some components are important in the formation of films and coatings like plasticizers, which are non-volatile substances with high fusion points when added to another material change its physical and, or mechanical properties. Inedible films and coatings, the most used plasticizers are glycerol and sorbitol, that act on the hydrogen bond reducing the intermolecular strength along the polymer chain, and improving the packaging its mechanical properties like flexibility, strength and resistance. McHugh et al. (1996) examined the effect of glycerol and sorbitol on whey protein isolated-based films and found that these plasticizers reduced the internal hydrogen bonds increasing, thereby, the flexibility and permeability to water vapor. In the same condition, deconcentration of plasticizer and relative humidity, the films with sorbitol present lesser permeability to water vapor than the ones with glycerol.

Edible films and coatings properties

Optical properties

For a good visual presentation of the product, it is desirable that the packaging (films and/or coating) presents great brightness and high transparency. On the other hand, many times the protection against incidence of light is necessary (low or null transparency), like in the storage of products sensible to reactions of deterioration catalyzed by light. Therewith opacity and color are the most important optical properties of films and coatings applied on the packaging. Color is considered an important parameter in the categorization of films because it is related to the raw material used in their elaboration (Vicentini, 2003).

For use as food packaging, it is necessary that the films and coatings have high transparency and great brightness, so the original characteristics of the product like color, for example is maintained, thus leaving a good visual presentation of the product (Yang and Paulson, 2000).

Regarding opacity of a material, one must take into consideration its highest or lowest transparency. The low transparency (or high opacity) of a material is characterized by the blocking of the light passage (Davaco, 2006). The degree of transparency or opacity of a polymer is a consequence of the chemical structure related to the molecular mass of the polymer used and its thickness (Crippa, 2006).

The coloring and the opacity of the film must not be modified during the storage of the material, this change can spoil the acceptance of the product (Sakanaka, 2002).

Barrier property: permeability to water vapor

According to Sarantópoulos (2002), the property of the barrier is defined as the capacity to resist absorption or evaporation of gas and vapor, resist permeation of lipids and the passage of light.

Edible films and coatings should have the potential to offer improvements in a combination of a factor when applied to food, factors such as humidity, oxygen, flavor, aroma, color or lipid barrier for a portion of food, with a consequent increase in quality and shelf life (Krochta, 2002).

In food with low activity of water, receiving humidity favors microbial growth, the oxidation of fats, enzymatic action and non-enzymatic darkening. In food of high-water activity, the loss of humidity to the environment represents the loss of weight and results in physical, chemical and sensorial alterations. Thus, the use of packaging with a good barrier to water vapor for humidity of reasonable products, allows that the intrinsic quality is maintained for a longer time (Sarantópoulos, 2002). Due to the great importance of humidity control to maintain the quality of food, the mass transference of vapor through edible films has been increasingly studied by methods and utilization of equipment that allows one to measure quite accurately.

The rate of permeability to water vapor is a relevant analysis in the field that includes polymeric materials. Krochta (2002) defined permeability as the concentration or difference of partial pressure between the phases adjacent to the two sides of the film, so that a property that describes the degree to which a substance that permeates dissolves, and then the rate at which the permeating diffuses through a film, with a motor force related with the difference of permeating concentration between both sides of the film.

This water vapor permeability through the polymeric films occurs in four stages: the first is the water vapor adsorption on the surface of the polymer; the second, the solubilization of water vapor in the polymeric matrix; the third, water vapor diffusion through the polymer; and the fourth, the water vapor desorption on the other surface of the polymer (Roy et al., 1999).

The knowledge of water vapor permeability is essential to define the possible applications of the films in packaging, however, it is not a restrictive property. One very permeable water vapor material can be indicated in fresh vegetable packaging, while a film barely permeable can be indicated in dehydrated products for example (Sobral, 2000).

Mechanical properties

One of the most relevant characteristics of edible films and coatings for their application in food are the mechanical properties, due to the handling where the products are susceptible during distribution and commercialization. Films and coatings must have as features rupture and abrasion resistance attributes that are important to reinforce the structure of the product and facilitate its handling, they must also have a flexibility that allows adaptation to possible deformations without causing ruptures (Sarantópoulos et al., 2002).

Some mechanical properties of films can be assessed by tensile tests that are useful for the identification and characterization of flexible films, with applications in the development, specification and quality evaluation of these materials. They are manifested by the capacity of these materials to develop reversible and irreversible deformations, these characteristics are generally evaluated by technical tests through answers that these materials present when submitted to various stress-strain dependencies (Sebio, 2003).

Resistance to traction and elongation are mechanical properties that must be presented in films and coatings for food. The elongation is measured by the malleability of the films and can be considered as a characteristic that defines the ability of the films in deforming before occurring its rupture. The resistance to traction is the maximum stress supported by the film until the moment of its rupture (Bertan, 2003).

It is important to know the differences between the tests and the mechanical stress in practice, related to the external variable that affects the mechanical behavior of the polymeric materials: temperature, time or deformation speed, the level of mechanical stress, the type of stress and environmental conditions, like for example the effect of the relative humidity over the properties of a hydrophilic polymer (Agnelli and Mascheroni, 2001).

Applications of edible films and coatings in food

Application in products of plant origin

Currently, people have shown a growing concern with health and with it the search for healthy and nutritive food that brings health benefits, with that the demand for fruits and vegetables in natural and minimally processed has been increasing, however, they are perishable food with a short shelf life (Ciolacu et al., 2014).

The quality factors of fresh food are important to ensure their acceptability, vegetables and fruits are highly perishable and during the process of post-harvest, there are considerable losses, due to injuries by insects, microorganisms, respiration, transportation and transpiration. The post-harvest losses are a big problem due to their quick deterioration by handling, transportation and storage resulting in financial losses (Barbosa et al., 2014).

Having said that, fresh fruits and vegetables need technologies post-harvest to preserve their quality and stability, one option would be coating by edible film that offers a semipermeable barrier to gas and water vapor and reduce the respiration rate, which offers a solution to extend the lifespan of fresh fruits and vegetables (Jung and Zhao, 2016). In the case of fruits and vegetables, the use of edible films and coatings has several objectives, which can be quoted as important: to reduce the loss of humidity that helps in the decrease of weight loss; to inhibit microbial growth, to control the gas permeability; to preserve the structural integrity; to slow the aerobics respiration increasing the appearance, controlling the maturation, reducing the penetrated oxygen in the fruit, controlling their metabolic activities, softening changes of the fruits and vegetables and then to enable the gradual release of the flavor and of antioxidants increasing the shelf life of these (Valencia-Chamorro, 2011).

Application in products of animal origin

Meat distinguishes itself by the structure of the tissue and is frequently treated as a different process that favors the development of microorganisms, with that there is a need for different methods to manage pathogens of food origin and to prolong its life span (Guo et al., 2014).

With the advance of technology, edible packaging emerged to develop mechanical properties, to avoid superficial dehydration of meats, as gas, humidity and microbial control barriers, decrease the lipids oxidation and preservation of the sensorial characteristics of meat products, beyond being used to reduce the absorption of fat while frying meat (Galus and Kadzińska, 2015).

Among the methods that can be used in the process of edible coatings of foods, one can highlight the method of foam, immersion, pulverization, electrostatic pulverization, casting, brushing an in individual packaging or winding in the case of films (Ustunol, 2018).

Fresh fish and minimally processed are extremely perishable and during storage the quick growth of microorganisms that are naturally found in fish or contaminated can occur and occasionally result in economical or health problems. For conservation, some strategies can be used, as the use of edible films and coatings to increase the lifespan, reduce the water loss, slow the lipid oxidation, to inhibit the microorganism growth and to help in fish products safety (Lacey et al., 2014).

Milk and dairy products are a significant source of necessary nutrients for childrens' growth and for the health maintenance of adults (Cardador and Gallego, 2016). Cheese is the most diverse group of dairy products, with a lifespan constrained by the action of microorganisms. Antimicrobial coatings applied to the surface of cheese is a way to limit the negative changes caused by the action of microorganisms that can be acquired during inappropriate handling and storage (Galus and Kadzińska, 2015).

For the use of food coatings in dairy products some important aspects must be assessed for the application of the method, since the choice will depend on the type and size of the cheese where the coating will be applied, these choices will affect the cost and efficiency of the coating and must be adapted to the production process. The most adequate method will be the one to guarantee the best description of dairy, so it will contribute to an extension of the expiration date and preserve its organoleptic characteristics (Zhong et al., 2014).

Conclusions

With the advancement of technology, new edible films and coatings have been developed, so that it can meet the market's needs, either by requiring biodegradable coatings or by reducing pollution, as well as increasing the shelf life and ensuring food safety. Thus, the production of films based on biopolymers has established itself as an alternative in the scope of various food products, compared to conventional plastics that are produced from petroleum.

References

Agnelli, M.E. and Mascheroni, R.H. 2001. Cryomechanical feezing. A model for the heat transfer process. J. Food Eng. 47(4): 263–270.

Allen, L. Nelson, A.I., Steinberg, M.P. and McGill, J.N. 1963. Edible carbohydrates food coating II: Evaluation on fresh meat products. Food Technol. (17): 1442–6.

Aloui, H. and Khwaldia, K. 2016. Natural antimicrobial edible coatings for microbial safety and food quality enhancement. Compr. Rev. Food Sci. Food Saf. 15(6): 1080–1103.

Alves, R.M.L., Grossmann, M.V.E. and Silva, R.S.S.F. 1999. Gelling properties of extruded yam (*Dioscorea alata*) starch. Food Chem. 67: 123–7.

Arvanitoyannis, I., Nakayama, A. and Aiba, S. 1998. Edible films made from hydroxypropyl starch and gelatin and plasticized by polyols and water. Carbohy. Pol. 36: 105–19.

Azeredo, H.M.C. 2012. 14 edible coatings. Advances in Technology and Fruit Processing, 345.

Baldwin, E.A., Nisperos-Carriedo, M.O. and Baker, R.A. 1995. Edible coatings for fruit and lighty processed vegetables. HortScience 30(1): 35–38.

Banker, G.S. 1966. Film coating theory and practice. J. Pharm Sci. 55(1): 81–9.

Barbosa, P.L., Angulo, I., Lagarón, J.M., Paseiro-Losada, P. and Cruz, J.M. 2014. Development of new films for active packaging containing bioactive Nano composite. Innov. Food Sci. Emerg. Technol. 26: 310–318.

Bertan, L.C. 2003. Development and characterization of simple and compound films based on gelatin, fatty acids and white pitch. (sn).

Bourtoom, T. 2008. Edible films and coatings: characteristics and properties. Int. F. Research J. 15(13): 237–248.

Campos, C.A., Gerschenson, L.N. and Flores, S.K. 2011. Development of edible films and coastings with antimicrobial activity. Food Bioproc. Tech. 4(6)L 849–875.

Cardador, M.J. and Gallego, M. 2016. The origin of haloacetics acids in milk and dairy products. Food Chem. 196: 750–75.

Chen, H. 1995. Functional properties and applications of edible films made of milk proteins. J. Dairy Sci. 78(11): 2563–83.

Ciolacu, L., Nicolau, A.I. and Hoorfar, J. 2014. Edible coatings for fresh and mini-fruits and poorly processed vegtables. Global Security of Fresh Products Produce.

Crippa, A., Sydenstricker, T.H.D. and Amico, S.C. 2006. The influence of the thermoforming process on co-extruded plastic films for sausage packaging. WORLD POLYMER CONGRESS - Macro 2006, 41st Int. J. Biol. Macromol., Rio de Janeiro. Anais.

Davaco, D. 2006. Development and Characterization of biofilms based on gelatin, triacetin, stearic acid or capricious acid and surfactants. Doctoral thesis, 2006. Faculty of Food Engineering. Universidade Estadual de Campinas, SP, p. 130.

De Lacey, A.L., López-Caballero, M.E. and Montero, P. 2014. Agar films containing green tea extract and probiotic bacteria for extending fish shelf-life. LWT-Food Sci. Technol. 55(2): 559–564.

Debeaufort, F., Quezada-Gallo, J.A. and Voilley, A. 1998. Edible films and coatings: tomorrow packaging: a review. Cr. R. F. Sci. 38(4): 299–313.

Debeaufort, F. and Voilley, A. 2009. Lipid-based edible films and coatings. pp. 135–168. *In*: Edible Films and Coatings for Food Applications. Springer, New York, NY.

Falguera, V., Quintero, J.P., Jiménez, A., Muñoz, J.A. and Ibarz, A. 2011. Edible films and coatings: Structures, active functions and trends in their use. Trends Food Sci. Technol. 22(6): 292–303.

Galus, S. and Kadzińska, J. 2015. Food applications of emulsion-based edible films and coatings. Trends Food Sci. Technol. 45(2): 273–283.

Garcia, M.A., Martino, M.N. and Zaritzky, N.E. 1999. Edible starch films and coatings characterization: scanning electron microscopy, water vapor, and gas permeabilities. Scanning 21(5): 348–353.

Garcia, M.A., Martino, M.N. and Zaritzky, N.E. 2000. Lipid addition to improve barrier properties of edible starch-based films and coatings. J. Food Sci. 65(6): 941–944.
Gennadios, A. 1990. Edible films and coatings from wheat and corn proteins. Food Techn. 44: 63–69.
Gennadios. A. and Weller, CL. 1990. Edible films and coatina from wheat and corn proteins. Food Technol. 44flOl: 63.
Gennadios, A., Brandenburg, A.H., Weller, C.L. and Testin, R.F. 1993. Effect of pH on properties of wheat gluten and soy protein isolate films. J. Agr. Food 41(11): 1835–1839.
Gennadios, A., Hanna, M.A. and Kurth, L.B. 1997. Application of edible coatings on meats, poultry and seafoods: a review. LWT-Food Sc. Tec. 30(4): 337–350.
Guilbert, S., Gontard, N. and Gorris, L.G. 1996. Prolongation of the shelf-life of perishable food products using biodegradable films and coatings. LWT- Food Sc. Tec. 29(1-2): 10–17.
Guo, M., Jin, T.Z., Wang, L., Scullen, O.J. and Sommers, C.H. 2014. Antimicrobial films and coatings for inactivation of Listeria innocua on ready-to-eat deli turkey meat. Food Contro. 40: 64–70.
Han, J.H. 2000. Antimicrobiano Food Packaging. Food Technology 54(3): 56–65.
Han, J.H. 2003. Antimicrobial food packaging. Novel Food Packaging Techniques 8: 50–70.
Han, J.H. (ed.). 2005. Innovations in Food Packaging. Elsevier Sci.
Han, J.H. and Gennadios, A. 2005. Edible films and coatings: A review. pp. 239–259. *In*: Han, J. (ed.). Innovations in Food Packaging.
Jung, D.S., Bodyfelt, F.W. and Daeschel, M.A. 1992. Influence of fat and emulsifiers on the efficacy of nisin in inhibiting Listeria monocytogenes in fluid milk. J. Dairy Sci. 75(2): 387–393.
Jung, J. and Zhao, Y. 2016. Antimicrobial packaging for fresh products and minimally processed fruits and vegetables. pp. 243–256. Em J. Barros-Velazquez (ed.). Embalagem de alimentos antimicrobianos. Cambridge, MA: Academic Press.
Kader, A.A. 2002. Postharvest Technology of Horticultural Crops (Vol. 3311). University of California Agriculture and Natural Resources.
Kester, J.J. and Fennema, O.R. 1986. Edible films and coatings: a review. Food Techn. Chicago 40(12): 47–59.
Khan, M.K.I., Schutyser, M., Schroën, K. and Boom, R. 2012. The potential of electrospraying for an hydrophobic films of food. J. Food Eng. 108(3): 410–416.
Khan, M.K.I., Boom, R., Maan, B.A.A., Schutyser, M. and Schroën, K. 2013. Electrospraying of water in oil emulsion for thin skin coating. J. Food Eng. 4 ed. 119: 776–780.
Klahorst, S. 1999. Applications: Credible edible films. Food Product Design 9(1): 1–6.
Krochta, J.M. 2002. Proteins as raw materials for films and coatings: definitions, current status, and opportunities. pp. 1–32. *In*: Gennadios, A. (ed.). Protein-based Films and Coatings. Boca Raton: CRC Press.
Krochta, Baldwin, E.A. and Nisperos-Carriedo, M. 1994. Edible Coatings and Films to Improve Food Quality. Pennsylvania, Technomic, 379 p.
Krochta, M. 1997. Edible and biodegradable polymer films: challenges and opportunities. Food Techn. 51: 61–74.
Krochta, J.M. and DeMulder-Johnston, C. 1997. Edible and biodegradable polymer challenges and opportunities. Food Techn. 51(2): 61–74.
Kuorwel, K.K., Cran, M.J., Orbell, J.D., Buddhadasa, S. and Bigger, S.W. 2015. Review of mechanical properties, migration, and potential applications in active food packaging systems containing nanoclays and nanosilver. Comprehensive Reviews in Food Science and Food Safety 14(4): 411–430.
Labuza, T.P. and Breene, W.M. 1989. Applications of "active packaging" for improvement of shelf-life and nutritional quality of fresh and extended shelf-life foods 1. J. Food Proc. 13(1): 1–69.
Lawton, J.W. 1996. Effect of starch type on the properties of starch containing films. Carbohy. Pol. 29(3): 203–208.
Lerdthanangkul, S. and Krochta, J.M. 1996. Edible coating effects on postharvest quality of green bell peppers. J. Food Sci. 67(1): 176, 9.
Maté, J.I. and Krochta, J.M. 1994. β-Lactoglobulin separation from whey protein isolate on a large scale. J. Food Sci. 59(5): 1111–1114.
Mchugh, T.H. and Krochta, J.M. 1994. Milk-protein-base edible films and coatings. Food Techn. 48(1): 97–103.

McHugh, T.H., Huxsoll, C.C. and Krochta, J.M. 1996. Permeability properties of fruit puree edible films. J. Food Sci. 61(1): 88–91.
Miller, K.S. and Krochta, J.M. 1997. Oxygen and aroma barrier properties of edible films: A review. Trends in Food Sci. Tec. 8: 228–37.
Morel, M.H., Bonicel, J., Micard, V. and Guilbert, S. 2000. Protein insolubilization and thiol oxidation in sulfite-treated wheat gluten films during aging at various temperatures and relative humidities. J. Agr. Food 48(2): 186–192.
Perez-Gago, M.B., Rojas, C. and DelRio, M.A. 2002. Effect of lipid type and amount of edible hydroxypropyl methylcellulose-lipid composite coatings used to protect postharvest quality of mandarins cv. fortune. J. Food Sci. 67(8): 2903–2910.
Petersen, K., Nielsen, P.V., Bertelsen, G., Lawther, M., Olsen, M.B., Nilsson, N.H. and Mortensen, G. 1999. Potential of biobased materials for food packaging. Trends Food 10(2): 52–68.
Rojas-Graü, M.A., Raybaudi-Massilia, M., Soliva-Fortuny, R.C., Avena-Bustillos, J.T.H. and Mchugh, O. Martin-Belloso. 2007. Apple puree-alginate edible coating as carrier of antimicrobial agents prolongs shelf-life of flesh-cut aples. Postharvest Biol. Technol. 45: 254–264.
Roy, S., Weller, C.L., Gennadios, A., Zeece, M.G. and Testin, R.F. 1999. Physical and molecular properties of wheat gluten films cast from heated film-forming solutions. J. Food Sci. 64(1): 57–60.
Sakanaka, L.S. 2002. Study in the Conditions of Storage Over Mechanical and Barrier Properties of Reticulated Gelatin. Masters dissertation. Faculty of Food Engineering, Universidade Estadual de Campinas, Campinas, 175p.
Salleh, N.S.M. 2013. Development of Starch and Soy Protein Edible Coating and its Effect on the Postharvest Life of Mango (*Mangifera indica* L.) Doctoral Dissertation. Universiti Teknologi MARA, Shah Alam. Malaysia.
Sarantópoulos, C.I.G.L., Oliveira, L.D., Padula, M., Coltro, L., Alves, R.M. and Garcia, E.E. 2002. Embalagens plásticas flexíveis: principais polímeros e avaliação de propriedades. Campinas: CETEA/ITAL, 1: 267.
SCOPES, R.K. 1994. Protein Purification: Principles and Practice. [s.l.] Springer, 1994. p. 380.
Sebio, L. 2003. Development of biodegradable plastic based on corn starch and gelatin by the process of extrusion: evaluation of mechanical, thermal and barrier properties. Doctoral thesis. Faculty of Food Engineering, Universidade Estadual de Campinas. Campinas.
Sobral, P.J.A. 2000. Influência Influence of the thicknesse of biofilms based on fibrofibrillar proteins over their funcional properties. Pesqui. Agropecu. Bras. Brasília 35(6): 1251–1259.
SOTERO, A.P. 2000. Biodegradable plastics bring environmental improvement. Jornal de plásticos. Unicamp.
Sothornvit, R. and Krochta, J.M. 2001. Plasticizer effect on mechanical properties of β-lactoglobulin films. J. Food Eng. 50(3): 149–155.
Ustunol, Z. 2018. Filmes e revestimentos comestíveis para food applications. pp. 245–268. Em: Huber, C.K. and Embuscado, E.M. (eds.). New York, NY: Springer.
Valencia-Chamorro, S.A., Pérez-Gago, M.B., Del Río, M.A. and Palou, L. 2010. Effect of antifungal hydroxypropyl methylcellulose-lipid edible composite coatings on penicillium decay development and postharvest quality of cold-stored "ortanique" mandarins. J. Food Sci. 75(8): S418–S426.
Valencia-Chamorro, S.A., Palou, L., Del Río, M.A. and Perez-Gago, M.B. 2011. Antimicrobial edible films and coatings for fresh and minimally processed fruits and vegetables: a review. Cr. R. F. Sci. 51(9): 872–900.
Vásconez, María B. et al. 2009. Atividade antimicrobiana e propriedades físicas de filmes e revestimentos comestíveis à base de amido de quitosana-amido de tapioca. Food Research Int. 42.7: 762–769.
Vicentini, N.M. 2003. Elaboration and Characterization of Edible Films Based on Cassava Starch for Post-Harvest Use. Doctoral thesis. Faculty of Agronomic Science, Universidade Estadual Paulista, Botucatu, 198 p.
Yang, L. and Paulson, A.T. 2000. Effects of lipids on mechanical and moisture barrier properties of edible gellan film. Food Res Int. 33(7): 571–578.
Zhong, Y., Cavender, G. and Zhao, Y. 2014. Investigation of different coating application methods on the performance of edible coatings on Mozzarella cheese. LWT-Food Sci. Technol. 56(1): 1–8.

Chapter 7

Perspectives For Using the Fruits of Cacti from the Caatinga of Piauí For Nutritional and Biotechnological Purposes

Tamiris Ramos Silva,[1] *Tiago Soares,*[1] *Mariany de Alencar,*[1] *Maria Nilka de Oliveira,*[3] *Boris Timah Acha,*[4] *Ana Cibele Pereira Sousa,*[2] *Alessandro de Lima,*[5] *Joilane Alves Pereira-Freire*[1] and *Stella Regina Arcanjo Medeiros*[1,*]

Introduction

Brasil stands out as one of the largest fruit producers in the world, due to its continental extension and consequent climatic and soil diversity, which enables the occurrence of many fruit species. The production is extremely varied, ranging from fruits typical to temperate to tropical climates. Many of these species are endemic to the country and represent a promising source of nutrients and bioactive compounds, which are increasingly becoming the object of study and interest in the search for prevention and treatment of diseases. However, some fruits still have no commercial exploitation, thus generating great waste that is perpetuated beyond the loss of raw

[1] Research Laboratory II, Federal University of Piaui, Campus Senator Helvídio Nunes de Barros - Junco, 64607-670, Picos - PI.

[2] Department of Nutrition, Centro Universitário Paraíso - UniFAP, 63010-465, Juazeiro do Norte, CE.

[3] Department of Food Engineering, Center for Agricultural Sciences, Federal University of Ceara, 60020-181, Fortaleza-CE.

[4] Medicinal Plants Research Center, Federal University of Piauí, Campus Ministro Petrônio Portella, SG-15, Ininga, 64049-550 Teresina, PI.

[5] Federal Institute of Education, Science and Technology of Piauí - IFPI, 64000-040, Teresina, PI.

* Corresponding author: stellaarcanjo@ufpi.edu.br

material, such as income and labor that could be aggregated (OECD-FAO, 2015; Infante et al., 2016).

Among these little-explored fruits, there are the fruits of cactaceae plants, such as mandacaru (Brazilian rosewood) (*Cereus jamacaru* DC.), facheiro (*Pilosocereus pachycladus* subsp. *pernambucoensis* (F. Ritter) Zappi), xiquexique (*Pilosocereus gounellei* (F.A.C. Weber) Byles & G.D. Rowley subesp. Gounellei) and Quipá (*Tacinga inamoena* (K. Schum) NPTaylor & Stuppy), whose use reported in literature refers mainly to the use in periods of drought, the plant as a whole, or its fruits for animal feed. In addition, human consumption of the fruits of these species is not widespread, and at present, their use is scarce. Some of these are reported as "hunger food" because they are used mainly in the northeastern dry season by the low-income population that suffers from the scarcity of financial resources and food (Almeida and Corrêa, 2012; Nascimento et al., 2012; Chaves and Barros, 2015; Neto et al., 2015).

However, some studies have revealed the nutritional and bioactive potential of tropical and exotic fruits from Northeast Brazil, highlighting their high antioxidant capacity (Omenaa et al., 2012; Nóbrega et al., 2015). The interest in these substances focuses on the ability to reduce oxidative damage in cellular metabolism, through the neutralization and elimination of free radicals (Siqueira et al., 2013; Paz et al., 2015). Thus, they can prevent premature aging, non-communicable chronic diseases, by exhibiting anti-inflammatory action (Silva et al., 2014). Moreover, these fruits have the potential to promote sustainable development with regard to the development of new products in the food industry and new compounds in the pharmaceutical industry that is increasingly seeking to invest in this market of natural substances (Souza et al., 2014; Gonçalves et al., 2015).

Food: new perspectives

Recently, food and nutrition issues have become internationally relevant and are present on the environmental, social, political and academic agendas due to the increasing urbanization of populations, accompanied by climate change. According to the United Nations (UN), by 2050 there will be approximately 9.8 billion people in the world. This will imply a greater demand for aspects related to food and food security, health and housing, among others. With regard to food security, it is essential to reduce food waste and use sustainable practices throughout the production chain in order to guarantee this resource for the populations of today and the future (Food and Agriculture Organization, 2011; United Nations, 2017; Mesquita and Bursztyn, 2018; Preiss et al., 2018).

Based on this perspective, new food aspects have been sought, and here the Non-Conventional Food Plants (NCFPs), which have one or more edible parts, not being routinely included in our daily menu, but that are great sources of nutrients and bioactives for human nutrition, although they have not yet been widely studied by the technical-scientific community and/or explored by all populations, of regional/local consumption, therefore they face the difficulty of acceptance and consumption by other regions of the country. Thus, modifications in the way of food consumption have been suggested and explored to include new plant species in the human diet, as well as to search for a sustainable system of food production, as well as to conserve, use and spread the vegetable biodiversity of each region. Leveraging research funding and periodic population surveys as strategies for food and nutrition surveillance in the country (Brasil, 2010; Kelen et al., 2015; Mesquita and Bursztyn, 2018; Liberato et al., 2019; Campos and Fonseca, 2021).

According to Kinupp and Lorenzi (2014), the cactaceae species are part of the Non-Conventional Food Plants (NCFPs) group, being an important food source in the caatinga. They have been consumed as a supplement to the main food plants. Chaves and Barros (2015) and Bezerril (2017) reported that the cactaceae are consumed during situations of food insecurity, for example during drought. This has a negative effect, possibly influencing the creation of a taboo, for their consumption. Brazil is very rich in biological diversity, with more than 45,000 native species in its territory, and in relation to non-conventional food plants, it is estimated that approximately three thousand species are known, but little used. This is the case of umbu (*Spondias tuberosa* Arruda), which in most cities of the South is considered an (NCFP), but in the Northeast, this plant is part of the daily menu (Chandra et al., 2016; Bezerra and Brito, 2020; Jacob, 2020).

Several projects have been designed and executed in search of solutions for food for various people, among these is the "Biodiversity for Food and Nutrition - BFN", implemented in four countries (Brazil, Kenya, Sri Lanka and Turkey) aimed at contributing to improving food and nutrition security of the population, seeking to conserve and promote the sustainable use of biodiversity, searching to enhance neglected and underutilized species, thus increasing the number of native species used as food. As a result of this project, more than 500 species with economic potential have already been cataloged in our country (Ministério do Meio Ambiente 2016).

Ordinance No. 284 of May 30, 2018, established the list of sociobiodiversity species, for the purposes of marketing *in natura* or their derivative products, in the context of operations carried out by the Food Purchase Program - PAA and in its various modalities, according to the Policy for Guaranteeing Minimum Prices for Sociobiodiversity Products - PGPMBio and the National Program for School Feeding - PNAE, among the species included, 63 are of interest to this study, and among these is the mandacaru (Ministry of the Environment, 2016; Brasil, 2018).

Brazil has the greatest diversity of plant species in the world, many of them under-exploited. Similar to what occurs in other regions of the country, the caatinga ecoregion, located in the Northeast of the country, despite the rich plant biodiversity, has many fruits that are still little-known despite being safe for consumption. There is still a great variety of underused fruits, especially when it comes to native species. In parallel to this, in recent years, several Brazilian fruits have been attracting attention for their potential auxiliary effect in the prevention and treatment of chronic diseases due to their high content of bioactive compounds and consequent antioxidant, antimicrobial, anti-inflammatory and anticancer potentials (Almeida et al., 2016; Infante et al., 2016; Li et al., 2016).

The caatinga biome

According to Antunes et al. (2018), caatinga is the vegetation that predominates in the Northeast of Brazil, with a semi-arid climate. The name caatinga comes from the first inhabitants of the region, the Indians, because in the dry season, most plants lost their leaves, and the trunks of the trees had a clear and whitish appearance. Hence the name Caatinga (caa: forest and tinga: white), which means "white forest" in Tupi. However, in the rainy season, the trees acquire varying shades of green.

Caatinga is the only exclusively Brazilian biome, occupying an area of 844,453 Km^2 in the Northeast region, extending over the entire state of Ceará and covering most of the states of Bahia, Paraíba, Pernambuco, Piauí and Rio Grande do Norte, almost half of the states of Alagoas and Sergipe, and small portions of the states of Minas Gerais and Maranhão (Fig. 7.1) (Brazilian Institute of Geography and Statistics, 2015).

This biome is under the influence of two well-defined seasons, the dry season, which lasts at least seven months and is marked by high-temperatures that can reach 45°C, and the rainy season is marked by high irregularity (Trentin et al., 2011; Kavamura et al., 2013).

The Caatinga has a rich biodiversity that offers primary goods such as plants (including those rich in bioactives) that can be used for human and animal food, medicinal use, construction, timbers and energy uses, among others. As the largest and most continuous area of the FATSS biome, it has typical vegetation called Seasonally Dry Tropical Forests and Shrubs (*FATSS* or *SDTFW*). This biome encompasses tropical vegetation rich in succulent plants with few or no grasses (Queiroz et al., 2017; Fernandes and Queiroz, 2018; Nascimento et al., 2018).

Thus, its vegetation can be described as shrubby-arboreal, with deciduous leaves in the dry season, present with thorns, and, in its massive extension, with the presence

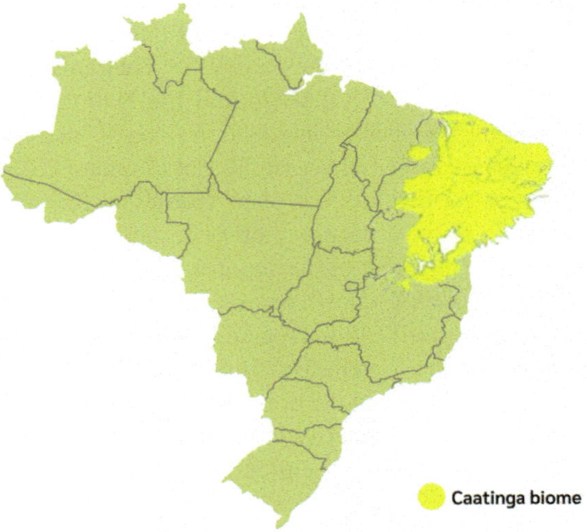

Figure 7.1. Map of Brazil highlighting the caatinga area in the Northeast region of the country. Source: WWF, 2015.

of cacti and bromeliads. It has a high capacity for physiological adaptations to the stressful conditions of the semi-arid ecosystem, due to its water deficit, with drought predominating most of the year; high temperatures and light intensity, which causes a high evaporative demand with consequent soil desiccation. Despite this, some plant species manage to establish themselves in this environment, often considered inhospitable and unviable for the survival of many species. The plants of FATSS have developed adaptations to survive the adverse conditions that include irregular precipitation and recurrent droughts (Trovao et al., 2007; Fernando and Queiroz, 2018).

Due to the climatic and soil characteristics of the caatinga region, combined with deforestation, burning and livestock practices, besides the lack of sustainable management, such as overgrazing and inadequate exploitation of natural resources, the loss of native vegetation occurs, making this biome vulnerable to desertification, and as a consequence, there is the threat of extinction of several endemic species, whether of the flora or fauna of this ecosystem (Araújo and Sousa, 2011; Nascimento et al., 2018).

Cactus family (Cactaceae)

Brazil is home to the greatest biodiversity of fauna and flora on the planet, due to the many climatic zones, leading to several ecological variations, in consequence, rich biogeographical zones (biomes). This is reflected in a unique condition, which makes Brazil a protagonist and responsible for the protection of its biodiversity (Joly et al., 2019; Moraes et al., 2021; Ministério do Meio Ambiente, 2021; Wohner et al., 2021).

Among this well divulged wealth, is the *Cactaceae* Juss, a family which is composed of approximately 124 genera and 1440 species, distributed almost exclusively in the tropical and dry regions of the Americas, with the exception of the species *Rhipsalis baccifera,* which can occur in Africa, Madagascar and Sri Lanka. These cacti can appear as trees, shrubs, climbers, epiphytes, geophytes and stems (stalks) with features that can vary from columnar to round, globular or tuberculate, rib-shaped, winged, flattened, generally segmented without leaves and with thorns. Its propagation can be carried out sexually or asexually, the former being a method still used only a little (Hunt et al., 2006; Abud et al., 2010).

Unlike other plants that have flowers, the species of the *Cactaceae* family have their own morphological characteristics, such as the presence of spines, the ovary below their floral receptacle and organization of the apical meristem divided into four different zones. In the semi-arid (dry areas), the species (*Pilosocereus gounellei* (A. Weber ex K. Schum.) Bly. ex Rowl.) occurs in a low to medium size, branching close to the base, green spines, tubular-shaped flowers and a mauve color (Barros et al., 2021), while the species *Pilosocereus pachycladus* F. Ritter, exists in the caatinga, with a shrubby, erect trunk, with few lateral branches, large tubular flowers, golden thorns, generally of a yellowish color (Silva et al., 2005; Pereira et al., 2021).

The cactaceae family can be divided into four subfamilies: *Maihuenioideae, Pereskioideae, Cactoideae* and *Opuntioideae*. In Brazil, it is distributed in all phytogeographic domains of the country and is represented by the occurrence of 277 species grouped into 39 genera, 14 of which are endemic (Zappi and Taylor,

2020). This wide range of species places Brazil as the holder of the third largest diversity of cacti in the world (Silveira, 2015; Simões et al., 2020; Zappi and Taylor, 2020).

Despite the abundant distribution throughout the national territory, the most important regions, in terms of biodiversity, are in the east of the country, mainly in the states of Bahia and Minas Gerais. The Northeast region is home to about 110 species of cacti; and in this semi-arid environment, the caatinga is one of the ecosystems that have the greatest diversity of these species, with about 95 species being part of this ecosystem (Zappi et al., 2011; Zappi et al., 2015).

They have a high adaptive power to high daytime temperatures, high radiation, low nighttime temperatures and low water content. Physiologically they are characterized by having the acid metabolism of the CAM (or MAC) crassulaceans. CAM plants (*Crassulacean Acid Metabolism*) are plants specially adapted to arid and semiarid regions, which allows water retention inside and its use in a different way, they open their stomata during the night and close them during the day, which generates a high efficiency in water use (Pereira et al., 2013).

The use of these and other caatinga species is reported mainly for the composition of the diet of ruminants during the dry period. It is estimated that 70% of the botanical species are used for this purpose. Among the Cactaceae species, *Pilosocereus gounellei* and *Cereus jamacaru* are described as the most used for feeding livestock and birds (Araújo et al., 2010; Lucena et al., 2012; Souza et al., 2013).

In human nutrition, cacti are described in literature as emergency food, used in situations where there is a shortage of food, however several studies have shown that the fruits and various other parts of these cacti are resources that have the potential to be used in everyday food either in natura or in the form of sweets, brown sugar, jellies, couscous, among others. Lucena et al. (2012) also pointed out that the whole plant is used in construction (living fences), has medicinal and veterinary use, and can also be used in ornamental gardens, among others (Nascimento et al., 2012; Chaves and Barros, 2015).

The species of the *Cactaceae* family are emerging sources of food for human and animal nutrition, possessing a supply of nutrients and bioactive substances (e.g., phenolic compounds, carotenoids, betalains, phytosterols, tocopherols), so they have industrial technological capabilities for use in human food (e.g., cakes, yogurts, bread, ice cream and juices) and others (pharmaceutical industry and natural dyes, biotechnology, sources of pectins, water treatment) (Araújo et al., 2021).

Recent research has shown that edible parts of cacti, especially fruits, have the potential for consumption in natura and application in technological processes aiming at the diversity of products such as sweets, jellies, beverages, and flours (Santos et al., 2019; Gonçalves et al., 2021).

Mandacaru (*Cereus jamacaru* DC.)

Cereus jamacaru DC (Fig. 7.2) is a species native to the caatinga vegetation, popularly known as *mandacaru, mandacaru-de-boi, mandacaru-facheiro, mandacaru-de-faix, cardeiro, jamacaru, jamaracurú, jumucurú, jumarucú, cumbeba and urumbeba.* It grows in rocky soils and rocky areas and can reach a height of 3 to 7 meters. The

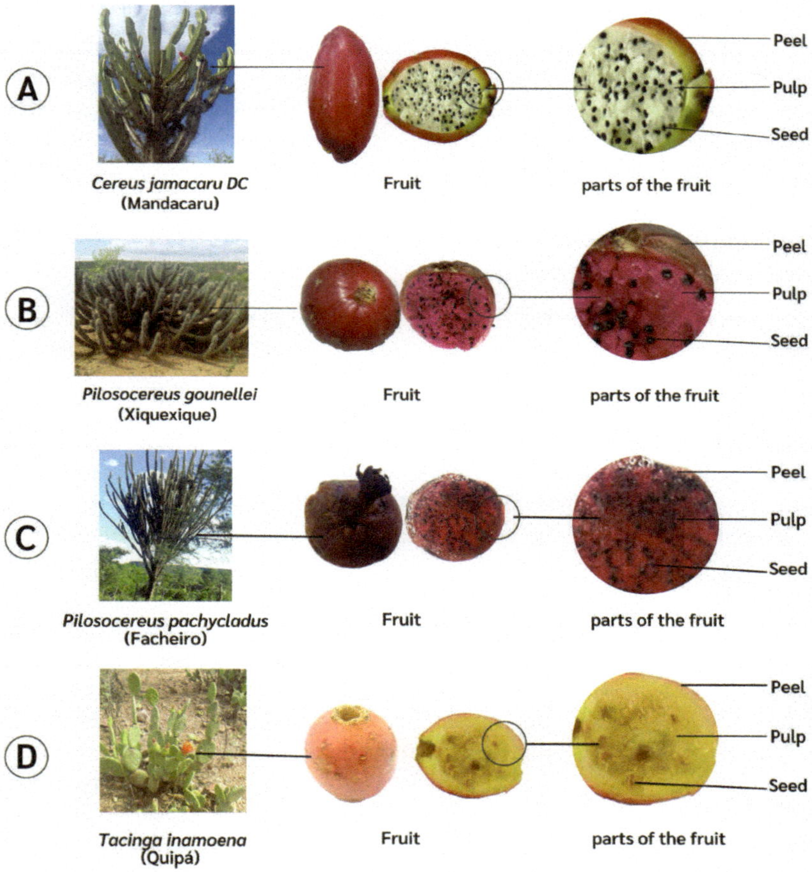

Figure 7.2. Cacti from Caatinga vegetation, their fruits and their respective parts. Source: Own Authors, 2022.

fruit (Fig. 7.2A) is an ovoid-shaped berry with an outer color ranging from pinkish to red, with dimensions 10–13 × 5–9 cm. It's funicular and mucilaginous pulp is light-colored with several black seeds with dimensions of 1.5–2.5 mm (Zappi and Aona, 2007; Mota et al., 2019).

Although edible, as with the other cactaceous fruits, there are no statistical reports of annual cultivation. Its fruit is fragile and perishable, which leads to a short shelf life, that can be solved with technological processing such as drying, and making sweets and jellies, thus constituting a biotechnological potential for dry regions (Oliveira et al., 2015). Most of the reports of fruit consumption in literature, as cited by Cavalcanti and Resende (2007) are for feeding birds and wild animals. Thus, as with other cacti, in periods of drought, the mandacaru is widely used for animal fodder.

In popular medicine, the roots are used in teas (decoction or infusion) to treat kidney, liver, respiratory, stomach problems, inflammation and sinusitis. The ethanolic extract of its stem has antibacterial activity. The decoction and maceration are shown for treating stomach problems, kidney complications and snake bites. As for general use, its wood is used to make wooden spoons, and the entire plant is used to shade spaces in the ornamentation of gardens. In addition, there are reports of magical and religious aspects attributed to this species (Davet et al., 2009; Lucena et al., 2012; Cordeiro and Félix, 2014; Saraiva et al., 2015; Mota et al., 2019).

Xique-xique (*Pilosocereus gounellei* (F.A.C. Weber) Byles & G.D. Rowley subesp. gounellei)

Pilosocereus gounellei (F.A.C. Weber) Byles & G.D. Rowley subsp. *gounellei,* popularly known as xique-xique (Fig. 7.2B), is found exclusively in areas of the Caatinga biome. It is a columnar shrub-like species with multi-articulated cladodes in candelabra-shaped branches with spines of various sizes, and its height varies between 0.8 and 1.5 m. Its fruit (Fig. 7.2B) is a 3–6 × 4–6 cm berry, juicy, purple in color; mucilaginous funicular pulp, present with several 2 mm long seeds, which are often exposed in the ripe fruit (Rocha and Agra, 2002; Zappi et al., 2015).

From the cladodes, flour can be produced for the preparation of couscous and cakes. Moreover, this food can also be consumed boiled or baked (Nascimento et al., 2012).

The study by Monteiro et al. (2015) reported that the xique-xique has attractive fruits both for human consumption and for birds, bats and other animals. Moreover, it is still one of the species used for livestock feeding, animals forage flowers, fruits, and the cladodes, despite its prominent thorns (Santos et al., 2012). Its medicinal and veterinary use refers to its pulp, used to treat infections and inflammations. Its stems, roots and flowers are used to treat prostate inflammation, jaundice, hyperglycemia and skin lesions (Maciel et al., 2016; Bezerril, 2017). It can also be used in ornamental gardens, and the people of the Sertanejos also use it as a bioindicator of rain from its flowering (Lucena et al., 2012). The pulp of xique-xique resembles that of a green papaya, which is used in the production of flour and couscous and is also good in the treatment of infections and inflammations (Almeida et al., 2007; Maciel et al., 2016).

Facheiro (*Pilosocereus pachycladus* subsp. *pernambucoensis* (F. Ritter) Zappi)

The *facheiro* (*Pilosocereus pachycladus* subsp. *pernambucoensis* (F. Ritter) Zappi) is a perennial species, shrubby, robust, erect trunk with lateral branches, but only a little branched, dark green, which has sharp thorns, and large, white and isolated flowers (Fig. 7.2C) (Braga, 1976). It occurs exclusively in the states of northeastern region of Brazil in the states of Alagoas, Bahia, Ceará, Paraíba, Pernambuco, Piauí and Rio Grande do Norte (Zappi et al., 2015).

The *facheiro* (*Pilosocereus pachycladus* F. Ritter subsp. *pernambucoensis* (F. Ritter) Zappi) is a columnar cactaceous endemic to the semi-arid region, it is very important for the maintenance, growth and conservation of the caatinga biome because its fruits are edible and during droughts, it is consumed a great deal (Santana et al., 2016; Batista et al., 2018).

It is a valued species from the ornamental point of view. The average length of the *facheiro* fruit (Fig. 7.2C) is 38.13 mm, diameter 50.53 mm, and 3,786 seeds, which are consumed by birds and the seeds are dispersed by birdlife. Besides the plant as a whole can be used for the production of animal fodder, the fruits can be used as human food, as well as the stem, which with appropriate treatment, can be used in the production of sweets, cakes, cookies, coconut cookie, among others (Lima et al., 2007; Abud et al., 2010).

Quipá (*Tacinga inamoena* (K. Schum) NPTaylor & Stuppy)

Tacinga is a genus of the subfamily Opuntioideae, composed of eight species endemic to Brazil, restricted to the Brazilian semiarid region. *Tacinga inamoena* (K. Schum) NPTaylor & Stuppy is a species popularly known as *quipá, cumbeba* or *gogóia.* It is a subshrubs cactus, with a round to oblong pads, solitary subapical flowers at the top of the branches, measuring 4–6 × 3.5–4 cm, with green globose pericarp (10 mm) and fleshy, acute, reddish-orange bract scales subtending the bristles (Fig. 7.2D). Its flowers are large, bright, have erect stamens forming a column around the stigma, provide pollen and nectar and are pollinated by hummingbirds. Quirino (2006) reported that its flowering starts at the end of the rainy season and throughout the dry season (September/October) (Lambert, 2009; Taylor and Zappi, 2004; Peixoto et al., 2016; Zappi and Taylor, 2020; Paixão et al., 2021).

Its fruits are ovoid to subglobose berries, with 3.0–4.0 × 2.4–3.5 cm longitudinal and transversal diameter, respectively. The color of its shell varies from yellow to matte orange, with a reddish basal portion or all matted red; its seminiferous chamber occupying almost the entire internal space, and is filled with a light peach-colored fleshy mass, consisting of the funicles and seeds (Fig. 7.2D) (Andrade-Lima, 1981).

Its fruits are suitable for fresh consumption and show potential for industrial utilization. Besides containing considerable values of minerals, among which calcium, magnesium and potassium stand out. The fruits and cladodes have been used in rural areas for animal feed, despite the mild flavor, not very sweet and presenting a higher proportion of fleshy portion (62.87%) than the shell, there are few reports of its insertion in human diet (Souza et al., 2007; Leal and Manfrin, 2017).

Nutritional profile of cactus fruits

Cactus fruits are widely used for animal feed, because they resist the various climatic changes throughout the year, and in human food they are only consumed when there is no other option, or to supplement the diet. However, literature has shown the great nutritional potential of these fruits, with high levels of nutrients, which are usually rejected due to lack of knowledge (Almeida and Corrêa, 2012; Neto et al., 2015).

The fruits of some of these cacti are composed mostly of carbohydrates, as shown in Table 7.1, in addition, they have low energy density and can be included in the human diet mainly due to their high content of minerals, nutrients that are essential for biological processes (Sousa, 2017).

Regarding minerals, there is a high content of manganese (Mn), especially in *xiquexique* with a value of 19.66 ± 0.85 mg/100 (Table 7.2), which is higher than

Table 7.1. Centesimal composition and total energy value - TEV (Kcal.100 g^{-1} sample) of the pulp and bark of *mandacaru* (*Cereus jamacaru* DC.), xiquexique (*Pilosocereus gounellei* (F.A.C. Weber) Byles & G.D. Rowley subsp. Gounellei), *facheiro* (*Pilosocereus pachycladus* subsp. *pernambucoensis* (F. Ritter) Zappi) and *quipá* (*Tacinga inamoena* (K. Schum) NP Taylor & Stuppy).

Constituents	Part of the fruit	Mandacaru	Xiquexique	Facheiro	Quipá
Moistness	Pulp (%) Shell (%)	$86{,}91^{a} \pm 0{,}33$ $87{,}53^{b} \pm 0{,}69$	$83{,}47^{c} \pm 0{,}41$ $88{,}85^{b} \pm 0{,}71$	$86{,}06^{b} \pm 0{,}23$ $93{,}93^{a} \pm 0{,}19$	$77{,}57^{d} \pm 0{,}06$ $87{,}87^{b} \pm 0{,}23$
Ash	Pulp (%) Shell (%)	$0{,}48^{c} \pm 0{,}02$ $1{,}35^{a} \pm 0{,}01$	$0{,}66^{b} \pm 0{,}01$ $1{,}45^{a} \pm 0{,}04$	$0{,}68^{b} \pm 0{,}02$ $1{,}35^{a} \pm 0{,}09$	$0{,}85^{a} \pm 0{,}11$ $1{,}40^{a} \pm 0{,}21$
Lipids	Pulp (%) Shell (%)	$1{,}41^{c} \pm 0{,}01$ $0{,}10^{b} \pm 0{,}01$	$1{,}64^{b} \pm 0{,}01$ $0{,}06^{c} \pm 0{,}01$	$3{,}16^{a} \pm 0{,}00$ $0{,}03^{d} \pm 0{,}01$	$1{,}13^{d} \pm 0{,}02$ $0{,}16^{a} \pm 0{,}01$
Proteins	Pulp (%) Shell (%)	$1{,}64^{a} \pm 0{,}01$ $1{,}91^{a} \pm 0{,}08$	$1{,}08^{c} \pm 0{,}02$ $0{,}60^{b} \pm 0{,}01$	$1{,}26^{b} \pm 0{,}06$ $0{,}66^{b} \pm 0{,}01$	$0{,}46^{d} \pm 0{,}01$ $0{,}28^{c} \pm 0{,}01$
Carbohydrates	Pulp (%) Shell (%)	$9{,}74^{c} \pm 0{,}28$ $8{,}91^{a} \pm 0{,}92$	$12{,}80^{b} \pm 0{,}39$ $9{,}03^{a} \pm 1{,}04$	$8{,}18^{d} \pm 0{,}23$ $4{,}07^{b} \pm 0{,}15$	$19{,}79^{a} \pm 0{,}04$ $10{,}37^{a} \pm 0{,}76$
The energy (Kcal/100 g)	Pulp (%) Shell (%)	$58{,}03^{c} \pm 1{,}00$ $44{,}22^{a} \pm 3{,}44$	$72{,}31^{b} \pm 1{,}72$ $39{,}16^{a} \pm 4{,}15$	$68{,}62^{b} \pm 1{,}17$ $19{,}30^{b} \pm 0{,}51$	$91{,}81^{a} \pm 0{,}76$ $43{,}64^{a} \pm 3{,}12$

Values expressed as mean ± standard deviation
Means followed by the same letter, in the same column (by parameter and fruit part), do not differ statistically, by Tukey's test ($p < 0.05$).
Source: Adapted from Sousa, 2017.

the maximum tolerable intake (UL) for this mineral, which is 11 mg per day, and its composition is higher than the food indicated with the highest value of this mineral by the Brazilian Table of Food Composition (TACO), which is the canned palm heart of juçara with 10.82 mg/100 g. Studies make a positive association of the adequate intake of Mn, relating it to the prevention of metabolic syndrome in men and with types 1 and 2 diabetes (TACO, 2011; Brazil, 2012; Forte et al., 2013; Zhou et al., 2016; Sousa, 2017).

Another mineral abundant in cactus fruits is magnesium (Mg), an important mineral for the proper functioning of the body, this is positively associated with the prevention of several diseases, such as type 2 diabetes, migraine, depression and hypertension, and also, its deficiency is related to oxidative stress markers and increased lipid peroxidation (Cocate et al., 2014; Baaij et al., 2015; Morais et al., 2016).

Phytochemical profile and biological activities

Living organisms, such as plants, undergo numerous chemical transformations along their biogenetic pathway. These give rise to natural compounds, such as phytochemicals, which act as a defense against diseases and pest attacks. These substances, in turn, play important biological roles, such as antioxidant, anti-inflammatory, antifungal, antibacterial and antiviral activity, and are able to add numerous benefits for maintaining health (Filho, 2010).

The alterations of the phytochemical profiles are given according to (1) solvent and the extraction processes, (2) growth stage, (3) storage conditions and (4) the plant species. They can also be classified according to their structure into phenolic

Table 7.2. Mineral content (mg.100 g^{-1} dry sample) of the fruit peels of *mandacaru* (*Cereus jamacaru* DC.), *xiquexique* (*Pilosocereus gounellei* (F.A.C. Weber) Byles & G.D. Rowley subsp. Gounellei), *facheiro* (*Pilosocereus pachycladus* subsp. *pernambucoensis* (F. Ritter) Zappi) and *quipá* (*Tacinga inamoena* (K. Schum) NPTaylor & Stuppy).

Minerais	Mandacaru	Xiquexique	Facheiro	Quipá
Cálcio (Ca)	694,80 ± 0,07	412,60 ± 0,11	820,17 ± 0,12	1.696,93 ± 0,10
% de adequação	69,48	41,26	82,01	169,69
Cobre (Cu)	0,72 ± 0,02	0,38 ± 0,10	0,98 ± 0,35	0,58 ± 0,16
% de adequação	80,54	42,69	109,30	64,70
Ferro (Fe)	3,71 ± 0,11	5,27 ± 0,73	17,17 ± 1,45	24,769 ± 0,90
% de adequação	28,55	40,51	132,11	189,89
Fósforo (P)	119,40 ± 0,02	60,27 ± 0,01	99,33 ± 0,08	51,84 ± 0,01
% de adequação	17,05	8,60	14,19	7,40
Potássio (K)	3.894,43 ± 0,25	5.042,84 ± 0,50	5.890,37 ± 0,92	2.433,50 ± 0,34
% de adequação	82,86	107,29	125,32	51,77
Magnésio(Mg)	496,13 ± 0,20	622,33 ± 0,11	646,17 ± 0,09	806,07 ± 0,06
% de adequação	139,75	175,30	182,01	227,06
Zinco (Zn)	1,74 ± 0,47	3,53 ± 0,25	3,79 ± 0,14	3,49 ± 0,26
% de adequação	18,29	37,12	39,92	36,72
Manganês (Mn)	12,24 ± 1,40	11,65 ± 0,61	55,24 ± 1,00	14,85 ± 1,46
% de adequação	597,23	568,45	2694,53	724,51
Selênio (Se)	Traços	Traços	Traços	Traços
% de adequação				

Values expressed as mean ± standard deviation
Source: Adapted from Sousa, 2017.

compounds, alkaloids, nitrogen compounds, organosulfur compounds, phytosterols and carotenoids (Baena, 2015; Cabañas-García, 2019).

Plant metabolism can be divided into two phases. Primary metabolism generates substances that are responsible for making the organisms able to live, grow and reproduce. The substances that come from secondary metabolism, on the other hand, express the individuality of the species and comprise the compounds that play some important role in the well-being of the species that produce it, such as acting in defense against predators (Dewick, 2002).

The phytochemical composition of the cactus fruits has a great diversity of secondary metabolites, representing an important tool capable of contributing towards food innovation and may act as raw materials for pharmaceutical industry, as can be seen in Table 7.3.

These bioactive compounds are responsible for the biological activity of this genus, which in turn, has been gaining prominence due to its antioxidant, anti-inflammatory, antibacterial and antifungal potentials. These are determined not only by a substance but by the sum capacity of each of its components (Madrigal-Santillán et al., 2013).

Table 7.3. Phytochemicals composition fruits of *Facheiro* (*Pilosocereus pachycladus*) and *mandacaru* (*Cereus jamacaru* DC.).

Species/Genus	Isolated compound	Plant tissue	Reference
Facheiro (***Pilosocereus pachycladus***)	Ascorbic Acid; Yellow Flavonoids; Betalains; Betacyanins; Betaxanthins.	Fruit	Rodrigues et al., 2019
Mandacaru (***C. jamacaru***)	Flavonoids	Fruit	Dutra et al., 2019
Coroa de frade (***M. zehntneri***)	Ascorbic Acid; Carotenoids; Flavonoids; Phenols.	Epidermis	Nunes et al., 2016

Source: Own Authors, 2022.

Studies such as that of Chahdoura et al. (2017), pointed out that the natural oil of *Opuntia macrorhiza* Engelm has a rich phytochemical profile, as well as, showed relevant antioxidant activity, especially in lipid peroxidation inhibition assays. Inhibitory activity of α-glucosidase, cytotoxicity against human tumor cell lines, as well as anti-inflammatory and analgesic activities.

In accordance with these findings, Rodrigues and colleagues (2019), evaluated fruits of *Pilosocereus pachycladus* Ritter, at different levels of seasonality, where it could be observed that the higher the concentration of betalains and the enzymatic activity of peroxidase (POD), the greater the antioxidant action and the lesser the quantity of pulp required to eliminate the free radical 2,2-diphenyl-1-picrylhydrazyl (DPPH). This implies that the antioxidant activity is higher when the fruit is ripe. However, the ripe state of the fruits increases the concentration of Ascorbic Acid (AA), which negatively influences the enzymatic activity.

The potential of betalain is attributed to its ability to inhibit lipid peroxidation, eliminate superoxide anion radicals, neutralize lipoperoxidases and prevent DNA cleavage. Also, it can modulate the imbalance between ROS and the antioxidant defense system (Sakihama et al., 2011; Canadanovic-Brunet et al., 2011; Esatbeyoglu et al., 2014).

Conclusion and future perspectives

In addition to high nutritional potential, the fruits of the caatinga have shown characteristics of industrial interest that are still little explored, such as hydrocolloid properties, which can be used to prepare thickeners, gelling agents and stabilizers, in addition to antioxidant and coloring properties, which are used not only in the food industry, but also in engineering, as adhesives in paints to improve the quality of house painting products, to increase water infiltration, for plaster mortar, lime or cement and in several other applications (Hamerski et al., 2013; Magalhães, 2009; Queiroga, 2017).

Mandacaru is widely used in the preparation of sweets and jellies, because it presents 9.82% of total sugars in its pulp, an important characteristic for alcoholic fermentation and other technological processes, making it attractive in the development of new products in the food industry (Almeida et al., 2011).

In a study by Lucena et al. (2013), technological applications were also found for *Xique-xique*, where it was used in the preparation of flour for making couscous, besides being used for making cookies and sweets. *Xique-xique* flour showed medium ash content, which corresponds to the minerals present, a favorable characteristic indicative of refining for flours, and a significant percentage of protein and vitamin C, similar results were reported by Deodato (2012) for the flour made from the *facheiro*.

The *Quipá* flour also showed satisfactory results for the food industry, it showed high water absorption capacity, which can be used for application in meat products, bread and cakes in order to maintain their moistness, its insoluble fractions also proved suitable to be used as ingredients in viscous products such as soups and pasta (Becker, 2010; Porte et al., 2011; Dionisio, 2017).

In this way, promising perspectives are observed regarding the use of these fruits in human food, besides their technological exploitation due to their nutritional and bioactive properties. They exhibit low cost, which is an alternative to generate income for families in the Northeast, leading to the valorization of the use of regional ingredients that are still little explored and the use of foods that are normally underutilized or undesired, being a great alternative for the industry for having sensory characteristics similar to the products available in the market.

Further studies are needed to investigate indepth and confirm the bioactive and non-toxic properties of these fruits through *in vivo* assays.

References

Abud, H.F., Gonçalves, N.R., Reis, R.G.E., Pereira, D.S. and Bezerra, A.M.E. 2010. Germinação e expressão morfológica de frutos, sementes e plântulas de *Pilosocereus pachycladus Ritter*. Cienc. Agron. 41: 468–474.

Almeida, C.A., Figueiredo, R.M.S., Queiroz, A.J.M. and Oliveira, F.M.N. 2007. Características físicas e químicas da polpa do xique-xique. Cienc. Agron. 38: 440–443.

Almeida, M.M., Conrado, F.L.H., Conrado, L.D.S., Mota, J.C. and Freire, R.M.M. 2011. Estudo cinético e caracterização da bebida fermentada do *Cereus jamacaru* D.C. Rev. Verde Agroecologia Desenvolv. Sustent. 6: 10–12.

Almeida, M.E.F.A. and Corrêa, A.D. 2012. Utilização de cactáceas do gênero Pereskia na alimentação humana em município de Minas Gerais. Cienc Rural 42: 751–756.

Almeida, M.L.B., Freitas, W.E.S., Morais, P.L.D., Sarmento, J.D.A. and Alves, R.E. 2016. Bioactive compounds and antioxidant potential fruit of *Ximenia americana* L. Food Chem. 192: 1078–1082.

Andrade-Lima, D.A. 1981. The caatingas dominium. Acta bot. Bras. 4: 149–163.

Antunes, A., Sena, L.M.M., Moura, L.M., Nascimento, M.A. and Silva, S.M. 2018. Conheça e Conserve a Caatinga - Atividades de educação ambiental. Fortaleza: Associação Caatinga, 104p. 2 ed., Fortaleza, Br.

Araújo, F.F., Farias, D.P., Neri-Numa, I.A. and Pastore, G.M. 2021. Underutilized plants of the Cactaceae family: Nutritional aspects and technological applications. Food. Chem. 362.

Araujo, K.D., Dantas, R.T., Andrade, A.P., Parente, H.N. and Silva, E.E. 2010. Uso de espécies da caatinga na alimentação de rebanhos no município de São João do Cariri – PB. Rev. Ra'e Ga 20: 157–171.

Araujo, C.S.F. and Sousa, A.N. 2011. Estudo do processo de desertificação na Caatinga: uma proposta de educação ambiental. Ciência & educação 17: 975–986.

Baena, R.C. 2015. Muito além dos nutrientes: o papel dos fitoquímicos nos alimentos integrais. Diagn. tratamento 20: 17–21.

Barros, E.S., Costa, V.S., Fonseca, W.B., Nero, J.D.P., Costa, P.M.A., Souza, V.C. et al. 2021. Sucesso reprodutivo da cactácea nativa, xique-xique (*Pilosocereus Gounellei*), em população natural. Braz. J. Dev. 7: 2980–299.

Batista, F.R.C., Almeida, E.M., Alves, L.I.S., Silva, P.K., Neves, J.A.L. and Freitas, J.G. 2018. Cactário Guimarães Duque: espécies da coleção botânica do INSA. Campina Grande, Br.

Bartlhott, W. and Hunt, D.R. 1993. Cactaceae. pp. 161–197. *In*: Kubiztki, K., Rohwer, J.G. e Bittrich, V. (eds.). The Families and Genera of Vascular Plants, V II, Flowering Plants – Dicctyledons. Berlin.

Becker, F.S. 2010. Caracterização de farinhas cruas e extrusadas obtidas a partir de grãos quebrados de diferentes genótipos de arroz. 79 f. Dissertação (Mestrado em Ciência e Tecnologia de Alimentos) – Escola de Agronomia e Engenharia de Alimentos, Universidade Federal de Goiás, Goiânia, Goiás, Br.

Bezerra, J.A. and Brito, M.M. 2020. Potencial nutricional e antioxidantes das Plantas alimentícias não convencionais (PANCs) e o uso na alimentação: Revisão. Res., Soc. Dev. 9.

Bezerril, F.F. 2017. Caracterização nutricional e de compostos bioativos do xique-xique (*Pilosocereus gounellei* (A. WEBER ex. k. Schum.) *Bly. Ex Rowl.*). Dissertação de Mestrado – Universidade Federal da Paraíba, Paraíba, Br.

Braga. 1976. Plantas do Nordeste, especialmente do Ceará. Mossoró: Escola Superior de Agricultura de Mossoró, 3. ed. pp. 510. Mossoró, Br.

BRASIL. 2010. Ministério da Agricultura, pecuária e Abastecimento. Secretaria de Desenvolvimento Agropecuário e Cooperativismo. Manual de hortaliças não convencionais. Brasília: MAPA/ACS.

BRASIL. 2018. Ministério do Meio Ambiente. Portaria Interministerial Nº 284, de 30 de maio de 2018. Institui a lista de espécies da sociobiodiversidade, para fins de comercialização in natura ou de seus produtos derivados, no âmbito das operações realizadas pelo Programa de Aquisição de Alimentos-PAA. Diário oficial da União, Brasília, Ed. 131, Seção: 1, página: 92.

Brasil, Instituto Brasileiro de Geografia e Estatistica - IBGE. 2019. Biomas e Sistemas Costeiro-Marinho do Brasil 45: 114–180.

Cabañas-García, E., Areche, C., Jáuregui-Rincón, J., Cruz-Souza, F. and Balch, E.P.M. 2019. Phytochemical profiling of *Coryphantha macromeris* (*Cactaceae*) growing in greenhouse conditions using ultra-high-performance liquid chromatography–tandem mass spectrometry. Molecules 24: 1–18.

Campos, D.S.L. and Fonseca, P.C. 2021. A vigilância alimentar e nutricional em 20 anos da Política Nacional de Alimentação e Nutrição. Caderno Saúde Pública, v. 7.

Čanadanović-Brunet, J.M., Savatović, S.S., Ćetković, G.S., Vulić, J.J., Djilas, S.M., Markov, S.L. et al. 2011. Antioxidant and antimicrobial activities of beet root pomace extracts. Czech. J. Food Sci. 29: 575–585.

Cavalvanti, N.B. and Resende, G.M.R. 2007. Efeito de diferentes substratos no desenvolvimento de mandacaru (*Cereus Jamacaru p. dc.*), facheiro (*Pilosocereus Pachycladus* Ritter), xiquexique (*Pilosocereus Gounellei* (A. Webwr Ex K. Schum.) *bly. ex rowl.*) e coroa-de-frade (*Melocactus Bahiensis Britton & Rose*). Rev. Caatinga 20: 28–35.

Chahdoura, H., Barreira, J.C.M., Barros, L., Santos-Buelga, C., Ferreira, I.C.F.R. and Achour, L. 2015. Seeds of *Opuntia* spp. as a novel high potential by-product: Phytochemical characterization and antioxidant activity. Ind. Crop. Prod. 65: 383–389.

Chandra, K.S., Dimple, G., Gautam, K.H. and Handique, A.K. 2016. Nutritive values of some non-conventional leafy vegetables and scarcity food plants of north east India. Afr. J. Food Sci. 10: 340–343.

Chaves, E.M.F. and Barros, R.F.M. 2015. Cactáceas: recurso alimentar emergencial no semiárido, Nordeste do Brasil. Gaia Scientia 9: 129–135.

Cocate, P.G., Natali, A.J., Oliveira, A. et al. 2014. Fruit and vegetable intake and related nutrients are associated with oxidative stress markers in middle-aged men. Nutrition 30: 660–665.

Cordeiro, J.M.P. and Félix, L.P. 2014. Conhecimento botânico medicinal sobre espécies vegetais nativos da caatinga e plantas espontâneas no agreste da Paraíba, Brasil. Rev. Bras. Pl. Med. 16: 685–692.

Davet, A., Virtuoso, S., Dias, J.F.G., Miguel, M.D., Oliveira, A.B. and Miguel, O.G. 2009. Atividade antibacteriana de *Cereus jamacaru* DC, Cactaceae, Braz. J. Pharm. Sci. 9: 561–564.

De Baaij, J.H.F., Hoenderop, J.G.J. and Bindels, R.J.M. 2015. Magnesium in Man: Implications for Health and Disease. Physiol Rev. 95: 1–46.

Deodato, J.N.V. 2012. Produção de farinha *Cereus squamosus* (Facheiro) e utilização como aditivo em biscoitos tipos cookies e barras de cereais. Trabalho de Conclusão de Curso (Bacharelado em Tecnologia de Alimentos). Centro de Ciências e Tecnologia de Alimentos - Universidade Federal de Campina Grande, Pombal, Br.

Dewick, P.M. 2002. Medicinal natural products. NewYork: Jonh Wiley & Sons Ltd.

Dionísio, M.T.M. 2017. Avaliação do potencial tecnológico da polpa do fruto do Quipá (*Tacinga inamoena*) em diferentes estádios de maturação. Trabalho de Conclusão de Curso (Graduação em Tecnologia de Alimentos). Centro de Tecnologia e Desenvolvimento Regional - Universidade Federal da Paraíba, João Pessoa, Br.

Dutra, J.C.V., Oliveira, J.B., Santos, V.S., Pereira, P.R.C., Ferreira, J.M. and Batitucci, M.C.P. 2019. Fruiting increases total content of flavonoids and antiproliferative effects of *Cereus jamacaru* D.C. cladodes in sarcoma 180 cells *in vitro*. J. Trop. Biomed. 9: 66–72.

Esatbeyoglu, T., Wagner, A., Motafakkerazad, R., Nakajima, Y., Matsugo, S. and Rimbach, G. 2014. Free radical scavenging and antioxidant activity of betanin: electron spin resonance spectroscopy studies and studies in cultured cells. Food Chem. Toxicol. 73: 119–126.

Fernandes, M.F. and Queiroz, L.P. 2018. Vegetação e flora da caatinga. Cienc. Cult. 70: 51–56.

Filho, R.B. 2010. Phytochemical contribution to development of a emergent country. Rev. Quim. Nova 33: 229–239.

Food and Agriculture Organization. 2011. Intergovernmental group on bananas and tropical fruits. CCP:BA/TF 11/Inf.4. Fifth Session Yaoundé, Cameroon. Disponível em: <http://www.fao.org/docrep/meet ing/028/ma937e.pdf>. Acesso em: 23 set. 2016.

Forte, G., Bocca, B., Peruzzi, A. et al. 2013. Blood metals concentration in type 1 and type 2 diabetics. Biol. Trace Elem. Res. 156(1–3): 79–90.

Gentry, A.H. 1995. Diversity and floristic composition of neotropical dry forests. pp. 146–194. *In*: Bullock, S.H., Mooney, H.A. and Medina, E. (eds.). Seasonally Dry Tropical Forests. Cambridge: Cambridge University Press.

Gonçalves, A.S.M., Peixe, R.G., Sato, A., Muzitano, M.F., Souza, R.O.M.A., Machado, T.B. et al. 2015. Pilosocereus arrabidae (Byles & Rowley) of the Grumari sandbank, RJ, Brazil: Physical, chemical characterizations and antioxidant activities correlated to detection of flavonoids. Int. Food Res. J. 70: 110–117.

Gonçalves, J.L.C., Santos, J.F., Carnelossi, M.A.G. and Moreira, J.J.S. 2021. Potencialidades das cactáceas brasileiras na tecnologia de alimentos: uma revisão integrativa. Capítulo 45, pp. 621–638. *In*: Cordeiro, C.A.M., Silva, E.M. and Evangelista-Barreto, N.S. (eds.). Ciência e tecnologia de alimentos: pesquisa e práticas contemporâneas. Volume 2. Guarujá, Br.

Hamerski, L., Rezende, M.J.C. and Silva, B.V. 2013. Usando as Cores da Natureza para Atender aos Desejos do Consumidor: Substâncias Naturais como Corantes na Indústria 50 Alimentícia. Rev. Virtual Quim. 5: 394–420.

Hunt, D.R., Taylor, N. and Charles, G. 2006. The New Cactus Lexicon. Text. dh Publications, Milborne Port.

Infante, J., Rosalen, P.L., Lazarini, J.G., Franchin, M. and Alencar, S.M. 2016. Antioxidant and anti-inflammatory activities of unexplored Brazilian native fruits. PLoS ONE 11: 1–13.

Instituto Brasileiro de Geografia e Estatística - IBGE, 2004. Mapas de Biomas e Vegetação.

Jacob, M.C.M. 2020. Biodiversity of underutilized food plants in a Community-based learning garden. Demetra: Food Nutr. Res. 15: 1–17.

Joly, C.A., Scarano, F.R., Seixas, C.S., Metzger, J.P., Ometto, J.P., Bustamante, M.M.C. et al. 2019. 1° Diagnóstico brasileiro de biodiversidade e serviços ecossistêmicos. Editora Cubo, São Carlos, Br.

Kavamura, V.N., Taketani, R.G., Lançoni, M.D., Andreote, F.D., Mendes, R. and Melo, I.S. 2013. Water regime influences bulk soil and rhizosphere of *Cereus jamacaru* bacterial communities in the Brazilian caatinga biome. J. Ethnopharmacol. 8: 1–10.

Kelen, M.E.B., Nouhuys, I.S.V., Kehl, L.C., Brack, P. and Silva, D.B. 2015. Plantas alimentícias não convencionais (PANCs): hortaliças espontâneas e nativas (1ª ed.). UFRGS, Porto Alegre, Br.

Kinupp, V.F. and Lorenzi, H. 2014. Plantas alimentícias não convencionais (PANCs) no Brasil: guia de identificação, aspectos nutricionais e receitas ilustradas. Instituto Platarum de Estudos da Flora, Br.

Lambert, S.M. 2009. Tacinga, the hummingbird pollinated prickly pear. Cactus Succul. J. 81: 156–161.

Leal, D.Y.B. and Manfrin, M.H. 2017. Reconstrução da história evolutiva de Tacinga inamoena (*K. Schum*) N. P. Taylor & Stuppy (*Cactaceae*) no bioma Caatinga. Universidade de São Paulo, Ribeirão Preto, Br.

Li, Y., Zhang, J.J., Xu, D.P., Zhou, T., Zhou, Y., Li, S. and Li, H.B. 2016. Bioactivities and health benefits of wild fruits. Int. J. Mol. Sci. 17: 1–27.

Liberato, P.S. Lima, D.V.T. and Silva, G.M.B. 2019. PANCs - Plantas alimentícias não convencionais e seus benefícios nutricionais. Environmental Smoke 2: 102–111.

Lima, E.E., Figueirêdo, R.M.F. and Queiroz, A.J.M. 2007. Cinética de secagem de polpa de facheiro. Rev. Bras. Produtos Agroindustriais 9: 17–28.

Lucena, C.M., Costa, G.G.S., Carvalho, T.K.N., Guerra, N.M., Quirino, Z.G.M. and Lucena, R.F.P. 2012. Uso e conhecimento de cactáceas no município de São Mamede (Paraíba, Nordeste do Brasil). Rev. Biol. Farm. 8: 121–134.

Lucena, C.M., Lucena, R.F.P., Costa, G.M., Carvalho, T.K.N., Costa, G.G.S., Alves, R.R.N. et al. 2013. Use and knowledge of *Cactaceae* in Northeastern. Brazil. J. Ethnobiol. Ethnomed. 62: 1–11.

Maciel, J.K.S., Chaves, O.S., Brito Filho, S.G., Teles, Y.C.F., Fernandes, M.G., Assis, T.S. et al. 2016. New alcamide and anti-oxidant activity of *Pilosocereus gounellei* A. Weber ex K. Schum. *Bly ex Rowl.* (*Cactaceas*). Molecules 21.

Madrigal-Santillán, E., García-Melo, F., Morales-González, J.A., Vázquez-Alvarado, P., Muñoz-Juárez, S., Zuñiga-Pérez, C., Sumaya-Martínez, M.T., Madrigal-Bujaidar, E. and Hernández-Ceruelos, A. 2013. Antioxidant and Anticlastogenic Capacity of Prickly Pear Juice. Nutrients 5: 4145–4158.

Magalhães, A.C.T.V. 2009. Estudo de fibras vegetais, mucilagem de cacto e gesso em componentes multi construtivos. Dissertação (Mestrado em Arquitetura e Urbanismo). Universidade de Brasília, Brasília, Br.

Melgar, B., Dias, M.I., Barros, L., Ferreira, I.C.F.R., Rodriguez-Lopez, A.D. and Garcia-Castello, E.M. 2019. Ultrasound and microwave assisted extraction of *Opuntia Fruit* peels biocompounds: optimization and comparison using RSM-CCD. Rev. Molecules 24: 1–20.

Mesquita, P.S. and Bursztyn, M. 2018. Alimentação e mudanças climáticas: percepções e o potencial de mudanças comportamentais em prol da mitigação. Desenvolvimento Meio Ambiente 49: 1–16.

Ministério da Saúde. Agência Nacional de Vigilância Sanitária (ANVISA). Resolução – RDC n° 54, de 12 de novembro de 2012. Aprova o Regulamento Técnico para Informação Nutricional Complementar. Diário Oficial [da] União, Poder Executivo, Brasília/DF, 12 de novembro de 2012.

Ministério do Meio Ambiente. 2016. Biodiversidade para Alimentação e Nutrição. 2016.

Ministério do Meio Ambiente - MMA. 2021. Biodiversidade.

Monteiro, E.R. et al. 2015. Genetic diversity and structure of populations in Pilosocereus gounellei (F.A.C.Weber ex K.Schum.) (Cactaceae) in the Caatinga biome as revealed by heterologous microsatellite primers. Biochemical Systematics and Ecolog 58: 7–12.

Moraes, L.A., Araújo, M.F.V. and Conceição, G.M. 2021. Levantamento florístico das angiospermas do Parque Estadual Cânion do rio Poti, Buriti dos Montes – PI. Rev. Bras. Geogr. Fís. 14: 987–2014.

Morais, J.B.S., Severo, J.S., Santos, L.R.d. et al. 2017. Role of Magnesium in Oxidative Stress in Individuals with Obesity. Biol. Trace Elem. Res. 176: 20–26.

Mota, T.R., Linhares, H.V.S., Araújo-Filho, J.H., Veras, D.M., Costa, H.P.S., Souza, C.M.P. et al. 2019. Protein extract from *Cereus jamacaru* (DC.) inhibits Colletotrichum gloeosporioides growth by stimulating ROS generation and promoting severe cell membrane damage. Microb. Pathog. 130: 71–80.

Nascimento, A.M.L., Bento-Silva, J.S. and Ramos, E.M.N.F. 2018. Conhecimento e uso das plantas da caatinga por agricultores locais moradores de uma comunidade rural do Município de Pesqueira Estado de Pernambuco. CIENTEC – Rev. Ciência, Tecnologia e Humanidades do IFPE 10: 75–91.

Nascimento, V.T., Vasconcelos, M.A.S., Maciel, M.I.S. and Albuquerque, U.P. 2012. Famine foods of Brazil's seasonal dry forests: ethnobotanical and nutritional aspects. Econ. Bot. 66: 2234.

Neto, J.A.S., Filho, E.S.C. and Araújo, H.B. 2015. Potencial das cactáceas como alternativa alimentar para ruminantes no semiárido. Rev. Eletrônica Nutr., v.12.

Nóbrega, E.M., Oliveira, E.L., Genovese, M.I. and Correia, R.T.P. 2015. The impact of hot air drying on the physical-chemical characteristics, bioactive compounds and antioxidant activity of acerola (Malphigia emarginata) residue. J. Food Process. Preserv. 39: 131–141.

Nunes, E.M., Lemoss, D.M., Silva, S.F., Rocha, A.P.T., Lucena, C.M., Meiado, M.V. et al. 2016. Quantification Physicochemical in Melon Cactus [*Melocactus zehntneri* (Britton & Rose) Luetzelburg - Cactaceae]. Rev. Bras. Pl. Med. 18: 81–88.

OECD-FAO. OECD-FAOAgricultural Outlook 2015-2024.OECD/FAO, 2015.

Oliveira, A.S., Figueirêdo, R.M.F., Queiroz, A.J.M. and Brito, J.G. 2015. Estabilidade da polpa do *Cereus jamacaru* em pó durante o armazenamento. Rev. Bras. Eng. Agric. Ambient. 19: 147–153.

Omenaa, C.M.B., Valentim, I.B., Guedes, G.S., Rabelo, L.A., Mano, C.M., Bechara, E.J.H. et al. 2012. Antioxidant, anti-acetylcholinesterase and cytotoxic activities of ethanol extracts of peel, pulp and seeds of exotic Brazilian fruits: Antioxidant, anti-acetylcholinesterase and cytotoxic activities in fruits. Int. Food Res. J. 49(1): 334–344.

Paixão, V.H.F., Gomes, V.G.N. and Venticinque, E.M. 2021. Florivory by lizards on Tacinga inamoena (*K.Schum.*) N.P.Taylor & Stuppy (*Cactaceae*) in the Brazilian Caatinga. Biota Neotrop. 21.

Paz, M., Gúllon, P., Barroso, M.F., Carvalho, A.P., Domingues, V.F., Gomes, A.M. et al. 2015. Brazilian fruit pulps as functional foods and additives: Evaluation of bioactive compounds. Food Chem. 172: 462–468.

Peixoto, M.R., Zappi, D.C., Silva, S.R., Costa, G.M. and Aona, L.Y.S. 2016. Cactus survey atthe floresta nacional of Contrndas do Sincorá, bahia, brazil. Brasleya 34: 38–54.

Pereira, F.C., Lima, V.L.A., Moreira, A.A.D., Rocha, C.S. and Lima, A.K.V.O. 2013. Fenologia do xiquexique (*Pilosocereus gounellei,* A. Weber ex K. Schum) cultivados em áreas degradas no seridó Paraibano. Revista Educação Agrícola Superior 28: 85–91.

Pereira, M.R.S., Silva, T.G., Ramos, G.J.A. and Correia, C.C. 2021. Visitantes florais em duas espécies do gênero Pilosocereus (*Cactaceae Juss.*) em área de Caatinga. Diversitas J. 6: 584–600.

Porte, A., Silva F.A. and Almeida, S.D.V. 2011. Propriedades funcionais tecnológicas das farinhas de sementes de mamão (Caricapapaya) E DE ABÓBORA (*Cucurbita* sp.). Rev. Bras. Produtos Agro. Campina Grande 13: 91–96.

Preiss, P., Vasconcelos, F.C.F. and Schneider, S. 2018. Relatório Verde 2018 - Agricultura e alimentação para o século 21 - novas referências, desafios e perspectivas. Agricultura familiar, produção de alimentos saudáveis e preservação ambiental: relatório verde, 37–58.

QUEIROGA, B.A.L. 2017. Dossiê Espessantes-foodingredientes-Brasil n°40–2017.

Queiroz, L.P., Cardoso, D., Fernandes, M. and Moro, M. 2017. Diversity and evolution of flowering plants of the Caatinga domain. pp. 23–63. *In*: Da Silva, J.C., Leal, I. and Tabarelli, M. (eds.). Caatinga: The Largest Tropical Dry Forest Region in South America. Cambridge: Springer.

Quirino, Z.G.M. 2006. Fenologia, síndromes de polinização e dispersão e recursos florais de especies vegetais do Cariri Paraibano. Recife: Universidade Federal de Pernambuco. Ph.D. Thesis.

Rocha, E.A. and Agra, M. 2002. F. Flora do pico do Jabre, Paraíba, Brasil: *Cactaceae Juss.* Acta bot. bras. 16: 15–21.

Rodrigues, T.R., Sousa, A.S.B., Silva, M.C.A., Nascimento, R.S. and Sousa, F.A.R.M. 2019. Quality, antioxidant and enzymatic activities of Facheiro (*Pilosocereus Pachycladus Ritter*) fruits during maturation. Rev. Caatinga 32.

Saad, A.B., Dalel, B., Rjeibi, I., Smida, A., Ncib, S. and Zouari, N. and Zourgui, L. 2017.. Phytochemical, antioxidant and protective effect of cactus cladodes extract against lithium-induced liver injury in rats. Pharm Biol. 55: 516–517.

Sakihama, Y., Maeda, M., Hashimoto, M., Tahara, S. and Hashidoko, A. 2011. Beetroot betalain inhibits peroxynitrite-mediated tyrosine nitration and DNA strand cleavage. Rev. Free Radical 46: 93–99.

Santana-Neto, J.A., Castro Filho, E.S. and Araújo, H.R. 2015. Potencial das cactáceas como alternativa alimentar para ruminantes no semiárido. Nutritime Rev. Eletr. 12: 4426–4434.

Santos, F.S.E., Maciel, J.R. and Siqueira Filho, J.A. 2012. Impacto da herbivoria por caprinos sobre as populações naturais de Bromelia laciniosa Mart. ex Schult. f. (*Bromeliaceae*). Rev. Árvore, 36: 143–149.

Santos, J.F., Gonçalves, J.L.C., Barros, V.M., Vasvary, H.S.C.E., Garcez Júnior, S.S. and Moreira, J.J.S. 2019. Prospective study of *Cereus jamacaru* DC Cactus. Rev. INGI 3: 564–574.

Saraiva, M.E., Ulisses, A.V.R.A., Ribeiro, D.A., Oliveira, L.G.S., Macêdo, D.G., Sousa, F.F.S. et al. 2015. Plant species as a therapeutic resource in areas of the savanna in the state of Pernambuco, Northeast Brazil. J. Ethnopharmacol. 171: 141–153.

Silva, J.G.M., Silva, D.S., Ferreira, M.A., Lima, G.F.C., Melo. A.A.S. and Diniz, M.C.N.M. 2005. Xiquexique (*Pilosocereus gounellei* (A. Weber ex K. Schum.) Bly. ex Rowl.) em substituição à silagem de sorgo (Sorghum bicolor L. Moench) na alimentação de vacas leiteiras. R. Bras. Zootec. 34: 1408–1417.

Silva, L.M.R. et al. 2014. Quantification of bioactive compounds in pulps and by-products of tropical fruits Brazil. Food Chemistry. 143: 398–404.

Silva, L.R. and Alves, R.E. 2009a. Avaliação da composição físico-química de frutos de mandacaru (*Cereus jamacaru P.*). Acta Agron. 58: 245–250.

Silva, L.R. and Alves, R.E. 2009b. Avaliação da composição físico-química de frutos de mandacaru (*Cereus jamacaru* D.C.). Acta Agron, Palmira, 58: 245–250.

Silveira, R.A. 2015. Ecologia de Discocactus pseudoinsignis e Discocactus placentiformis simpátricos e endêmicos da Serra do Espinhaço, MG, Brasil. Dissertação (Mestrado em ecologia). Programa de Pós-Graduação Ecologia de Biomas Tropicais, Universidade Federal de Ouro Preto, Ouro Preto, Br.

Simões, S.S., Zappi, D.C. and Aona, L.Y.S. 2020. A família *Cactaceae* no Parque Nacional de Boa Nova, Estado da Bahia, Brasil. Hoehnea 47.

Siqueira, E.M.A., Rosa, F.R., Fustinoni, A.M., Sant'ana, L.P. and Arruda, S.F. 2013. Brazilian savanna fruits contain higher bioactive compounds content and higher antioxidant activity relative to the conventional red delicious apple. Plos One 8: 1–7.

Sousa, A.C.P. 2017. Frutos de cactáceas da caatinga piauiense: potencial bioativo e tecnológico. Dissertação (Mestrado) – Programa de Pós-Graduação em Alimentos e Nutrição, Universidade Federal do Piauí, Teresina-PI, Br.

Souza, A.C.M.S., Gamarra-Rojas, G., Andrade, S.A.C. and Guerra, N.B. 2007. Características físicas, químicas e organolépticas de quipá (*Tacinga Inamoena, Cactaceae*). Rev Bras. Frutic. 29: 292–295.

Souza, C., Barreto, H.F., Gurgel, V. and Costa, F. 2013. Disponibilidade e valor nutritivo da vegetação de caatinga no semiárido norte riograndense do Bras. Holos 3: 196–204.

Souza, K.O., Moura, C.F.H., Brito, E.S. and Miranda, M.R. 2014. Antioxidant compounds and total antioxidant activit y in fruits of acerola from cv. Flor Branca, Florida sweet and BRS 366. Rev Bras Frutic. 36(2): 294–304.

TACO. Tabela Brasileira de Composição de Alimentos/NEPA. UNICAMP. Campinas. NEPA-UNICAMP. 2011. 164p.

Taylor, N. and Zappi, D. 2004. Cacti of Eastern Brazil. 1 ed. Royal botanical gardens, London.

Trentin, D.S., Giordani, R.B., Zimmer, K.R., Da Silva, A.G., Da Silva, M.V., Correia, M.T. et al. 2011. Potential of medicinal plants from the Brazilian semi-arid region (Caatinga) against Staphylococcus epidermidis planktonic and biofilm lifestyles. J. Ethnopharmacol. 137: 327–335.

Trovao, D.M.B.M., Fernandes, P.D., Andrade, L.A. and Neto, J.D. 2007. Variações sazonais de aspectos fisiológicos de espécies da Caatinga. Rev. Bras. de Eng. Agricola e Ambient. 11: 307–311.

United Nations (UN). 2017. Department of Economic and Social Affairs, Population Division. World Population Prospects: The 2017 Revision, Key Findings and Advance Tables. Working Paper No.

Wallace, R.S. 1995. Molecular systematic study of the Cactaceae: using chloroplast DNA variation to elucidate Cactus phylogeny. Bradleya 13: 1–12.

Wohner, C., Ohnemus, T., Zacharias, S., Mollenhauer, H., Ellis, E.C., Klug, H. et al. 2021. Assessing the biogeographical and socio-ecological representativeness of the ILTER site network. Ecol. Indic. 127: 1–11.

Zappi, D. and Aona, L. 2007. Cactaceae in flora brasiliensis revisitada.

Zappi, D., Taylor, N., Ribeiro-Silva, S., Machado, M., Moraes, E.M., Calvente, A. et al. 2011. Plano de ação nacional para a conservação das cactáceas. Suelma Ribeiro Silva (Orgs). Brasília: instituto chico mendes de conservação da biodiversidade, ICMBIO, pp. 112.

Zappi, D.C., Santos, M.R. and Larocca, J. 2015. *Cactaceae*. IN. Jardim Botânico do Rio de Janeiro. Lista de Espécies da Flora do Brasil.

Zappi, D.C. and Taylor, N. 2020. *Cactaceae*. *In*: Flora do Brasil 2020 em construção. Instituto de Pesquisas Jardim Botânico do Rio de Janeiro, Br.

Chapter 8

Cashew's Industry

Products, Co-products and its Applications

Rafael Marinho Bandeira,[2] *Lindalva de Moura Rocha,*[4] *Leanne Silva de Sousa,*[5] *Gabriela Almeida de Paula,*[6] *Josélia Borges de Moura Furtado,*[7] *Paulo Afonso Furtado Filho,*[8] *Raimundo Oliveira Lima Júnior,*[9] *Thamires Pereira dos Santos Lima,*[10] *Inês Maria de Souza Araújo*[3,*] and *José Ribeiro dos Santos Júnior*[1]

Introduction

The cashew tree (*Anacardium occidentale*) is a plant native to Brazil that was introduced in other regions of the globe in the 16th century with the aim to conserve the soil (Sharma et al., 2020). This tree has great importance in the world and its major growers are Brazil, India, Vietnam and some African countries.

[1] Full Professor at Federal University of Piauí. Doctor in Chemistry (Physicochemical) from University of São Paulo.

[2] Visiting Researcher at Federal University of Piauí. Doctor in Science from University of São Paulo.

[3] Assistant Professor at Federal University of Piauí. Doctor in Biotechnology from Northeast Network of Biotechnology/UFPI, Teresina-PI.

[4] Doctoral Student in Biotechnology from Northeast Network of Biotechnology/UFPI, Teresina-PI, Brazil.

[5] Professor at Federal Institute of Education, Science and Technology of Piauí. Master in Chemistry from Federal University of Piauí.

[6] Assistant Professor at Federal University of Piauí. Doctor in Biotechnology from Northeast Network of Biotechnology/UFC, Fortaleza-CE.

[7] Professor at Federal University of Piauí. Master in Chemistry from Federal University of Piauí.

[8] Professor at Federal University of Piauí. Master in Physics from Federal University of Piauí.

[9] Professor at Federal Institute of Education, Science and Technology of Piauí. Master in Chemistry from Federal University of Piauí.

[10] Doctoral Student in Veterinary Medicine from Federal University of Piauí, Teresina-PI, Brasil.

* Corresponding author: inesmaria@ufpi.edu.br

The cashew tree has crooked trunks, originating from the North and Northeast regions of Brazil, and belongs to the family *Anacardiaceae*, which have approximately 80 genera and more than 700 species, that can be found in tropical and subtropical regions of the planet (Gazzola et al., 2006; Muianga et al., 2016).

In Brazil, there are 15 genera and 70 species. In regions with dry weather and sandy soil and low fertility, the cashew trees have a small size, large and crooked trunks, long and sinuous branches, leading to the formation of irregular crowns (Serrano and Oliveira, 2013).

The cashew tree reaches between 5 and 1 meters of height, reaching up to 20 meters. It is also known by different names: acaju, acajaíba, caju-comum, cajuil, caju-manso, cajuzeiro and ocaju (Gazzola et al., 2006).

Its period of fruiting is in spring and summer. The harvest occurs during the months of January and February, and is done manually. The cashew contains the peduncle and the nut, as shown in the Fig. 8.1. The peduncle is called pseudo fruit, and it can be consumed in its natural or industrialized state, and composes 90% of the fruit's total weight, with sizes that varies from 5 to 11 cm. The cashew nut is the main product, considered the true fruit with small size, about 2 cm, it has rigid structure and with grayish-green color, the inner nut is the edible part (Lima and Duarte, 2006; Muianga et al., 2016; Neves et al., 2016).

The cashew nut and the peduncle have independent growth. The nut grows fast, taking about 30 d to reach its maximum size after flower fertilization. The peduncle grows slowly until reaching 30 d. After that time, the growth accelerates, taking about 50 d to reach its maximum size, with a color that ranges from yellow, orange or red (Muianga et al., 2016). The complete maturation of the cashew occurs approximately between 49 to 56 d after flower fertilization (Serrano and Oliveira, 2013).

Figure 8.1. Cashew fruit image containing the fruit (cashew nut) and the peduncle.

The peduncle

The peduncle is the fresh edible part of the cashew, where one gets juices, cajuína and food fibers that represent 90% of the total weight (Brasil, 2010). The peduncle is also the soft and juicy, from where the cashew pulp is extracted. In terms of commercialization the natural consumption of the cashew's peduncle as fresh fruit is an accomplishment of the Brazilian cajaculture (Moura et al., 2013). The physical characteristics of the cashew peduncle are of fundamental importance to the definition of post-harvest handling techniques, as well as for the good acceptance of the product in the market (Moura et al., 2001).

The cashew peduncle presents a high nutritive value, containing about three to five times more vitamin C than oranges, besides sugars, minerals (calcium, iron and phosphor), phenolic compounds (tannins, carotenoids and anthocyanins) among other nutrients (Quirino, 2019). Tannins are also present in the proportion of 0,12 to 0,37% (p/p) in its composition, being responsible for the sensation of "locking" of the taste when consuming the natural cashew, this sensation is caused by a property present in some fruits, known as astringency. Tannins also contribute to antioxidant activity, being capable of helping in the prevention of cardiovascular diseases and cancer (Pinho, 2009).

With the extraction of the peduncle juice, bagasse is obtained as a coproduct of the extraction. The bagasse consists of the skin and remaining pulp, being used for the production of several products. However, the characterization of derived products of the cashew peduncle is still a little-explored field as a new form of consumption (Barros and Moura, 2020). The production of the cashew peduncle in Brazil is estimated at around 1.8 million tons per year, concentrating basically in the Northeast region, with industrial exploitation of just 15% of the total (Kiss, 2005). This scheme can be related to its perishability. Furthermore, it is estimated that around 90% of the peduncles are wasted as a result of the low post-harvest conservation, which does not exceed 48 hr. Thus, there is a high rate of post-harvest loss, which demands studies to minimize these impacts. For this reason, several researches in the area of conservation of the fruit have been initiated, like dehydration, drying, freezing, handling, among others (Machado et al., 2011).

The cashew nut

The nut coincides to the smallest portion of the cashew, presenting approximately 10% of the cashew weight, with a grayish-green color. The fruit is formed by the peel (pericarp), skin and nut. The pericarp is made of three layers: epicarp, mesocarp and endocarp. The pericarp has a leathery consistency corresponding to the outermost layers. In percentual terms, the pericarp presents 65 to 70% of the cashew nut total weight. The middle layer is the mesocarp and different from the pericarp has a spongy feature where it can be found in the Cashew Nut Liquid (CNL), which has inflammable properties. The endocarp is the innermost layer, being hard and therefore characterized by the purpose of protecting the nut (Serrano and Oliveira, 2013; Muianga et al., 2016).

The nut is formed by two ivory-colored edible cotyledons. It is the main product used by cashew trees worldwide. As it is an oilseed, an oil can be extracted from it, and can be used as a substitute for olive oil.

Besides, the cashew nut is nutritive and shows cholesterol content that does not compromise human health, it is rich in proteins, lipids, phosphorus, iron, carbohydrates, calcium, vitamin C, vitamin A, B complex, fiber, polyunsaturated fatty acids (around 60% oleic acid and 21% linoleic acid) and other minerals (Gazzola et al., 2006; Lima et al. 2018; Muianga et al., 2016; Neves et al., 2016). The main components of the nut are lipids, proteins and starch (Soares et al., 2015). After the nut, the main product of the cashew is the CNL, followed by the juice of the pseudo fruit (Chaves et al., 2010).

The countries with the highest production of nuts are India, Brazil, Mozambique, Tanzania, Kenya and Vietnam. The Brazilian states that show the highest production of nuts are Ceará, Piauí, Bahia and Rio Grande do Norte. In addition to nuts, cashews provide other products accepted by the market, including pulp, sweets, etc. (Muianga et al., 2016).

Cashew nut constituents

The cashew nut is a real nut, a source of proteins, and essential amino acids, with the exception of lysine, methionine and cysteine, which has a high lipid content (Freitas and Naves, 2010). It consists of the shell, skin or nut integument and the nut, is the edible part (Paiva et al., 2000).

The nut, shown in Fig. 8.2, is surrounded by a film which during its processing is removed and alkaloids and tannins are extracted. The CNL is a caustic, viscous, reddish-brown liquid, consisting of long-chain saturated and unsaturated phenols such as anacardic acid, cardanol, cardol and 2-methyl cardol, representing about 15 to 30% of the cashew nut's total weight, considered a by-product of the cashew production chain, sold at a very low cost (Amorati et al., 2001; Mazzetto et al., 2009; Andrade et al., 2010; Mgaya et al., 2019).

The CNL is classified according to the extraction method, the technical CNL and solvent extracted CNL. The material obtained by solvent extraction mainly contains anacardic acid (60–65%), cardol (15–20%), cardanol (10%) and traces of 2-Methyl-cardol (Fig. 8.3). On the other hand, technical CNL, got by roasting the husks at high temperatures, has cardanol (60–65%) as its main constituent.

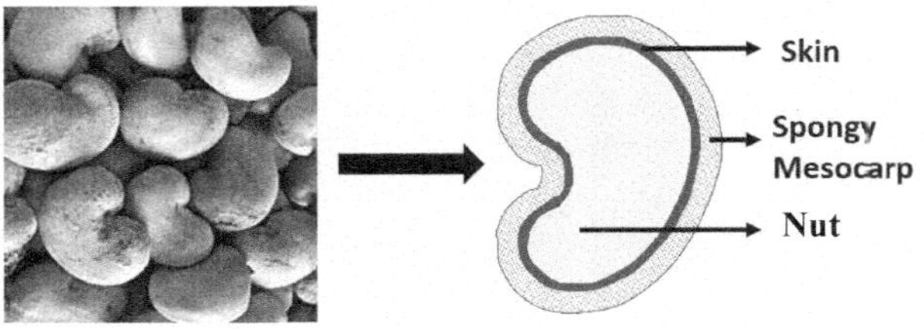

Figure 8.2. Cashew nuts and their parts. The CNL accumulates in the spongy mesocarp, being released during the processing of the nuts for the subsequent nut removal.

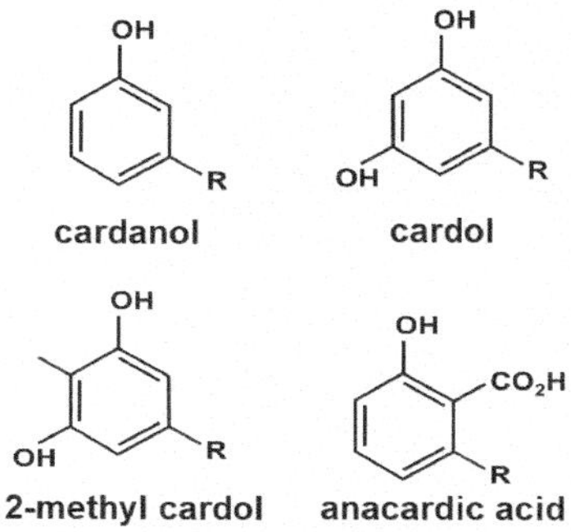

Figure 8.3. CNL chemical constituents.

Cashew nuts are made up of unsaturated fatty acids, such as oleic and alpha-linolenic acids, tocopherols, phytosterols, and phenolic compounds that act as antioxidants (Zanqui et al., 2020).

Due to the presence of phenolic components, CNL has several biological properties such as anti-inflammatory, antitumor potential, molluscicide activity, antioxidant and insecticide activity against *Aedes aegypti*, showing excellent therapeutic potential and acting as a mushroom tyrosinase inhibitor (Bastos et al., 2019).

The industry of the cashew nut

The cashew industry (cashew nut and peduncle) are specific industries, although both are food. Thus, the peduncle industry is focused on the production of cajuína, sweets, juices and other food products. The cashew nut industry has three products; the raw or roasted nuts, the CNL and the solid residue of the nut shell. In these nut industries, the solid residue of the shell is used in the generation of heat used in the nut processing, cooking, drying and others applications. To illustrate the production process, the flowchart in Fig. 8.4 is presented. The process starts with the harvest of cashews in the field and then the transportation to the production unit (Lima, 1988). Cashew is determined as a 'fruit' for sale in the market, in processing to produce cajuína and in the manufacture of sweets and other products from the peduncle. On the other hand, the nuts, when separated, are meant to produce nuts and the obtaining of CNL (Medina, 1978). For each processing, the steps are different, for example, to produce cajuína it is necessary to select the cashews, in obtaining nuts, it is necessary to select the nuts by size. The cashew nut industry has stages that require completely different equipment from those used in the processing of the peduncle. While the production of cajuína is artisanal, the production of nuts is a small, medium or large-scale industry (Paiva et al., 2000).

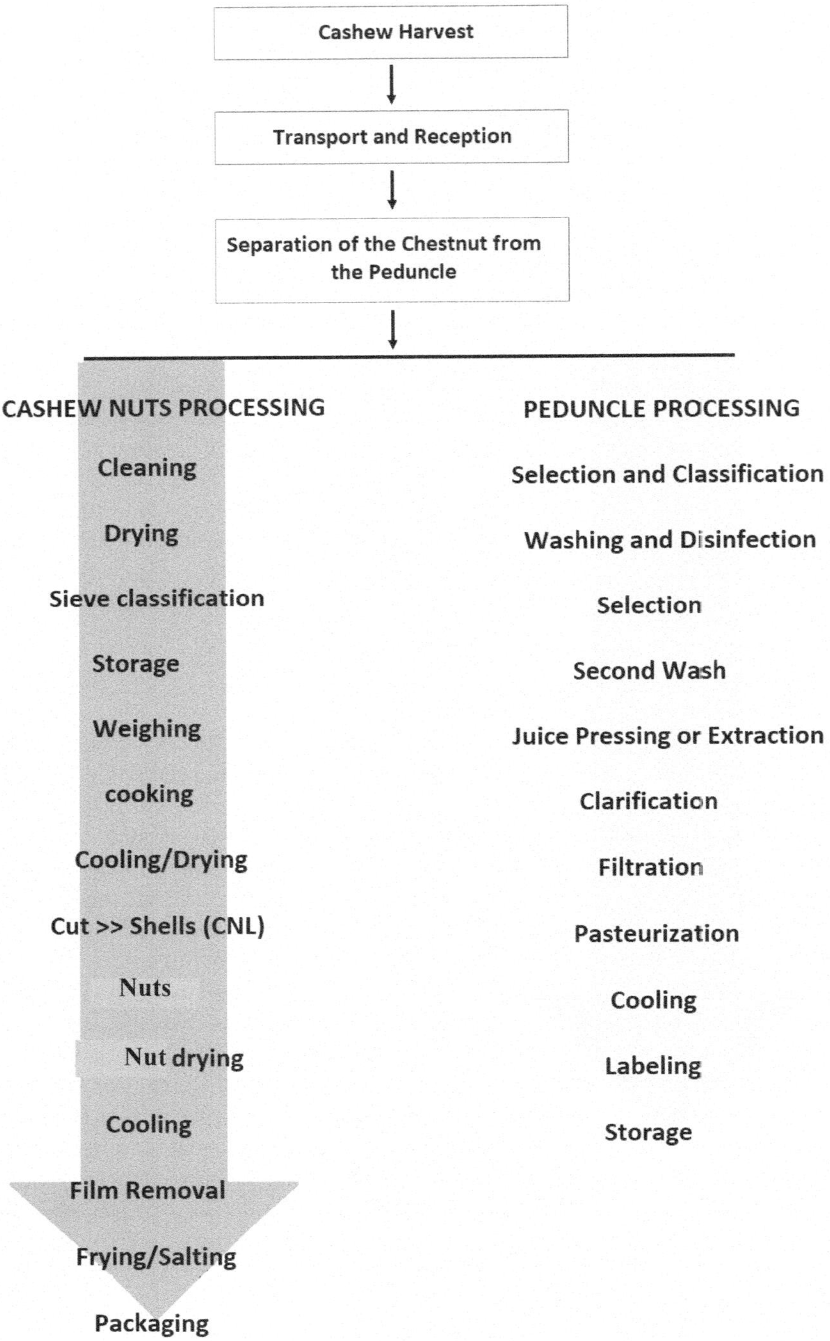

Figure 8.4. Flowchart of industrial processing steps of cashew nuts and peduncles.

The product of the cashew industry

The agroindustry related to the cashew fruit and pseudo-fruit plays an important role in the Brazilian economy, especially in the Northeast region of the country, with the nut, which is the fruit, corresponding to 10% of cashews in mass, being the main product and even entering the export agenda at some states, such as Ceará and Piauí (Silva, 2015).

The peduncle, pseudo fruit, corresponds to 90% of cashews in mass, being used for natural consumption or for the production of juices and many other products that are developed from this part of the cashew, as is the case of clarified juices (cajuína) concentrates of juices, liqueurs, soft drinks, wines, spirits, in addition to the production of freeze-dried sweets and pieces of peduncle for consumption. However, most of these products have only Brazilians as a consumer market, since this portion of cashew is perishable and suffers from production seasonality (Araújo, 2013; Costa, 2015).

There are many incentives in the states that have large cashew production, aimed at improving production conditions, processing and, consequently, increasing the added value of products from cashew farming. In 2000, Embrapa launched a manual on the production of various products in the cashew industry, emphasizing the production methods of whole juice, nectar, cajuína, wine, jam in syrup, jam, jam in mass, cashew jelly, candied cashews, plum-cashew, clarified cashew honey, cashew pulp and cashew brown sugar. Many of these products manage to solve the problem of the seasonality and of the rapid perishability of the cashew stalk (Paiva et al., 2000).

In turn, cashew nuts are a product aimed at both the domestic and the foreign market, becoming the third most consumed nut in the world (Araújo, 2013). This product usually goes through a large production chain, since the producer usually does not have a large volume for commercialization, requiring that the processing be done in cooperatives and industries that buy the nuts from small producers, making the production chain economically viable and strengthened (Guanziroli, 2009).

Beyond the food consumption of cashew nuts, this fruit also provides the CNL, as mentioned earlier, resulting from the industrial processing of nuts in chemical industries. This liquid is of foremost importance, as it can be used in various applications as can be seen next. Among the applications, its use as a monomer to produce polymer with antioxidant effect for biodiesel is highlighted. Figueiredo (2009) studied the production of resin for application purposes as a concrete waterproofing agent to protect against corrosion. The implications of the development of the cashew industry are positive, as it allows for a source of income for the population, having an important social and economic effect for low-income people in northeastern Brazil (Silva Neto et al., 2009).

Cashew nut liquid (CNL) and its applications

As mentioned earlier, CNL is a by-product of the processing of nuts. The cashew nuts are roasted in an oven and a flammable liquid is released, consisting mainly of cardanol (60–65%), cardol (15–20%), polymeric material (~ 10%), in addition to showing traces of 2-methylcardol (Lopes et al., 2008). The CNL obtained from the industrial processing of nuts is called technical CNL. The CNL by cold extraction is called natural, consisting of anacardic acid (60–65%), cardol (15–20%), cardanol (~ 10%) and traces of 2-methylcardol (Maia et al., 2012).

CNL is a natural source of long-chain unsaturated phenols and has been used in several applications such as; antioxidant activity (Lopes et al., 2008; Maia et al., 2012), corrosion protection (Da Silva et al., 2020; Furtado et al., 2019; Tewary et al., 2019; Wazarkar et al., 2018), matrix for mechanical reinforcement of composites (Dashtizadeh et al., 2020; Kumar and Venkatesh Babu, 2020), plasticizer and composites (Gomes et al., 2018; Guna et al., 2018), antibacterial and antifungal activity (Garcia et al., 2018; Koteich Khatib et al., 2020; Mashraf and Rathinasamy, 2018), anticancer (Mashraf and Rathinasamy, 2018; Reddy et al., 2018), technologies with applications in agriculture (Pandiyan et al., 2020), fuel and energy (Anas et al., 2019; Krishnamoorthi et al., 2019; Kumar et al., 2018; Sawadogo et al., 2018; Scaldaferri and Pasa, 2019), etc. These applications are economically viable because CNL has a low cost, is renewable, in addition to the important environmental factor characterized by the need to apply this residue (Sharma et al., 2020). According to Sharma et al. (2020) the global production of cashew nuts in 2005 and 2013 was 3.21 and 4.44 million tons, respectively. As the demand for cashew nuts increases in the world, CNL becomes an environmental problem, requiring studies to be carried out in order to make the use of this by-product viable.

In Brazil, cashew production is concentrated in the Northeast region, accounting for 98.8% of national production (Brainer and Vidal, 2018). In 2017, the production of cashew nuts in Brazil was 133,465 tons, a number that can vary significantly from one year to another due to the climatic conditions in the region (Brainer and Vidal, 2018). Efforts to increase cashew production have grown, due to the increased demand for the nut, cashew juice and clarified cashew juice (cajuína). As a result, the accumulation of CNL stands out, being characterized as a low-cost by-product, but with several potential technological applications as mentioned above.

Therefore, the cashew production chain is increasingly highlighted, being an agribusiness sector that generates income for families and for small cashew and nut processing companies, which can expand income by valuing the CNL against new technologies related to the use of the constituents of this by-product.

Food products

In addition to the economic aspect, cashew-derived products have high nutritional importance. Cashew contains about 156 mg to 387 mg of vitamin C, 14.70 mg of calcium, 32.55 mg of phosphorus and 0.575 mg of iron per 100 ml of juice (Codevasf, 2012).

Cashew farming plays a particularly important role in the rural economy of the Northeast. The main exploration agenda for this culture is the cashew nut (Filho et al., 2006). The castanha-de-caju, as Brazilians know it, has become an expensive and luxury spice. Among its uses, the cashew nut has been consumed salted, as a snack or in its natural form, and can be consumed and used in the production of cereal bars, flour, vegetable milk, pastes, sweets and confectionery, among others.

Table 8.1. Products derived from the Cashew Peduncle (Paiva et al., 2000).

Products obtained from the liquid fraction	**Products obtained from the fibrous fraction**	**Products obtained from the cashew fiber**
Juice	Candy in mass	Basic cashew meat
Nectar	Candied cashew	Baked pastries
Liquor	Flour	Shrimp cashew fry
Concentrated juice	Candy in syrup	Strawberry cashew
Soft drink	Brown sugar	White sauce
Brandy	Pies	Cashew stew
Sweetened juice	Jam bread fillings	Cashew bread
Soda	Biscuit pulp	Indian cashew
Alcohol	Pizza	Cashew omelet
Cajuína	Cashew plum	Mousse
Wine	Cake	Burger
Syrup	Fried meatballs	Cake
Sparkling wine	Cashew raisin	Fried
Honey	Ketchup	meatballs
Vinegar	Hot dishes	Provençal cashew
Jam	Pickles	Oriental rice
Ice creams	Cold dishes	Roulade
		Cashew pate

Source: EMBRAPA; SEBRAECE (2000)

Cashew nuts have excellent nutritional values, being rich in vitamins, minerals, monounsaturated fatty acids and approximately 25% of proteins (Melo et al., 1998).

The industrialization of the peduncle through its best use remains a challenge to be faced (Oliveira, 2009). However, the peduncle can be used in the production of various food products, as shown in Table 8.1.

Applications

Besides the applications mentioned above, there are several other operations characterized by the use of CNL and the cashew nut shell. The work developed by Paiva et al. (2000) summarizes the implementation of CNL and the nut shell. Among the practices mentioned in the work, is the use of CNL in paints, varnishes, insecticides, lubricating oils, rubbers, waterproofing, etc. They are part of the list of applications that CNL currently finds. In addition, the nut shell has applications in the manufacture of glues, fertilizers and can also act as fuel residue, etc.

Some fine applications of CNL were described earlier in this work. It is noteworthy that there are also fine applications aimed at the use of the cashew nut shell. Garg and Das (2019) obtained microporous carbon from the shell of cashew nuts and applied the material as a CO_2 adsorbent. More recently, Cai et al. (2020) published a "short communication" in which they obtained porous carbon from the cashew nut shell and used the carbon in the manufacture of electrodes with a high specific capacity, its use in high-performance supercapacitors, making it plausible. These applications are in line with what is expected from Green Chemistry in contributing to alternatives that can mitigate environmental impacts generated by residues from the processing of materials, as in the case of the cashew nut processing process.

Therefore, there are several application possibilities for both CNL and the cashew nut shell. Many of these applications are recent discoveries, which will allow the cashew industry to gain greater socio-economic and environmental visibility soon.

Conclusions

In summary, this work showed that the cashew tree (*Anacardium occidentale*) is a plant that has a broad spectrum of technological applications due to the commercial use of its products and co-products. Both peduncle and cashewnut are important to the maintenance and valorization of cashew products processing activity, bringing income to the producer families, creating a relevant socioeconomic impact. Moreover, there are other promising materials such as the CNL with fine technological applications—which encourages the improvement of technologies for products from cashew culture, product processing, co-products and wealth generation for producers. The CNL, for example, is applied in the protection of metallic corrosion, being an important way to use it instead of discarding it in the environment. Consequently, applications of CNL and other cashew-derived materials add to the cashew productive chain.

References

Amorati, R., Pedulli, G.F., ValgimiglI, L., AttanasI, O.A., Filippone, P., Fiorucci, C. and Saladino, R. 2001. Absolute rate constants for the reaction of peroxyl radicals with cardanol derivatives. J. Chem. Soc., Perkin Trans 2(11): 2142–2146, 10.1039/B105079F.

Anas, M., Jahiding, M., Ratna and Sudiana, I.N. 2019. Production and characterization of activated carbon from cashew nut shell using N_2 as activation agent. IOP Conference Series: Mat. Scien. and Eng. 550(1): 012035,10.1088/1757-899X/550/1/012035.

Andrade, T.d.J.A.d.S., Araújo, B.Q., Citó, A.M.d.G.L., Da Silva, J., Saffi, J., Richter, M.F. and Ferraz, A.B.F. 2010. Antioxidant properties and chemical composition of technical Cashew Nut Shell Liquid (tCNSL). Food Chem. 126(3): 1044–1048, 10.1016/j.foodchem.2010.11.122.

Anghel, A.G., Grumezescu, A.M., Chirea, M., Grumezescu, V., Socol, G., Iordache, F., Oprea, A.E., Anghel, I. and Holban, A.M. 2014. MAPLE Fabricated Fe_3O_4@Cinnamomum verum antimicrobial surfaces for improved gastrostomy tubes. Molecules 19(7): 8981–8994. 10.3390/molecules19078981.

Araújo, J.P.P. 2013. Cashew Agribusiness: Practices and Innovations. Brasília-DF: Embrapa.

Bastos, T.M., Russo, H.M., Moretti, N.S., Schenkman, S., Marcourt, L., Gupta, M.P., Wolfender, J.-L., Queiroz, E.F. and Soares, M.B.P. 2019. Chemical constituents of anacardium occidentale as inhibitors of Trypanosoma cruzi Sirtuins. Mol. 24(7): 1299, 10.3390/molecules24071299.

Barros, T. and Moura, R. 2020. Cashew technology will contribute to food education. 2020. Available at: https://www.embrapa.br/busca-de-noticias/-/noticia/18011524/tecnologia-do-caju-vai-contribuir-com-educacao-alimentar. Acesso em: 13 mar. 2020.

Brainer, M.S.C.P. and Vidal, M.F. 2018. Northeastern cashew farming in recovery. Sector Book ETENE, Year 3(54): 1–13.

Brasil, F.B.D. 2010. Sustainable regional development. Fruit Growing - Cashew. FBB. Brasília. 4.

Cai, N., Cheng, H., Jin, H., Liu, H., Zhang, P. and Wang, M. 2020. Porous carbon derived from cashew nut husk biomass waste for high-performance supercapacitors. J. of Elect. Chem. 861(15): 113933, 10.1016/j.jelechem.2020.113933.

Chaves, M.H., Citó, A.M.G.L., Lopes, J.A.D., Costa, D.A., Oliveira, C.A.A., Costa, A.F. and Júnior, F.E.M.B. 2010. Fenóis totais, atividade antioxidante e constituintes químicos de extratos de *Anacardium occidentale* L., Anacardiaceae. Rev. Bras. Farmacogn. 1(20): 106–112, 10.1590/S0102-695X2010000100021.

Codevas, F. 2012. The cashew production chain. CODEVASF, B. I. d. P. d. : Codevasf.

Costa, M. 2015. Cajuína project in Piauí. Cashew Growing: Basic Information. Implantation of Orchard and Use of Peduncle. Teresina: EMATER.

Da Silva, L.R.R., Avelino, F., Diogenes, O.B.F., Sales, V.d.O.F., Da Silva, K.T., Araujo, W.S., Mazzetto, S.E. and Lomonaco, D. 2020. Development of BPA-free anticorrosive epoxy coatings from agroindustrial waste. Prog. in Organic Coatings 139: 105449, 10.1016/j.porgcoat.2019.105449.

Dashtizadeh, Z., Abdan, K. and Jawaid, M. 2020. Mechanical properties enhancement of cardanol by hybridization with kenaf/recycled carbon. Mech. of Mater. 148: 103475, 10.1016/j.mechmat.2020.103475.

Figueiredo, F.C. 2009. Obtaining CNL polymers for application as an antioxidant in soy biodiesel. Master's Dissertation, Federal University of Rio Grande do Norte, Natal, BR.

Figueiredo, F.C. 2016. Resina de Líquido da castanha de caju (CNL) utilizada para redução da permeabilidade de superfícies de concreto. Ph.D. Thesis, Federal University of Piaui, Teresina, BR.

Filho, M.d.S., Aragão, A.d.O., Alves, R.E. and Filgueiras, H.A.C. 2006. Aspects of postharvest, harvesting and industrial processing of cashew peduncles (*Anacardium occidentale* L.). Technological, A. E. d. I.: Embrapa.

Freitas, J.B. and Naves, M.M.V. 2010. Chemical composition of edible nuts and seeds and their relationship to nutrition and health. J. of Nut. 23(2): 269–279, 10.1590/S1415-52732010000200010.

Furtado, L.B., Nascimento, R.C., Seidl, P.R., Guimarães, M.J.O.C., Costa, L.M., Rocha, J.C. and Ponciano, J.A.C. 2019. Eco-friendly corrosion inhibitors based on Cashew nut shell liquid (CNSL) for acidizing fluids. J. of Molec. Liq. 284: 393–404, 10.1016/j.molliq.2019.02.083.

Garcia, N.Z.T., Barbosa, G.F., Matias, R., Pedrinho, D.R., Bono, J.A.M. and Martini, D. 2018. Antifungal potential of cashew nut shell liquid in the control of plant pathogens. Biosc. J. 34(1): 95–103, 10.14393/BJ-v34n1a2018-36695.

Garg, S. and Das, P. 2020. Microporous carbon from cashew nutshell pyrolytic biochar and its potential application as CO_2 adsorbent. Biomass. Convers. and Bioref. 10(6): 1043–1061, 10.1007/s13399-019-00506-1.

Gazzola, J., Gazzola, R., Coelho, C.H.M., Wander, A.E. and Cabral. J.E.O. 2006. The cashew nut kernel: composition and importance of fatty acids - Production and world trade. *In*: XLIV Congresso Da Saber. Fortaleza, 23 to 27 July 2006.

Gazzola, J., Gazzola, R., Coelho, C.H.M., Wander, A.E. and Cabral, J.E.O. 2006. A amêndoa da castanha-de-caju: composição e importância dos ácidos graxos - produção e comércio mundiais. Proc. XLIV Congresso da Sober: "Questões Agrárias, Educação no Campo e Desenvolvimento". Fortaleza, BR.

Gomes, V.N.C., Carvalho, A.G., Furukava, M., Medeiros, E.S., Colombo, C.R., Melo, T.J.A., Araújo, E.M., Morais, D.D.S., Ueki, M.M., Paskocimas, C.A. and Santos, A.S.F. 2018. Characterization of wood plastic composite based on HDPE and cashew nutshells processed in a thermokinetic mixer. Polym. Comp. 39(8): 2662–2673, 10.1002/pc.24257.

Guanziroli, C.E. 2009. Cashew nut production chain: study of market relations. Fortaleza-CE: Banco of Northeast: IICA.

Guna, V., Ilangovan, M., Nataraj, D. and Reddy, N. 2018. Bioproducts from wheat gluten with high strength and aqueous stability using cashew nut shell liquid as plasticizer. J. of Appl. Polym. Scien. 135(43): 46719, 2018/11/15 2018.

Kiss, j. 2005. Reforma na casa. Revista Globo Rural. 233: 58–63.

Koteich Khatib, S., Bullón, J., Vivas, J., Bahsas, A., Rosales-Oballos, Y., Marquez, R., Forgiarini, A. and Salager, J.L. 2020. Synthesis, characterization, evaluation of interfacial properties and antibacterial activities of dicarboxylate anacardic acid derivatives from cashew nut shell liquid of *Anacardium occidentale* L. J. of Surfact. and Deterg. 23(3): 503–512, 10.1002/jsde.12384.

Krishnamoorthi, S., Prabhu, L., Sakthikumar, R., Prabakaran, K. and Pradeep, M. 2019. Performance and emission characteristics of cashewnut shell liquid as alternate fuel with five hole nozzle. Int. J. of Eng. and Adv. Tech. 8(6S3): 2117–2119, 10.35940/ijeat.F1407.0986S319.

Kumar, J.G.K. and Venkatesh BAabu, R. 2020. Mechanical behaviour and characterization of reinforced CNSL composite material. Mater. Today: Proceed. 22: 404–409, 10.1016/j.matpr.2019.07.398.

Kumar, S., Dinesha, P. and Rosen, M.A. 2018. Cashew nut shell liquid as a fuel for compression ignition engines: a comprehensive review. Energ. and Fuels 32(7): 7237–7244, :10.1021/acs.energyfuels.8b00579.

Lima, V. de P.M.S. 1988. The cashew crop in Northeast Brazil. Fortaleza: BNB-ETENE, 1988. 486p. (BNB-ETENE. Economic and Social Studies, 35).

Lima, J.R. and Duarte, E.A. 2006. Pastas de castanha-de-caju com incorporação de sabores. Pesq. Agropec. Bras., Brasília 8(41): 1333–1335, 10.1590/S0100-204X2006000800019.

Lima, L.V. 2018. Obtention, characterization and application in foods of cashew nut (*Anacardium occidentale* L.) protein isolate and concentrate, Master Dissertation, Federal University of Piauí, Teresina, BR.

Lima, D.P., Reddy, T.N., Beatriz, A., Lopes, R.S., Marques, M.R., Dos Santos, E. dcs A., Nantes, C.I., Matos, M. de F.C., Perdomo, R.T., Pereira, I.C. and Weber, S.S. 2018. Design, synthesis and structure-activity relationship of novel phenolic based pyrimidine hybrids from Cashew Nut Shell Liquid (CNSL) components as potential antitumor agents. Chem. Biol. Lett. 5(2): 41–54.

Lopes, A.A.S., Carneiro, E.A., Rios, M.A.S., Hiluy Filho, J.J., Carioca, J.O.B., Barros, G.G. and Mazzetto, S.E. 2008. Study of antioxidant property of a thiosphorated compound derived from cashew nut shell liquid in hydrogenated naphthenics oils. Braz. J. of Chem. Eng. 25: 119–127, 10.1590/s0104-66322008000100013.

Mashraf, S. and Rathinasamy, K. 2018. Antibacterial and anticancer activity of the purified cashew nut shell liquid: implications in cancer chemotherapy and wound healing. Nat. Prod. Res. 32(23): 2856–2860, 10.1080/14786419.2017.1380022.

Machado, A.V., Alves, F.M.S. and Queiroga, K.H. 2011. Food produced from cashew flour, obtained by drying. Green J. Agroec. and Sust. Devel. 6(3): 8.

Maia, F.J.N., Ribeiro, V.G.P., Lomonaco, D., Luna, F.M.T. and Mazzetto, S.E. 2012. Synthesis of a new thiophosphorylated compound derived from cashew nut shell liquid and study of its antioxidant activity. Ind. Crops and Prod. 36(1): 271–275, 10.1016/j.indcrop.2011.10.019.

Mazzetto, S.E., Lomonaco, D. and Mele, G. 2009. Cashew nut oil: opportunities and challenges in the context of industrial development and sustainability. New Chemistry 32: 732–741.

Medina, J.C. 1978. Cashew: From Culture to Processing and Marketing. Campinas: ITAL, 178p.

Melo, M.L.P., Maia, G.A., Silva, A.P.V., Oliveira, G.S.F. and Figuereido, R. 1998. Physicochemical characterization of raw and toasted cashew nuts (*Anacardium occidentale* L.). Food Scienc. and Techn. 18: 184–187.

Mgaya, J., Shombe, G.B., MasikanE, S.C., Mlowe, S., Mubofu, E.B. and Revaprasadu, N. 2019. Cashew nut shell: a potential bio-resource for the production of bio-sourced chemicals, materials and fuels. Green Chem. 21(6): 1186–1201, 10.1039/C8GC02972E.

Moura, C.F.H., Alves, R.E., Innecco, R., Filgueiras, H.A.C., Mosca, J.I. and Pinto, S.A.A. 2001. Physical characteristics of cashew peduncles for fresh marketing. Brazilian. J. of Fruits 23: 537–540, 10.1590/S0100-29452001000300017.

Moura, C.F.H., Alves, R.E. and Silva, E.d.O. 2013. Harvesting and postharvest conservation of cashew stalk. pp. 277–290. *In*: ARAÚJO, J.P.P. de (ed.). Cashew Agribusiness: Practices and Innovations. Brasília, DF: Embrapa.

Muianga, C.A., Muniz, J.A., Nascimento, M.D.S. and Fernandes, T.J. 2016. Description of the cashew fruit growth curve by nonlinear models. Brazil. J. of Fruit. 38: 22–32, 10.1590/0100-2945-295/14.

Neves, W.B.P., Carmo, S.K.S., Palácio, C.M. and Melo, R.P.F. 2016. Physicochemical characterization of cashew nuts obtained in the western region of Rio Grande do Norte. Chem.: Sci. Techn. and Soc. 5(2): 48–58, 10.1016/j.lwt.2011.08.018.

Oliveira, L.G.L. 2009. Integration of the cashew agribusiness production chain with sustainable development. Master's Dissertation, State University of Ceará, Fortaleza, BR.

Paiva, F.F.d.A., GarrutI, D.d.S. and Silva Neto, R.M.d. 2000. Industrial use of cashew. Fortaleza: Embrapa. Embrapa-CNPAT. Documentos.

Pandiyan, C.V., Shylaja, G., Srinivasan, G.R. and Saravanan, S. 2020. Studies on use of Cashew Nut Shell Liquid (CNSL) in biopesticide and biofertilizer. Nat. Environm. and Pollut. Techn. 19(1): 103–111.

Pinho, L.X. 2009. Use of cashew penduncle residue (*Anacardium occidentale* L.) for human consumption. Master's Dissertation, Federal University of Ceará, Fortaleza, BR.

Quirino, E.C.G. 2019. Obtaining cashew peduncle flour and its use in the formulation of fiber-rich cake. Advisor: ARBOS, K.A. 50 f. - Department of Food Technology, Federal University of Paraíba, João Pessoa-PB.

Reddy, K.J., Menon, K.R. and Thattil, A. 2018. Academic Stress and its Sources among University Students. Biomedical and Pharmacology Journal 11: 531–537. https://doi.org/10.13005/bpj/1404.

Rosa, M. de F. and Abreu, F.A.P. de. 2000. Coconut water: conservation methods. 40 p. (Embrapa - CNPAT - SEBRAE/CE. Documents 37) Fortaleza.

Sawadogo, M., Tchini Tanoh, S., Sidibé, S., Kpai, N. and Tankoano, I. 2018. Cleaner production in Burkina Faso: Case study of fuel briquettes made from cashew industry waste. J. of Clean. Prod. 195: 1047–1056, 10.1016/j.jclepro.2018.05.261.

Scaldaferrl, C.A. and Pasa, V.M.D. 2019. Green diesel production from upgrading of cashew nut shell liquid. Renew. and Sust. Energ. Rev. 111: 303–313, 10.1016/j.rser.2019.04.057.

Serrano, L.A.L. and Oliveira, V.H.d. 2013. Botanical aspects, phenology and cashew crop management. ARAÚJO, J.P.P.d. Brasília, DF: Embrapa: 77–165.

Sharma, P., Gaur, V.K., Sirohi, R., Larroche, C., Kim, S.H. and Pandey, A. 2020. Valorization of cashew nut processing residues for industrial applications. Indust. Crops and Prod. 152: 112550, 10.1016/j.indcrop.2020.112550.

Silva Neto, R.M.d., Abreu, F.A.P.d. and Paiva, F.F.d.A. 2009. Cajuína. Embrapa Tropical Agroindustry - Technical Circular (INFOTECA-E). Fortaleza-CE, p. 4.

Silva, V.V.d. 2015. Cashew: The producer asks, Embrapa answers. Brasília-DF: Embrapa.

Soares, C.M.S., Martins, G., Alves, D.G., Conte, C., Adorno, W.T. and Sousa, J.P. 2015. Optimization of the processing of toasted cashew nuts. Braz. Magaz. of Agroind. Prod. 17: 65–74, 10.1080/23311916.2018.1531457.

Tewary, N.K., Gupta, A., Mandal, G., Ghosh, S.K., Das, A., Datta, P. and Ghosh, A. 2019. Improvement of corrosion performance of DI pipe by newly invented CNSL based paint. Materials Today: Proceedings 18: 5202–5208, 10.1016/j.matpr.2019.07.520.

Wazarkar, K., Kathalewar, M. and Sabnis, A. 2018. Anticorrosive and insulating properties of cardanol based anhydride curing agent for epoxy coatings. React. and Func. Polym. 122: 148–157, 10.1016/j.reactfunctpolym.2017.11.015.

Zanqui, A.B., da Silva, C.M., Ressutte, J.B., Rotta, E.M., Filho, L.C. and Matsushita, M. 2020. Cashew nut oil extracted with compressed propane under different experimental conditions: Evaluation of lipid composition. J. of Food Proces. and Preserv. 44(6): 14599, 10.1111/jfpp.14599.

Chapter 9

Chemistry, Management, Functional Properties and Biotechnological Applications of *Ocimum* Essential Oils in the Food Industry

Celma de Oliveira Barbosa,[1,*] *Selene Maia de Morais,*[2] *Halisson Araújo de Sousa,*[3] *Vitor Carvalho Martins,*[4] *João Francisco Câmara Neto,*[4] *Icaro Gusmão Pinto Vieira,*[1] *Rita de Cássia Alves Pereira,*[5] *Ana Livya Moreira Rodrigues*[2] and *José Osvaldo Beserra Carioca*[1]

Introduction

Brazil is one of the main producers of food, but it presents the reality of waste in all stages of the production chain. The losses occur from harvest, transport and industrialization, and also due to inadequate culinary processing and bad eating habits. Besides these stages, losses occur in the supply centers all over the country and in supermarkets, with estimated total of 30%. According to the Food and Agriculture Organization of the United Nations (FAO), Brazil loses about 64% of all annual food production, causing an increase in the final price, quantity and quality of products (Roriz, 2012).

[1] PADETEC – Technological Development Park, Federal University of Ceará.
[2] Natural Products Chemistry Laboratory, Chemistry Department, Ceará State University.
[3] Master in Biochemistry, Federal University of Ceará.
[4] Doctoral Student in Postgraduate Chemistry, Chemistry Department, Federal University of Ceará.
[5] Embrapa Tropical Agroindustry.
* Corresponding author: celmaoliver@yahoo.com.br

Waste often occurs in the fruit and vegetable sector, because usually the food already arrives damaged for production, due to its collection, transport and handling until it reaches the stall (Parisoto et al., 2013). Twenty eight percent of wasted food is lost in the production and harvest processes; 22% is lost in transport, handling and storage; 6% in the processing stage; 28% is lost in supermarkets and in consumers' homes, the main factor being the validity of the products (Moraes and Souza, 2018).

The waste of raw materials is a very important issue for food industries. In this perspective, the food industry presents a growing demand for the use of essential oils due to their important applications as food preservatives, addition to food packaging and fight against pathogens that can generate food poisoning or infection (Asbahani et al., 2015) and may have biotechnological application in reducing waste, increase the shelf life of products, use in packaging and as dietary additives in animal species for the preservation of their chemical characteristics that also have interesting functional properties in the prevention of chronic diseases.

Essential oils are rich in biologically active compounds, insoluble in water, obtained by steam distillation or cold mechanical pressing or by organic solvent extraction, either being one of the main ones (Bizzo et al., 2009; Bizzo and Rezende, 2022) and can be extracted from all plant components such as roots, stems, leaves, flowers, fruits, bark and seeds, and applied to reduce food waste and add value to the food industry.

The genus *Ocimum* (Fig. 9.1), popularly known as basil, belongs to the family Lamiaceae and is characterized by great variability in morphology and chemotypes from tropical and subtropical regions of Asia, Africa, Central and South America, generating a large number of subspecies, varieties and forms for its ease of cross-pollination, for example: *Ocimum gratissimum* (alfavaca), *Ocimum basilicum* L., *Ocimum tenuiflorum*, *Ocimum selloi* Benth (paregoric elixir), species that produce essential oils rich in methyl-chavicol, methyl-cinnamate, eugenol, citral, linalool, thymol, camphor, pinene and tannins with great economic value in the development of pharmaceuticals, foods, perfumes and cosmetics (Özcan and Chalchat, 2002; Pereira and Moreira, 2011; Veloso, 2012; Borges et al., 2012), drawing attention from researchers due to their antioxidant, anti-inflammatory, anti-acetylcholinesterase,

Figure 9.1. *Ocimum* spp.

antimicrobial and larvicidal properties (Barbosa et al., 2021), which can be applied in food preservation and prevention of chronic diseases.

Thus, the aim of this chapter is to focus on the chemical analysis and functional properties of basil essential oils, discussing the biotechnological applications of the use of essential oils in the food industry.

Extraction of essential oils (EOs) in the food industries and the genus *Ocimum*

Many fruits and cereals contain essential oils which are important in the food, drug and cosmetic industries owing to their desirability and nutraceutical attributes (Gayas and Kaur, 2017). Herbs and spices are also gaining prominence for their chemical constitution and functional properties (Barbosa et al., 2021). Essential Oils (EOs) are complex mixtures of volatile compounds extracted from all parts of the plants including seeds, flowers, peels, stems, bark and the whole plants (Bhavaniramya et al., 2019). These low molecular mass compounds are insoluble in water (Dima and Dima, 2015). The method used to extract essential oils in plants is steam distillation in a Clevenger type apparatus. The extraction time is 4 hr (A.O.A.C., 1995). However, new extraction methods are being used in the food industry, such as: Ultrasound, microwave and supercritical fluid (Gayas and Kaur, 2017).

Essential oils (EOs) are secondary metabolites important for plant defense mechanisms with various medicinal properties including antimicrobial activity. Due to the aroma, flavors and natural antimicrobial contents, EOs are primarily used in the food industry for food preservation. For example, EOs which are extracted from citrus such as monoterpenes, sesquiterpenes and oxygenated derivatives have strong inhibitory activities against pathogenic bacteria; and, suggest their use as antioxidant and flavoring agents (Bhavaniramya et al., 2019).

Historically, essential oils were extracted from cinnamon, sassafras, vetiver and other natural sources due to the presence of aromatic compounds (Dhifi et al., 2016), and, hence, it was widely used in aromatherapy and the cosmetics industry. Inhalation and external application of the EOs significantly relieves stress, rejuvenates and regenerates the individual for the next day's work. It mainly works on the olfactory nerve from the nose to the brain and has proven efficient for microbial infections, cardiovascular diseases, cancer and Alzheimer's and labor pain in pregnancy (Faroogi and Sharma, 2000; Pery and Perry, 2006; Shina et al., 2007; Jimbo et al., 2009). There is an increasing trend to use aromatherapy for cancer and sleeping disorders (Bhavaniramya et al., 2019).

Essential oils are very complex in nature as they are composed of a mixture with more than 50 components at quite different concentrations. Mainly, EOs are located in the cytoplasms of certain plant cells, specifically secreted in trichomes or secretory hairs, epidermal cells, internal secretory cells and secretory pockets (Lee et al., 2015). The most common volatile compounds in essential oils are mixtures of aldehydes, ketones, alcohols, amines, amides, phenols and especially, terpenes (Silva et al., 2009).

In this perspective, fruits have F-nerolidol, floral (linalool) and herbals (γ-selinene) as major compounds. Volatile compounds such as hydrocarbons,

2-methyl, 4-heptanone and trimethylsilyl methanol extracted from bitter apples, are the major components responsible for pharmacological activities as antioxidant, antimicrobial, antidiabetic and antiinflammatory activity (Gurudeeban et al., 2011). Terpenes and terpenoids are present in most of the EOs and are responsible for various activities, including food preservation. EOs extracted from rosemary have several monoterpenes including 1,8-cineole and camphor as the major constituents, which act as antimicrobial biofilms in food production (Moore et al., 2016). Peels of four species, including lemon (*Citrus limon*), bitter orange (*Citrus aurantium*), Berkane clementine (*Citrus clementina*) and umbilicus, a species of oranges (*Citrus sinensis*), co-products in the food industries, have different types of essential oils with different yields, different composition and different antioxidant and antimicrobial activity (Brahmi et al., 2021).

Basil is the main commercially grown essential oil crop in many countries. Basil leaves contain essential oils that can be used fresh or dried to flavor various types of meals. Common basil (*Ocimum basilicum*) is an annual herb that grows in several regions around the world with more than 150 species of the genus *Ocimum* (Naidu et al., 2016). Barbosa et al. (2021) highlighted the following species of the genus *Ocimum*: *Ocimum x citriodorum*, *Ocimum gratissimum*, *Ocimum campechianum*, *Ocimum selloi* Benth, *Ocimum tenuiflorum* and *Ocimum basilicum* L., the latter being divided into varieties: ball, cinnamon, 'licorice', greek a palla, white and 'Maria Bonita'. Generally, the genus *Ocimum* is rich in terpenes and terpenoids, related to color variation in the predominantly purple varieties, usually differentiating in thymol or eugenol content (Fig. 9.2).

The average essential oil content of basil species was considered satisfactory by Barbosa et al. (2021), being from 0.03 to 2.00%, and these amounts may be associated with environmental factors, such as: temperature, solar radiation intensity, soil and others, as explained by Botrel et al. (2010). In addition, experimental factors must be considered, such as: the matrix used, the steam-dragging extraction performed by the distiller and the collection (Burt, 2004). Trevisan et al. (2006) attributed the highest yield in the extraction of essential oils to 'alfavaca' basil (*O. gratissimum*) (3.5%)

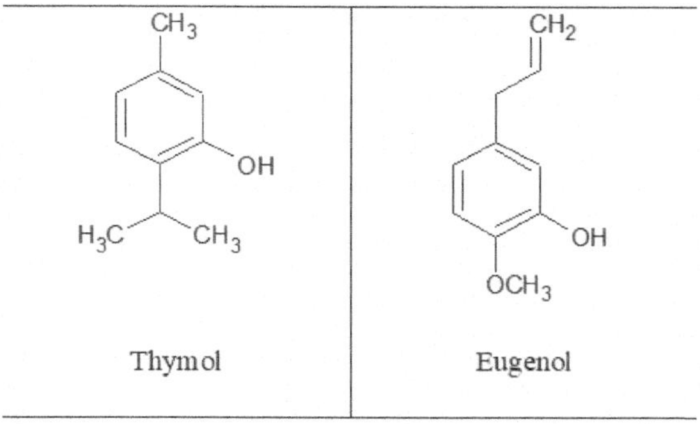

Figure 9.2. Representation of the structures of the main constituents present in basil.

and the lowest to *O. basilicum* purpurascens variety (0.5%), with *O. campechianum* being within this range in the study.

Chemical constitution of basil (*Ocimum* spp.) essential oils

The chemical constitution of essential oils varies by the presence of the following compounds: terpene hydrocarbons, terpenoids, simple terpene alcohols, aldehydes, ketones, phenols, esters, peroxides, furans, organic acids and coumarins (Venancio, 2006).

The terpenes may differ both in structure and form, is created by the condensation of isoprene (C5) units by the acetate-mevalonate pathway and sharing its origin with fatty acids. The difference between the latter occurs in the presence of branching and cyclic structure (Simões et al., 1994). The terpenes are classified by the number of units present in its carbon skeleton, such as: monoterpenes (C10), sesquiterpenes (C15) and diterpenes (C20) (Bakkali et al., 2008). Additionally, when they present oxygen, they are called terpenoids (Toscan, 2010), found in *Ocimum*.

In this view, the composition of essential oils extracted from basil leaves and stems may vary according to the genetic makeup of the plant and geographic location. However, this characteristic aroused the interest of researchers due to its biological potential (Santos, 2007; Martins, 2010).

The genus *Ocimum* had its chemical composition reported in several books. Table 9.1 shows the bibliographic survey regarding the chemical composition of *Ocimum* species in books from 2017 to 2021.

Avetisyan et al. (2017) showed that the chemical composition of the three essential oils of basil cultivars (*O. basilicum* var. thyrsiflora, *O. x citriodorum* and *O. basilicum* var. purpureum), grown in Armenia, had the main chemotypes: methyl-chavicol (estragol) (57.3%) in *O. basilicum* var. purpureum, nerol (23.0%) and citral (20.7%) in *O. x citriodorum* and linalool (68.0%) in *O. basilicum* var. thyrsiflora.

Ocimum basilicum essential oil (grown in the Republic of Srpska) indicated the presence of oxygen-containing monoterpenes (60.9%), sesquiterpene hydrocarbons (28.3%), oxygen-containing sesquiterpenes (3.8%) and monoterpene hydrocarbons (3.6%). Linalool (31.6%) and methyl chavicol (23.8%) have also been reported (Stanojevic et al., 2017).

Li et al. (2017) compared distillate fraction, residue fraction and crude oil from *Ocimum basilicum* grown in Fengsheng Herb Garden (Haikou, Hainan, China). In the study, the results for the distillate fraction indicated that the main components from *Ocimum basilicum* were estragole (13.00%), methyl eugenol (16.96%), α-cadinol (16.24%) and α-bergamotene (11.92%), and represented 63.87% of the total composition.

The essential oil from *O. tenuiflorum* was predominantly constituted by phenylpropanoids (84.8%), while *O. basilicum* presents high amounts of oxygenated monoterpenes (47.7%), phenylpropanoids (21.1%) and hydrocarbon sesquiterpenes (17.7%). Regarding the major compounds, *Ocimum tenuiflorum* essential oil was rich in methyl eugenol (84.7%) with β-caryophyllene as the second major compound

Table 9.1. Bibliographic survey regarding the chemical composition of *Ocimum* species in books from 2017 to 2021.

References	Species (Origin)	Main chemical constituents
Avetisyan et al. (2017)	*O. basilicum* var. *purpureum* (Armenia)	Methyl-chavicol (estragole) (57.3%)
	O. basilicum var. *thyrsiflora* (Armenia)	Linalool (68.0%)
	O. x citriodorum (Armenia)	Nerol (23.0%), citral (20.7%).
Stanojevic et al. (2017)	*O. basilicum* L. (northwestern Republic of Srpska)	Linalool (31.6%) and methyl chavicol (23.8%)
Li et al. (2017)	*Ocimum basilicum* L. (Haikou, Hainan, China)	Methyl eugenol (16.96%) α-cadinol (16.24%)
Piras et al. (2018)	*Ocimum tenuiflorum* L. (India)	Methyl eugenol (84.7%)
	Ocimum basilicum L. (India)	Linalool (35.1%) Eugenol (20.7%)
Ladwani et al., (2018)	*Ocimum basilicum* species (Saudi Arabia)	β-Linalool, Methyl Chavicol Methyl Eugenol, Methyl Cinnamate, trans Methyl Cinnamate, Eugenol, Methyl Eugenol, 1,8-Cineole
Ilić et al. (2019)	*Ocimum basilicum* L. (Serbia)	Linalool (13.68–40.97%)
Rezzoug et al. (2019)	*Ocimum basilicum* L. (Algeria)	Linalool (52.1%) Linalyl acetate (19.1%)
Ricarte et al. (2020)	*Ocimum campechianum* (Ceará, Brazil)	Eugenol (72.1%)
	Ocimum carnosum (Ceará, Brazil)	Linalool (52.1%)
Tacchini et al. (2020)	*O. campechianum* (Ecuador)	Eugenol (43.6%)
Barbosa et al. (2021)	*Ocimum x citriodorum* *Ocimum gratissimum* L. *Ocimum campechianum* *Ocimum selloi* Benth *Ocimum tenuiflorum* *Ocimum basilicum* L. (ball, cinnamon, 'licorice', greek a palla, white and 'Maria Bonita' varieties) (Ceará, Brazil)	Eugenol (42.15–81.91%) Methyl – chavicol (65.82–92.48%) Linalool (70.58%) Geranial (47.62%) Neral (36.10%)
Vasconcelos et al. (2021)	*Ocimum basilicum* (Goiás, Brazil)	Eugenol (74%) β-Linalool (16%)
	Ocimum gratissimum (Goiás, Brazil)	Eugenol (89%)

(7.4%). The essential oil isolated from *Ocimum basilicum* was mainly constituted of linalool (35.1%), eugenol (20.7%) and 1,8-cineole (9.9%) (Piras et al., 2018).

According to the study of Ladwani et al. (2018), the plants from *Ocimum basilicum* L. collected from different cities in the Kingdom of Saudi Arabia showed the following main chemical constituents in the essential oils: β-Linalool (51.93%) in the Abharegion samples, Methyl chavicol (44.9%) in the Dammam samples, β-Linalool (38.75%) in the Jeddah samples, β-Linalool (43.33%) in the Madina samples, trans methyl cinamate (46.69%) and β-linalool (31.58%) in the Mecca samples, methyl chavicol (38.95%) and β-linalool (29.16%) in the Riyadh samples, β-linalool (60.15%) in the Al-Hada samples, β-linalool (72.59%) in the Al-Shafa samples and β-linalool (59.61%) in the Yanbu samples. The results proved that there is an obvious difference in the chemical composition between samples regarding the difference in environmental and geographical factors which are known to have a significant influence on the essential oil composition of the plants.

Ilić et al. (2019) showed that the main classes of compounds from *Ocimum basilicum* L. (B1EO and B3EO) were sesquiterpene hydrocarbons (38.39 and 37.95%), oxygenated monoterpenes (25.44 and 28.04%) and phenylpropanoids (17.43 and 15.71%). Monoterpene alcohol linalool (13.68 and 15.38%), phenoyl derivate eugenol (10.83 and 8.97%) and sesquiterpene hydrocarbon α-bergamotene (8.12 and 9.25%) were the main constituents of both *Ocimum basilicum* EOs. In both EOs, epi-bicyclosesquiphellandrene was detected in considerable amounts (7.03 and 8.07%). The most abundant compound classes in B2EO were oxygenated monoterpenes (52.07%), sesquiterpene hydrocarbons (24.27%) and phenylpropanoids (10.95%). Linalool was the dominant compound (40.97%), followed by epi-bicyclosesquiphellandrene (8.70%) and methyl chavicol (7.92%).

In the essential oil from *O. basilicum*, 26 compounds were identified, accounting for 98.3% of the essential oil. Interestingly, the content of oxygenated monoterpenes (92.0%) was 23 times greater than that of monoterpene hydrocarbons (4.0%). Linalool (52.1%) and linalyl acetate (19.1%) were detected as the major compounds in the oil (Rezzoug et al., 2019).

The essential oil from *O. campechianum* had eugenol (72.1%), β-elemene (6.8%), (E)-caryophyllene (6.4%) and bicyclogermacrene (5.2%). Linalool (79.0%), α-epi-cadinol (5.4%), terpinen-4-ol (3.2%) and 1,8-cineole (2.8%) were the major constituents in the *O. carnosum* essential oil (Ricarte et al., 2020).

Tacchini et al. (2020) highlighted that the most abundant component was the monoterpene eugenol (43.6%), followed by 1,8-cineole (4.4%); among the sesquiterpenes. Tacchini et al. (2020) detected β-caryophyllene (10.8%), β-elemene (8.1%), δ-elemene (4.2%) and bicyclogermacrene (2.9%).

Barbosa et al. (2021) gathered a screening with 11 essential oils from Ocimum grown in Ceará (Brazil) as *Ocimum x citriodorum*, *O. selloi*, *O. gratissimum*, *O. tenuiflorum*, *O. basilicum* (ball, cinnamon, 'licorice', greek a palla, white and 'Maria Bonita' varieties). The results showed the following mains chemotypes: eugenol in *O. tenuiflorum* (81.91%), *O. campechianum* (68.74%), *O. basilicum* var. ball (56.53%), *O. gratissimum* (47.03%) and *O. basilicum* var. greek (42.15%); methyl–chavicol in *O. basilicum* var. 'licorice' (92.48%), *O. basilicum* var. cinnamon

(87.47%), *O. basilicum* var white (67.85%) and *O. selloi* (65.82%); linalool in *O. basilicum* var. 'Maria Bonita' (70.58%); and geranial (47.62%) and neral (36.10%) in *O. x citriodorum*.

The major compounds from *O. gratissimum* grown in Goiás (Brazil) were eugenol and eucalyptol, with the highest yield of 89% eugenol with leaves collected in the afternoon, and 9% eucalyptol for leaves collected in the morning with four hours of extraction. The major compounds of OE from *O. basilicum* were eugenol, β-linalool and muurolol, with the highest yield being 74% eugenol in plants collected in the afternoon and 16% β-linalool and 6% muurolol in plants collected in the morning, both with 4 hours of extraction (Vasconcelos et al., 2021).

In view of the above findings, geographic location may influence the chemical composition of essential oils, revealing higher contents of linalool, eugenol (Fig. 9.2), methyl-chavicol (estragol), methyl–eugenol and 1,8-cineol (Fig. 9.3).

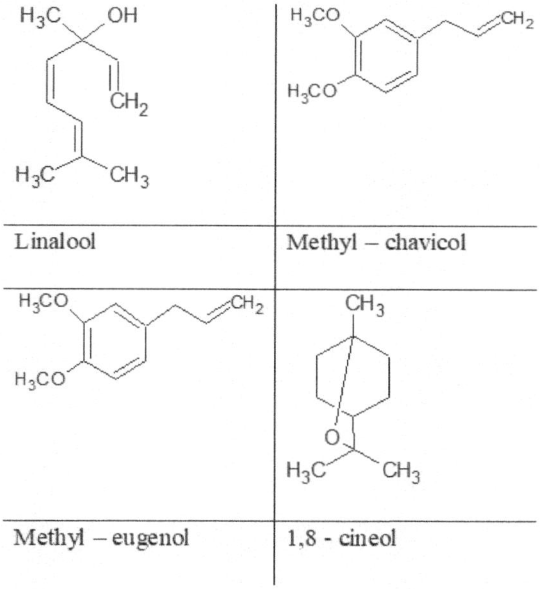

Figure 9.3. Representation of chemical structures of some major chemical constituents of basil (*Ocimum* spp.).

Variations in the chemical composition of essential oils, products of secondary metabolism of plants, occur due to changes in abiotic factors, as well as the geographical location of the plant and time of plant collection (Sobrinho et al., 2016).

Antioxidants and antimicrobial properties of Basil (*Ocimum* spp.) essential oils

Avetisyan et al. (2017) highlighted that the essential oil from *O. basilicum* var. thyrsiflora had the highest antioxidant activity (2.5 μL/mL). This oil has also exhibited the highest tyrosinase inhibition level, whereas the *O. x citriodorum* oil had the highest antimicrobial activity. The essential oil from *O. x citriodorum* was quite active against *B. subtilis* and *St. aureus*, with the MIC (Minimal Inhibitory Concentration) of 3.125 $\mu L.mL^{-1}$.

Stanojevic et al. (2017) showed that the *O. basilicum* essential oil had good antioxidant properties after 90 minutes of incubation with inhibitory concentrations a 50% (IC_{50}) value of 2.38 $mg.ml^{-1}$. The best antimicrobial activity on coagulase-positive *Staphylococcus* was reported.

According to Li et al. (2017), the distillate fraction from *Ocimum basilicum* indicated 1.729 ± 0.056 mg/mL for DPPH and 1.390 ± 0.058 mg/mL for ABTS. However, the residue fraction of sweet basil had a higher antioxidant activity, while the distillate fraction exhibited better anti-inflammatory.

Piras et al. (2018) analyzed the essential oil from *O. tenuiflorum* that showed a pre-eminent effect against *C. neoformans* (0.16 µL/mL) and dermatophytes (0.32 µL/mL). The effect on the germ tube formation of both essential oils was described for the first time showing that *O. tenuiflorum* decreases germ tube formation by more than 50% at values four times lower than MIC (Minimal Inhibitory Concentration) while *O. basilicum* decreased at values eight times lower than MIC. Furthermore, *O. basilicum* showed a more pre-eminent effect in inhibition of *C. albicans* biofilm formation and in disruption of preformed biofilm.

The *O. basilicum* essential oil showed the following IC_{50}: 16.296 ± 0.394 mg/mL by DPPH method, 0.6870 ± 0.0203 mg/mL by ABTS method, 0.003 ± 0.0007 mg/mL by AEAC-FRAP assay and 0.760 ± 0.001 mg/mL by AEAC-Phosphomolybdenum assay. Moderate antifungal activity was associated the essential oils from *O. basilicum* when examined against *Candida glabrata* and *Candida albicans* (Rezzoug et al., 2019).

In a study with *O. basilicum* collected from three locations in Egypt, Assiut, Minia and BeniSuef governorates, the essential oil from the Minia basil showed high activity in DPPH radical scavenging with IC_{50} (11.23 mg/mL) and contained the highest content of phenolic (41.3 mg PE/g) (Ahmed et al., 2019).

In particular, the essential oil from *O. campechianum* showed the highest radical scavenging activity in both assays with an IC_{50} of 7.7 ± 0.1 µg/mL against DPPH and 3.18 ± 0.29 µg/mL against ABTS. EO and eugenol had significant activity against *Pseudomonas syringae pv. syringae* and a moderate effect against *Candida* strains, with possible synergism in association with fluconazole against *Candida* strains (Tacchini et al., 2021).

According to Barbosa et al. (2021), the antioxidant activities by DPPH method of basil essential oils were in the following decreasing orders: *O. basilicum* var. greek a palla > *O. gratissimum* > *O. basilicum* var. bola > *O. campechianum* > *O. tenuiflorum*. Statistically, the IC_{50} value of *O. tenuiflorum* (2.31 ± 0.02 µg/mL) was the best activity and equal to the pure eugenol (1.91 ± 0.06 µg/mL), with no statistically significant difference ($p < 0.05$). The IC_{50} values by DPPH method from *O. basilicum* var. greek (7.40 ± 0.34 µg/mL), *O. gratissimum* (6.53 ± 0.34 µg/mL) and *O. basilicum* var. ball (6.13 ± 0.13 µg/mL) and *O. campechianum* (4.93 ± 0.03 µg/mL) were statistically equal to each other, and statistically significantly different

($p < 0.05$) when compared to eugenol. Using the ABTS method, the antioxidant potentials of four essential oils of *Ocimum* did not differ statistically when compared to pure eugenol, showing excellent antioxidant activity of the following essential oils: *O. tenuiflorum* (2.22 ± 0.23 μg/mL), *O. campechianum* (2.51 ± 0.07 μg/mL), *O. basilicum* var. ball (2.69 ± 0.39 μg/mL) and *O. gratissimum* (2.84 ± 0.26 μg/mL). By the β-carotene/linolenic acid system, *Ocimum gratissimum* showed IC_{50} of 3.43 $\mu g/mL^{-1}$ with statistically significant difference when compared to eugenol ($p < 0.05$, $IC_{50} = 7.85\ \mu g/mL^{-1}$). There is a high correlation between antioxidant and eugenol in these basil essential oils.

Vasconcelos et al. (2021) pointed out that the antioxidant activity for both *O. gratissimum* and *O. basilicum* was 100% DPPH free radical scavenging at the highest essential oil concentrations 15.6 $\mu g\ mL^{-1}$. The content of total phenolic compounds in essential oils from *O. gratissimum* and *O. basilicum* was = 17.90 and 16.44 mg EAG 100 g^{-1}, respectively. The essential oils from *O. gratissimum* and *O. basilicum* presented 100 and 83% of bactericidal activity against *Staphylococcus aureus* and from 100 to 66% against *Escherichia coli*, respectively, thus highlighting the biotechnological potential of these essential oils.

The essential oils from *O. tenuiflorum, O. gratissimum, O. campechianum* and *O. basilicum* showed biotechnological potential with excellent yields and strong antioxidant activities attributed to eugenol and synergistic effect by several chemotypes present, which can be used in food preservation (intelligent packaging and stability of food products), in animal feed and the human diet as functional food, preventing non-transmissible chronic diseases such as obesity, diabetes mellitus, hypertension, dyslipidemias and cancer, and also neurological diseases such as Alzheimer's disease.

Biotechnological aspects of essential oils in the food industry and in waste reduction

Correlating the antioxidant effect of essential oils and food preservation, recent researches try to develop intelligent packages that add quality to products by incorporating essential oils in polymers. In this view, essential oils with antioxidant action incorporated or used in packaging can eradicate the action of oxidation by means of elimination and/or inhibition of free radicals, conferring longer shelf life to foods. For example, the use of edible coatings (carrageenan, wheat protein concentrate) in combination with plant antioxidants was efficient to extend the shelf life of minimally processed apples for two weeks when stored at 3°C (Lee et al., 2003).

The orange juice industry generates waste of around 50% of the fruit weight, alternatives such as the extraction of essential oils from *Citrus sinensis* are obtained by crushing the peel. D-limonene was the main component of the orange peel essential oil at 90.85%, followed by β-myrcene, γ-terpinene, linalool and α-pinene. Limonene has a high tendency to autoxidation and polymerization, and formation of off-flavors, making it a suitable industrial by product for bioconversions to high commercial value compounds. In addition, there are several applications for limonene, such as a solvent for resins, synthesis of other chemical compounds, applications in rubber, paints and dispersant agent for oil, besides the use in the chemical synthesis of menthol in the food industry (Maróstica Jr. and Pastore, 2007).

Another biotechnological application of essential oils, a commercial product based on spice blends showed better antioxidant activity of its essential oils than TBHQ (synthetic antioxidant) in potato flakes (Baardseth, 1989; Mariutti and Bragagnolo, 2007).

The antioxidant activity of essential oil from plants of the Lamiaceae family used in margarine acted as a natural antioxidant and flavoring agent in margarine (Ozkan et al., 2007).

It is also worth noting that fruit biofilms including that of clove EO, oregano and 1:1 binary combinations of EO of both spices had an antioxidant effect with a significant difference ($p < 0.05$) between treatments in relation to TBA (2-thiobarbituric acid) values at the end of the 15-day storage period under refrigeration (Ugalde, 2014).

Over the past two decades, several studies have reported positive results of using plant essential oils (EOs) as dietary additives in various animal species. Recently, these nutritional alternatives have been evaluated and reported in fish feed production to increase disease resistance and prevent outbreaks, as well as improve fish growth, animal welfare, better stability of chemical constituents and nutritional composition for use in the human diet (Sutili et al., 2017).

In a study conducted by Tofino-Rivera et al. (2017) in Colombia, aimed to replace the use of chemical preservatives in the production of artisanal sausages with OEs of thyme and clove, which had mostly, : 46% thymol and 26% para-cymene; and eugenol (84%). The authors first found that *Salmonella* spp., *Staphylococcus aureus* and *Escherichia coli* species were the most commonly found microorganisms in the contamination of these specific meat sausages, and it was found that the samples containing OEs showed *in vitro* inhibition of 28.3 mm and 27.3 mm, respectively, against *Salmonella* spp., *Staphylococcus aureus* and *Escherichia coli*.

Films obtained from renewable sources are considered promising due to their degradability and decreasing environmental impacts. Scudeler (2019) developed films based on tilapia protein isolate with different concentrations of gelatin, glycerol and sorbitol, and incorporated them with essential oils of cloves and oregano. The antimicrobial activity of the protein films with nanoclay and essential oils was verified by the disk diffusion method against *Staphylococcus aureus*, *Staphylococcus epidermidis*, *Escherichia coli*, *Pseudomonas aeruginosa* and *Salmonella typhymurium*. The films were found to be promising from the point of view of visual appearance, manageability and continuity. Films with nanoclay and essential oils showed high solubility and low water vapor permeability (PVA), tensile strength and elongation when compared to protein films with gelatin.

Singh et al. (2019) nanoencapsulated the essential oil from *Ocimum sanctum* to evaluate against aflatoxigenic fungi contaminating herbal raw materials and its novel mode of action as a methyglyoxal inhibitor. The chemically characterized nanoencapsulated *Ocimum sanctum* essential oil showed an effect against fungi contaminating herbal raw materials and aflatoxin B1 secretion, being important to extend the shelf life of stored herbal raw materials and the findings also led towards the development of aflatoxin and methylglyoxal resistant medicinal plants through green transgenics.

Effect of essential oil from *Ocimum gratissimum* (200, 400, 600, 800 or 1000 mg/L) was analyzed against six fungi including *Aspergillus flavus*, *Aspergillus tamarii*, *Fusarium poae*, *Fusarium verticillioides*, *Penicillium citrinum* and *Penicillium griseofulvum* isolated from a traditional cheese named wagashi. There was a significant fungistatic activity against all the examined species with MIC values ranging from 800–1000 mg/L. However, the concentration of these essential oils applied in cheeses should be considered carefully because of their possible negative impacts on organoleptic properties, but the microencapsulation of essential oil, incorporation into edible film or application of these oils along with other antimicrobial agents and preservatives can be more effecient in prevention of microbial growth and enhancing cheese safety (Philippe et al., 2012).

The solid residue that is also generated in large quantities during the extraction of essential oil from plants has added value, making it a candidate for raw material for sustainable products with the low environmental impact that act in the alternative control of phytopathogenic fungi (Kasper et al., 2020).

Conclusion

Ocimum basilicum is a plant of great interest with a variability in its chemical composition influenced by geographic location and other factors. The high contents of eugenol in *O. campechianum*, *O. tenuiflorum*, *O. basilicum* and *O. gratissimum* essential oils give them strong antioxidant potential and antimicrobial properties, being very interesting in future biotechnological applications for the food industry as the development of active packaging, the addition of antioxidants and antimicrobial substances in food preservation, functional value to various chronic diseases and increase the shelf life with natural products that do not harm the environment.

References

Ahmed, A.F., Attia, F.A.K., Liu, Z., Li, C., Wei, J. and Kang, W. 2019. Antioxidant activity and total phenolic content of essential oils and extracts of sweet basil (*Ocimum basilicum* L.) plants. Food Sci. and Hum. Wellness 8: 299–305.

Association of Official Analytical. AOAC. 1995. Official Methods of Analysis. 16th ed. AOAC International: Gaithersburg, MD.

Asbahani, A.El., Miladi, K., Badri, W., Sala, M., Aït Addi, E.H., Casabianca, H. et al. 2015. Essential oils: From extraction to encapsulation. Int. J. Pharm. 483: 220–243.

Avetisyan, A., Markosian, A., Petrosyan, M., Sahakyan, N., Babayan, A., Aloyan, S. et al. 2017. Chemical composition and some biological activities of the essential oils from basil *Ocimum* different cultivars. BMC Comp Alt. Med. 17: 1–8.

Baardseth, P. 1989. Effect of selected antioxidants on the stability of dehydrated mashed potatoes. Food Add. Cont. 6: 201–207.

Bakkali, F., Averbeck, S., Averbeck, D. and Idaomar, M. 2008. Biological effects of essential oils—a review. Food Chem. Tox 46: 446–475.

Barbosa, C.O., Morais, S.M., Sousa, H.A., Martins, V.C., Camara Neto, J.F., Vieira, I.G. et al. 2021. Chemical composition and antioxidant potential of essential oils from different *Ocimum* species (Basil). Braz J. Developm. 7: 24422–24442.

Bhavaniramya, S., Vishnupriya, S., Al-Aboody, M.S., Vijayakumar, R. and Dharmar Baskaran, D. 2019. Role of essential oils in food safety: Antimicrobial and antioxidant applications. Grain & Oil Sci. Technol. 2: 49–55.

Bizzo, H.R., Hovell, A.M.C. and Rezende, C.M. 2009. Óleos essenciais no Brasil: aspectos gerais, desenvolvimento e perspectivas. Quim. Nov 32: 588–594.

Bizzo, H.R. and Rezende, C.M. 2022. O mercado de óleos essenciais no Brasil e no mundo na última década. Quim. Nov. 45: 949–958.

Borges, A.M., Pereira, J., Cardoso, M.G., Alves, J.A. and Lucena, E.M.P. 2012. Determinação de óleos essenciais de alfavaca (*Ocimum gratissimum* L.), orégano (*Origanum vulgare* L.) e tomilho (*Thymus vulgaris* L.). Rev. Bras. Plan. Med. 14: 656–665.

Botrel, P.P., Pinto, J.E.B.P., Ferraz, V., Bertolucci, S.K.V. and Figueiredo, F.C. 2010. Teor e composição química do óleo essencial de *Hyptis marrubioides* Epl. Laminaceae em função da sazonalidade. ACTA Scient. Agron. 32: 533–538.

Brahmi, F., Mokhtari, O., Legssyer, B., Hamdani, I., Asehraou, A., Hasnaoui, I. et al. 2021. Chemical and biological characterization of essential oils extracted from citrus fruits peels. Mat. Tod. 45: 7794–7799.

Burt, S. 2004. Essential oils: their antibacterial properties and potential applications in foods—a review. Int. J. Food Microbiol. 94: 223–253.

Dhifi, W., Bellili, S., Jazi, S., Bahloul, N. and Mnif, W. 2016. Essential oils' chemical characterization and investigation of some biological activities: a critical review. Med (Basel, Switzerland) 3: 25.

Dima, C. and Dima, S. 2015. Essential oils in foods: extraction, stabilization, and toxicity. Curr. Opin. Food Sci. 5: 29–35.

Farooqi, A.H.A. and Sharma, S. 2000. Aromatherapy: a promising holistic system. J. Med. Aromat. Plant Sci. 22: 704–706.

Gayas, B. and Kaur, G. 2017. Novel oil extraction methods in food industry: A review. J. Oil seed Brassica 8: 1–11.

Gurudeeban, S., Ramanathan, T. and Satyavani, K. 2011. Characterization of volatile compounds from bitter apple (*Citrulllus colocynthis*) using GC-MS. Int. J. Chem. Anal. Sci .2: 108–110.

Ilić, A.S., Antić, M.P., Jelačić, S.C. and Šolević Knudsen, T.M. 2018. Chemical composition of the essential oils of three *Ocimum basilicum* L. cultivars from Serbia. Not Bot Horti Agro Cluj-Napoca. 47: 347–351.

Ilić, A.S., Antić, M.P., Jelačić, S.C. and Šolević Knudsen, T.M. 2019. Chemical composition of the essential oils of three *Ocimum basilicum* L. cultivars from Serbia. Not Bot Horti Agro Cluj-Napoca. 47: 347–351.

Jimbo, D., Kimura, Y., Taniguchi, M., Inoue, M. and Urakami, K. 2009. Effect of aromatherapy on patients with Alzheimer's disease. Psychoger. 9: 173–179.

Kasper, A.A.M., Sousa, S.F., Martin, B.S.S., Satoratto, A., Nunes, K.M., Sousa Júnior, J.J.V. et al. 2020. Aproveitamento dos resíduos de priprioca (*Cyperus articulatus* L.) no controle alternativo de fungos fitopatogênicos. Rev IberoAm Ciênc Amb 11: 80–88.

Ladwani, A.M.A., Salman, M. and Abdel Hameed, E.S. 2018. Chemical composition of *Ocimum basilicum* L. essential oil from different regions in the Kingdom of Saudi Arabia by using Gas chromatography mass spectrometer. J. Med. Plan. Stud. 6: 14–19.

Lee, S.H., Do, H.S. and Min, K.J. 2015. Effects of essential oil from *Hinoki cypress*, *Chamaecyparis obtusa*, on physiology and behavior of flies. PLoS ONE 10: e0143450.

Lee, J.Y., Park, H.J., Lee, C.Y. and Choi, W.Y. 2003. Extending shelf life of minimally processed apples with edible coatings and antibrowning agents. LWT – Leb Wiss Technol. 36: 323329.

Li, H., Ge, Y., Luo, Z., Zhou, Y., Zhang, X., Zhang, J. and Fu, Q. 2017. Evaluation of the chemical composition, antioxidant and antiinflammatory activities of distillate and residue fractions of Sweet basil essential oil. J. Food Sci. Technol. 54: 1882–1890.

Mariutti, L.R.B. and Bragagnolo, N. 2007. Revisão: antioxidantes naturais da família Lamiaceae - aplicação em produtos alimentícios. Braz J. Food Technol. 10: 96–103.

Marostica Junior, M.R. and Pastore, G.M. 2007. Biotransformação de limoneno: uma revisão das principais rotas metabólicas. Quim. Nov 30: 382–387.

Martins, A.G.L.A. 2010. Antibacterial activity of essential oils of basil (*Ocimum basilicum* Linnaeus) and ginger (Zingiber officinale Roscoe) front the line of enteropathogenic *Escherichia coli* isolated of grasses. Ph.D. Thesis, Federal University of Paraíba, João Pessoa, Br.

Moore, J., Yousef, M. and Tsiani, E. 2016. Anticancer effects of rosemary (*Rosmarinus officinalis* L.) extract and rosemary extract polyphenols, Nutrients 731. [30] R. Tisserand, R. Young, Essential Oil Safety: A Guide for Health Care.

Moraes, C.C. and Souza, T.A.S. 2018. Panorama mundial do desperdício e perda de alimentos no contexto de cadeias de suprimentos agroalimentares. Rev. Agroneg Meio Amb. 11: 901–924.

Naidu, J.R., Ismail, R.B. and Sasidharan, S. 2016. Chemical profiling and antioxidant activity of Thai basil (*Ocimum basilicum*). J. Ess. Oil Bear Plan. 19: 750–755.

Özcam, M. and Chalchat, J.C. 2002. Essential oil composition of *Ocimum basilicum* L. and *Ocimum minimum* L. in Turkey. Czech J. Food Sci. 20: 223–228.

Ozkan, G., Simsek, B. and Kuleasan, H. 2007. Antioxidant activities of *Satureja cilicica* essential oil in butter and *in vitro*. J. Food Engineer. 79: 1391–1396.

Parisoto, D.F., Hautrive, T.P. and Cembranel, F.M. 2013. Redução do desperdício de alimentos em um restaurante popular. R Bras Tecnol. Agroind 7: 1106–1117.

Pereira, R.C.A. and Moreira, A.L.M. 2011. Manjericão: cultivo e utilização. Fortaleza: Embrapa Agroindústria Tropical.

Perry, N. and Perry, E. 2006. Aromatherapy in the management of psychiatric disorders clinical and neuropharmacological perspectives. Psychogeriatrics 9: 257–580.

Philippe, S., Souaïbou, F., Guy, A., Sébastien, D.T., Boniface, Y., Paulin, A. et al. 2012. Chemical composition and antifungal activity of essential oil of fresh leaves of *Ocimum gratissimum* from Benin against six Mycotoxigenic Fungi isolated from traditional cheese wagashi. Int. Res. J. Biol. Sci. 1: 22–27.

Piras, A., Gonçalves, M.J., Alves, J., Falconieri, D., Porcedda, S., Maxia, A. et al. 2018. *Ocimum tenuiflorum* L. and *Ocimum basilicum* L., two spices of Lamiaceae family with bioactive essential oils. Ind. Crops Prod. 113: 89–97.

Rezzoug, M., Bakchiche, B., Gherib, A., Roberta, A., Guido, F., Kilinçarslan, O. et al. 2019. Chemical composition and bioactivity of essential oils and Ethanolic extracts of *Ocimum basilicum* L. and *Thymus algeriensis* Boiss. & Reut. from the Algerian Saharan Atlas. BMC Compl. Alt. Med. 19: 146.

Ricarte, L.P., Bezerra, G.P., Romero, N.R., Silva, H.C., Lemos, T.L.G., Arriaga, A.M.C. et al. 2020. Chemical composition and biological activitiesof the essential oils from Vitex-agnus castus, *Ocimum campechianum* and *Ocimum carnosum*. An Acad. Bras Cienc. 92: e20180569.

Roriz, R.F.C. 2012. Aproveitamento dos resíduos alimentícios obtidos das Centrais de Abastecimento do Estado de Goiás S/A para alimentação humana.

Santos, E.F. 2007. Seleção de tipos de *Ocimum basilicum* L. de cor purpura para o mercado de plantas ornamentais. Dissertação. Universidade de Brasília, Brasília, Br.

Scudeler, C.G.S. 2019. Desenvolvimento e caracterização de filmes biopoliméricos a partir de resíduos de tilápia (*Oreochromis niloticus*) com incorporação de óleos essenciais, nanoargila, gelatina e plastificantes. Ph.D. Thesis Federal University of Grande Dourados, MS, Br.

Shina, Y., Funabashi, N., Lee, K., Toyodo, T., Sekine, T., Honjo, S. et al. 2007. Relaxation effects of lavender aromatherapy improve coronary flow velocity reserve in healthy men evaluated by transthoracic Doppler echocardiography. Int. J. Cardiol. 129: 193–197.

Silva, C.B., Simionatto, E., Hess, S.C., Peres, M.T.L.P., Simionatto, E.L., Junior, W.A., Poppi, N.R., Faccenda, O., Cândido, A.C.S. and Scalon, S.P.Q. 2009. Composição química e atividade alelopática do óleo volátil de *Hydrocotyle bonariensis* Lam (Araliaceae). Quim. Nov. 32: 2373–2376.

Simões, C.M.O., Mentz, L.A., Schenkel, E.P., Irgang, B.E. and Stehmann, J.R. 1994. Plantas da medicina popular no Rio Grande do Sul. 4ª ed., Porto Alegre: Editora da UFRGS.

Singh, V.K., Das, S., Dwivedy, A.K., Rathore, R. and Dubey, N.K. 2019. Assessment of chemically characterized nanoencapuslated *Ocimum sanctum* essential oil against aflatoxigenic fungi contaminating herbal raw materials and its novel mode of action as methyglyoxal inhibitor. Posth Biol. Technol. 153: 87–95.

Sobrinho, A.C.N., Souza, E.B., Rocha, M.F.G., Albuquerque, M.R.J.R., Bandera, P.N., Santos, H.S. et al. 2016. Cytotoxicity, antifungal and antioxidant activities of the essential oil from *Eupatorium ballotifolium* Kunth (Asteraceae). African J. Pharmacy and Pharmacol. 10: 346–355.

Stanojevic, L.P., Marjanovic-Balaban, Z.R., Kalaba, V.D., Stanojevic, J.S., Cvetkovic, D.J. and Cakic, M.D. 2017. Chemical composition, antioxidant and antimicrobial activity of Basil (*Ocimum basilicum* L.) essential oil. J. Ess. Oil Bearing Pl. 20: 1557–1569.

Sutili, F.J., Gatlin, D.M., Heinzmann, B.M. and Baldisserotto, B. 2017. Plant essential oils as fish diet additives: benefits on fish health and stability in feed. Rev. Aquacult. 0: 1–11.

Tacchini, M., Guevara, M.P.E., Grandini, A., Maresca, I., Radice, M., Angiolella, L. et al. 2020. *Ocimum campechianum* Mill. from Amazonian Ecuador: chemical composition and biological activities of extracts and their main constituents (Eugenol and Rosmarinic Acid). Molecules 26: 84.

Tacchini, M., Guevara, M.P.E., Grandini, A., Maresca, I., Radice, M., Angiolella, L. et al. 2021. *Ocimum campechianum* Mill. from Amazonian Ecuador: chemical composition and biological activities of extracts and their main constituents (Eugenol and Rosmarinic Acid). Molecules 26: 84.

Tofiño-Rivera, A., Ortega-Cuadros, M., Herrera-Hinojosa, B.K., Fragoso-Castilla, P. and Pedraza-Claros, B. 2017. Conservación microbiológica de embutido carnico artesanal con aceites esenciales *Eugenia caryophyllata* y *Thymus vulgaris*. Rev. BioAgro. 15: 30–41.

Toscan, C.M. 2010. Atividade antimicrobiana e antioxidante de terpenóides. Dissertação. Universidade de Caxias do Sul, Caxias do Sul, Br.

Trevisan, M.T.S., Vasconcelos Silva, M.G., Pfundstein, B., Spiegelhalder, B. and Owen, R.W. 2006. Characterization of the volatile pattern and antioxidant capacity of essential oils from different species of the genus *Ocimum*. J. Agricultural Food Chem. 54: 4378–4382.

Ugalde, M.L. 2014. Biofilmes ativos com incorporação de óleos essenciais. Ph.D. Thesis. Universidade Regional Integrada do Alto Uruguai e das Missões, Rio Grande do Sul, Br.

Vasconcelos, S.C., Régis, L.A., Menezes Filho, A.C.P., Cazal, C.M., Pereira, P.S. and Christofoli, M. 2021. Composição química, atividade bactericida e antioxidante dos óleos essenciais das folhas de *Ocimum basilicum* e *Ocimum gratissimum* (Lamiaceae). Research, Society and Develop. 10: e51810817109.

Veloso, R.A. 2012. Divergência genética, análise do óleo essencial e bioatividade de acessos de manjericão (*Ocimum basilicum* L.) no estado do Tocantins. Dissertação. Universidade Federal do Tocantins, Gurupi - TO, Br.

Venancio, A.M. 2006. Toxicidade aguda e atividade antinociceptiva do óleo essencial do *Ocimum basilicum* L. (manjericão), em *Mus musculus* (camundongos). Dissertação. Universidade Federal de Sergipe, Aracaju, Br.

Chapter 10

Moringa oleifera Seed Cake
A By-product of Oil Extraction with Biotechnological Potential

Ardilles Juan Carlos Alves dos Santos,[1,*]
Eva Luana Almeida da Silva,[2] *Panait Abu Nidal Alves dos Santos,*[3]
Hanna Gracie Inez de Freitas Lima,[2]
Welton Aaron de Almeida,[2] *Romildo de Albuquerque Nogueira*[2] and
Emmanuel Viana Pontual[2]

Introduction

Moringa oleifera Lamarck (1785) (Fig. 10.1) is popularly known as "moringa", "drumstick tree", "horseradish tree", "ben oil tree" or "benzolive tree". It is a pantropical tree of the Moringaceae family, native to India, that survives for a long period in poor nutrient soils with low moisture content; because of this, *M. oleifera* thrives well in semi-arid regions (Parrota, 2009).

M. oleifera has aroused great interest worldwide because of its medicinal properties and use as a forage plant, as well as being a promising source of oils and biogas (Makkar and Becker, 1996; Foidl et al., 2001; Karadi et al., 2006). All of the parts of *M. oleifera* (Fig. 10.2), including its leaves, fruits, flowers and seeds, have been described as sources of compounds with the most varied applications.

The leaves of *M. oleifera* are rich in nutrients, including β-carotene, vitamin C, iron, potassium and essential amino acids (Makkar and Becker, 1996; Parrotta,

[1] Departamento de Nutrição, Universidade Federal do Piauí, Picos, Piauí, Brazil.
[2] Departamento de Morfologia e Fisiologia Animal, Universidade Federal Rural de Pernambuco, Recife, Brazil.
[3] Departamento de Ciência do Consumo, Universidade Federal Rural de Pernambuco, Recife, Pernambuco, Brazil.
* Corresponding author: ardillesjuan@ufpi.edu.br

Figure 10.1. *Moringa oleifera* tree.

2009). Several medicinal applications have been attributed to leaf preparations, such as antioxidant, hypolipidemic, hypoglycemic, anti-atherosclerotic, antibacterial and antifungal activities (Jaiswal et al., 2009; Rahman et al., 2009).

Flowers of *M. oleifera* contain calcium, potassium and the antioxidants α- and γ-tocopherol, and are consumed as raw food or after simmering (Makkar and Becker, 1996; Sánchez-Machado et al., 2006). There are reports of the use of

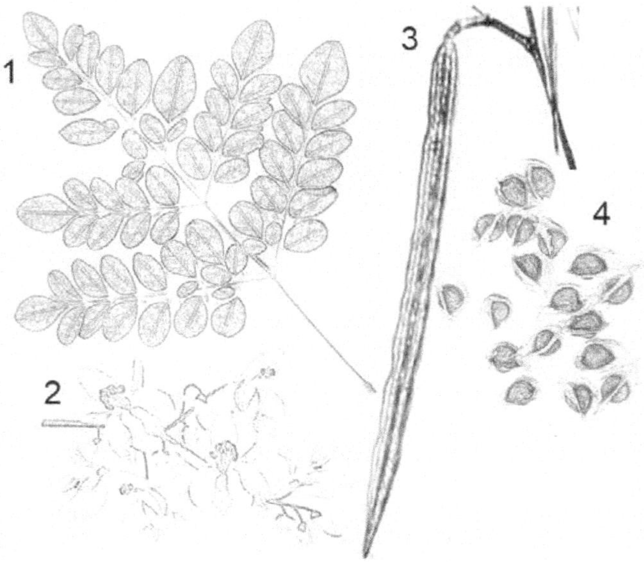

Figure 10.2. Aerial parts of the pantropical tree *Moringa oleifera*. Leaves (1), Flowers (2), Fruits (3) and Seeds (4).

M. oleifera flowers as collagogue, diuretic and hypoglycemic agents (Khare, 2007; Parrotta, 2009). *M. oleifera* flower preparations are cytotoxic to *Bacillus subtilis*, *Enterococcus faecalis*, *Escherichia coli*, *Proteus mirabilis*, *Salmonella enteritidis* and *Staphylococcus aureus* and reduce the growth of bacteria in ambient water, as well as cause lysis of *Trypanosoma cruzi* trypomastigotes and modulates the immune response *in vitro* of *T. cruzi*-infected human cells (Moura et al., 2011; Nova et al., 2020).

A water extract of *M. oleifera* flowers was able to kill *Aedes aegypti* (Diptera; Culicidae) larvae at second third and fourth instars (L2, L3 and L4, respectively) and the Trypsin inhibitor (MoFTI) purified from it damaged the survival and development of newly hatched larvae, as well as inhibited the growth or killed L4 midgut microbiota (Pontual et al., 2012a; Pontual et al., 2014). *M. oleifera* flowers also contain aspartic, cysteine and serine proteases with caseinolytic and milk clotting activities that give it the potential for use as a vegetable rennet for cheese production (Pontual et al., 2012b).

M. oleifera seeds contain 9.2% carbohydrates, 35.4% proteins, 43% lipids, 4.7% fibers and ions that include calcium, sodium, potassium and phosphorus at concentrations ranging from 86 to 602 mg/g of seeds. In addition, vitamins B3, C and E are also found at 0.2, 4.5, and 752 mg/100 g, respectively (Olagbemide and Philip, 2014; Jacques et al., 2020).

The seeds show coagulant properties due to the presence of proteins and a 3 kDa organic polyelectrolyte (Okuda et al., 2001; Ghebremichael et al., 2005; Santos et al., 2009). In developing countries, *M. oleifera* seeds are used by the population to treat water for human consumption. Among the seed proteins are lectins (carbohydrate-binding proteins) such as cMoL (Coagulant *M. oleifera* lectin) and WSMoL (Water-Soluble *M. oleifera* Lectin) that were able to remove particles suspended in kaolin solution, revealing their involvement in the mechanism of coagulation by the seeds (Santos et al., 2009; Ferreira et al., 2011).

Although the compounds of *M. oleifera* seeds are less efficient than some commercially used coagulants such as aluminum sulfate and ferric sulfate in removing turbidity, their biodegradability together with the fact that they do not interfere with the water pH, and their ability to remove colors make them an interesting candidate for this purpose (Cardoso et al., 2008; Balbinot et al., 2018). WSMoL was also able to reduce bacterial contamination in ambient water by killing or causing bacteria agglutination and settling (Ferreira et al., 2011). Rolim et al. (2011) demonstrated the use of *M. oleifera* seeds at 0.2 g/L, as recommended by the non-governmental organization Esplar (www.esplar.org.br), as well as WSMoL does not result in genotoxicity or mutagenicity.

M. oleifera seeds have been reported as a source of natural insecticides. WSMoL killed *A. aegypti* L4 by promoting morphological damages in its digestive tract including hypertrophy of segments, enlargement of the gut lumen and rupture of the layer that delimits the gut epithelium (Coelho et al., 2009). WSMoL also showed a stimulating oviposition effect for pregnant *A. aegypti* females and reduced egg hatchability by killing the embryos under laboratory or semi-field conditions (Santos et al., 2012; Santos et al., 2014). cMoL damaged the nutrition and delayed the development of *Ephestia kuehniella* (Lepidoptera; Pyralidae) larvae, as well as reduced pupae weight and survival (Oliveira et al., 2011).

M. oleifera seeds are a source of oil, popularly known as Ben oil, corresponding to about 19–47% of the total seed mass, and presenting a high content of palmitic, stearic and oleic acids (Ojiako and Okeke, 2013). Due of its chemical properties, the oil shows many different applications used for cosmetics manufacture and as a raw material for biodiesel production (Mofijur et al., 2014; Fernandes et al., 2015). In this chapter, a review of the biotechnological value of the *M. oleifera* seed cake, a by-product of oil extraction are provided.

Extraction of the *M. oleifera* seed oil and cake production

The extraction of *M. oleifera* seed oil can be chemically carried out, using solvents, which represent the oldest reported form (Bhutada et al., 2016). In this method, after collecting and washing the seeds, they can be soaked in ether solution for 4 hr to eliminate impurities. Then the seeds are dried, crushed and sieved in a 40 μm mesh. Next, the powder is placed in the paper thimble of a Soxhlet glass (Fig. 10.3) using chloroform and methanol (1: 1, v/v) as described by Ramluckan et al. (2014). Cold pressing, supercritical fluid extraction and enzymatic fluid extraction are other procedures used to extract the *M. oleifera* seed oil (Abdulkarim et al., 2005; Fakayode and Ajav, 2016).

The mechanical extraction by cold pressing consists of using a screw-type press for vegetable oil, in which the seeds are ground and then pressed; the oil is conducted through a tube, being filtered and stored at –20°C (Gharsallah et al., 2021). This method has the main advantages of not using organic solvents such as n-hexane,

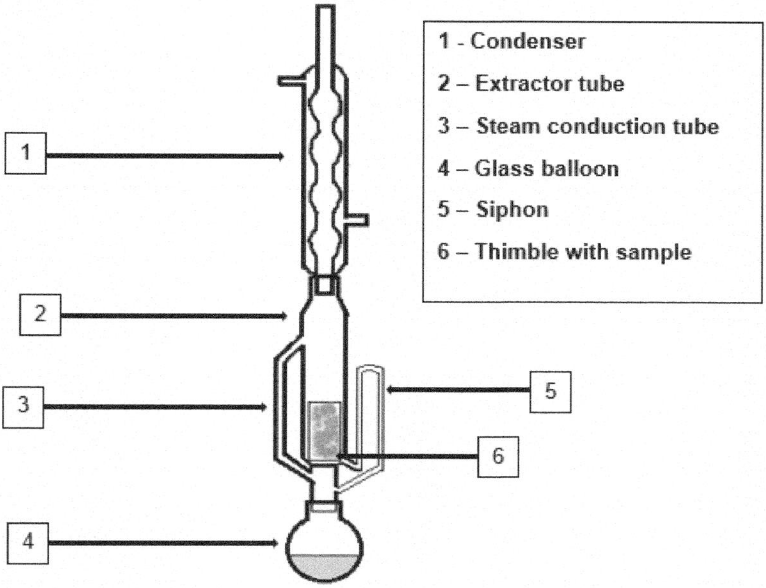

Figure 10.3. Schematic representation of the Soxhlet tube used in extraction of the *Moringa oleifera* seed oil. The dried material is placed at the paper thimble in the extraction tube. Next, the solvent in the glass flask is gradually heated, and goes to the condenser. The solvent heating gradually dissolves the sample, and the dried particles are extracted.

which are highly toxic and polluting to the environment, as well as can modify the structure and activity of the seed compounds (Rezig et al., 2018).

The extraction of *M. oleifera* seed oil comes as a residual material known as seed cake as a by-product. Oliveira et al. (2016) obtained the seed cake by chemical extraction. *M. oleifera* seeds were powdered and 100 g of the flour were taken to the Soxhlet extractor with n-hexane for 6 hr. After the oil extraction, the seed cake was treated with saline solution producing an extract with a high protein content (19.5 mg/ml) regarding the saline extract (4.1 mg/ml of protein) prepared with the powder of whole seeds. This result may be a consequence of the oil interference in the extraction environment when whole seeds were used.

M. oleifera seed cake biotechnological applications

Water treatment

Water turbidity is caused by the presence of negatively charged particles and suspended organic material. The synthetic substances currently used to destabilize and coagulate suspended matter can be harmful to human health and the environment, in addition to causing a sharp drop in pH, as occurs with aluminum sulfate (Saini et al., 2016). In regions with a dry climate, the restricted access to water makes its reuse almost mandatory. In this sense, natural coagulants may represent eco-friendly alternatives of low toxicity to replace conventional coagulants (Rodrigues and Barros, 2013).

According to the data available in international literature, the oil extraction procedures do not damage the coagulant ability of *M. oleifera* seeds, since the cake retains this effect. Shan et al. (2016) investigated the properties of *M. oleifera* seed cake to remove the turbidity and heavy metals from the water of the "Sungai Baluk" River (Malaysia), and from samples of sewage. The authors recorded a reduction of 85 to 93% in water turbidity and 96% in sewage samples after treatment with *M. oleifera* seed cake. Reduction rates of up to 70% in the concentration of lead (Pb), 90% of copper (Cu) and cadmium (Cd) and 100% of iron (Fe) were recorded. In the sewage samples, Fe was totally (100%) removed, and there was a reduction of up to 78.1% in the concentration of Pb and 98% in the concentrations of Cu and Cd. Similarly, promising results regarding water turbidity were found by Zaid and Ghazali (2019). They analyzed a source of drinking water, before or after being treated with *M. oleifera* seed cake. A detection of 96% reduction in turbidity and an increase of 93 and 77% in total soluble solids and electrical conductivity, respectively.

In another study, the best conditions for chromium removal in water using the seed cake were confirmed by Meneghel et al. (2013). The authors demonstrated that the largest (80%) chromium removal occurred at an acid pH value (close to 5.0), and that the time of contact between the metal ion and the adsorbent material increased the removal rate; however, after the first hour, this increase becomes subtle. *M. oleifera* seed cake was also able to remove 74% of benzene, 85% of toluene, 86% of ethylbenzene and o-xylene and 87% of p-xylene from water; all of the volatile organic compounds harmful to the atmosphere and health of animals (Almeida et al., 2012). Tie et al. (2014) used the seed cake to remove a benzidine-derived dye commercially known as Congo Red, which is toxic to humans and may cause allergic reactions. The seed cake removed 95.8% of the Congo Red present in the water and the authors contended that the removal mechanism involves the adsorption and neutralization of the dye charges by seed cake proteins, followed by flocculation.

Another study showed that *M. oleifera* seed cake was efficient in flocculating approximately 90% of *Chlorella vulgaris* cells from water (Teixeira and Teixeira, 2017). The authors stated that the addition of 1.0 M NaCl improves the *C. vulgaris* flocculation due to the effect known as "salting-in".

Animal feed and veterinary use

Seed cakes have great appeal as ingredients in the production of feed for ruminants, especially for causing a decrease in the release of methane gas, improving microbiological fermentation at the rumen and, consequently, the animal performance (Gonzaga-Neto et al., 2015; Ceruti et al., 2016).

Moreira et al. (2016) carried out a comparative study between seed cakes of cotton (*Gossypium hirsutum* L.), sunflower (*Helianthus annuus* L.), castor bean (*Ricinus communis*), jatropha (*Jatropha curcas* L.), sugar cane (*Saccharum* sp.) and *M. oleifera*. In this study, *M. oleifera* seed cake showed the highest amount of protein, *in vitro* increased food digestibility, as well as reduced the production of methane and acetate by the animals as regards the other seed cakes. The high protein content of *M. oleifera* seed cake seems to be linked to the animal well-being, since diets with less protein may reduce food degradability by microbiota from rumen and decrease nutrient absorption, which can lead to a decrease in animal production, as well as damage to the digestive, reproductive and immune systems.

Population control of *A. aegypti*

The hematophagous mosquito *Aedes aegypti* is the main vector of important tropical diseases such as dengue, chikungunya, Zika and yellow fever (WHO, 2020). Their life cycle (Fig. 10.4) begins when pregnant females lay their eggs in standing water, where the first instar larvae (L1) hatch, which sequentially develops into the second, third and fourth instar larvae (L2, L3 and L4, respectively). After the fourth stage, the larvae become pupa from which the adult mosquitoes emerge (Tsunoda et al., 2014). The population control of *A. aegypti* represents an important strategy to reduce the spread of the diseases transmitted by its bite, and because of problems that include the emergence of resistant insects, new insecticidal principles have been searched for in nature (WHO, 2020). In this sense, *M. oleifera* seed cake was used as a starting material for bioprospecting new larvicidal agents.

M. oleifera seed cake obtained by chemical extraction using n-hexane contains a lectin named WSMoLc. The protocol for extraction of this protein (Fig. 10.5) included extraction in saline solution, precipitation with ammonium sulfate (60% saturation) and affinity chromatography on chitin column, the same protocol used in the purification of WSMoL from whole seeds (Oliveira et al., 2016). A polyacrylamide gel electrophoresis under denaturing conditions revealed WSMoLc as two polypeptides of 15 and 20 kDa and this result was similar to that obtained for WSMoL under the same conditions. Mass spectrometry revealed similarities of WSMoLc with other *M. oleifera* seed proteins including a coagulant protein and a 2 S albumin precursor; Similar data were found for the sequence of WSMoL from whole seeds. WSMoLc showed an affinity for fructose, glucose, mannose,

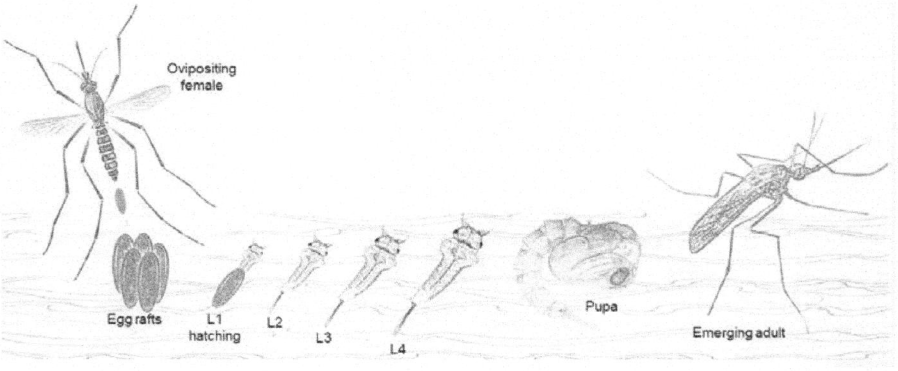

Figure 10.4. *Aedes aegypti* life cycle. After the oviposition by pregnant females, larvae hatch from the eggs and will pass through four instars (L1, L2, L3 and L4). Next, the larvae become pupae from which the adults will emerge, restarting the cycle.

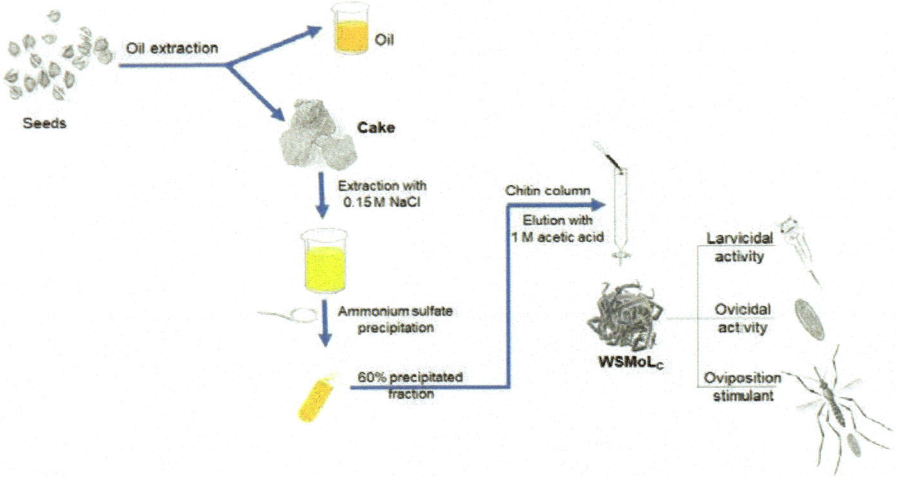

Figure 10.5. Purification and biological activities of a lectin from *M. oleifera* seed cake (WSMoLc). The chemical extraction of seed oil using n-hexane generates the seed cake as a by-product. After protein extraction from seed cake with saline solution followed by precipitation with ammonium sulfate, a protein-rich fraction is obtained, which is chromatographed on a chitin column to isolate WSMoLc. This lectin has the potential to be used as an insecticide against *A. aegypti* for killing larvae, reducing egg hatchability and acting as an oviposition stimulant cue for pregnant females.

N-acetylglucosamine and galactose, plus a characteristic shared with WSMoL. Together, these findings led the authors to conclude that the lectin recovered from *M. oleifera* seed cake (WSMoLc) resembles WSMoL and because of this, they studied the effects of WSMolc on *A. aegypti* larvae, eggs and pregnant females.

Oliveira et al. (2016) demonstrated that WSMoLc promoted mortality of *A. aegyti* L4 and this activity was 4.5 times greater than that detected for WSMoL. The lectin concentration required to kill 50% of larvae (LC50) was 0.89 [0.84–0.94] mg/mL for WSMoLc, and 0.99 [0.92–1.07] mg/mL for WSMoL (Coelho et al., 2009). WSMoLc (0.1 mg/mL) acted as a stimulating oviposition signal since it increased the capture of eggs laid by pregnant *A. aegypti* females, and reduced the hatching rate of eggs in a dose-dependent way. The lectin concentrations required to reduce egg hatching by 50% (EC50), 90% (EC90) and 99% (EC99) were 0.14 [0.13–0.15], 0.24 [0.22–0.26] and 0.32 [0.30–0.36] mg/mL, respectively (Oliveira et al., 2016). These data encourage the use of the *M. oleifera* seed cake as a source of larvicidal agent, oviposition stimulant for use in traps, and ovicidal against *A. aegypti*, as well as show its potential as a tool for control of dengue, chikungunya, Zika and yellow fever. Figure 10.5 outlines the potential of WSMoLc as an insecticidal agent.

Conclusion

It is clear from literature discussed here that the *M. oleífera* seed cake represents a by-product of oil production with a high biotechnological value in different areas of action.

M. oleifera seed cake can be used in water purification and decontamination by decreasing the concentration of heavy metals in water. It is also used in the production of animal feed with various activities in the animal microbiota and a higher protein concentration in its composition. Finally, *M. oleífera* a seed pie in the population control of insects that act as arbovirus vectors.

Acknowledgments

The authors express their gratitude to the *Conselho Nacional de Desenvolvimento Científico e Tecnológico* (CNPq; 408789/2016-6), as well as to the *Coordenação de Aperfeiçoamento de Pessoal de Nível Superior* and *Fundação de Amparo à Ciência e Tecnologia do Estado de Pernambuco* for financial support. WAA and ELAS would like to thank CAPES (finance code 001) for the graduate scholarships.

References

Abdulkarim, S.M., Long, K., Lai, O.M., Muhammad, S.K.S. and Ghazali, H.M. 2005. Some physico-chemical properties of *Moringa oleifera* seed oil extracted using solvent and aqueous enzymatic methods. Food Chem. 93: 253–263.

Almeida, I.L.S., Filho, N.R.A., Alves, M.I.R., Carvalho, B.G. and Coelho, N.M.M. 2012. Removal of BTEX from aqueous solution using Moringa oleifera seed cake. Environ. Technol. 33(11): 1299–1305.

Balbionot, G.S., Collares, F.M., Visioli, F., Soares, P.B.F., Takimi, A.S., Samuel, S.M.W. and Leitune, V.C.B. 2018. Niobium addition to sol-gel derived bioactive glass powders and scaffolds: *In vitro* characterization and effect on pre-osteoblastic cell behavior. Dental Materials 34(10): 1449–1458.

Bhutada, P.R., Jadhav, A.J., Pinjari, D.V., Nemade, P.R. and Jain, R.D. 2016. Solvent assisted extration of oil from *Moringa oleífera* Lam. Seeds. Ind. Crops Prod. 82: 74–80.

Cardoso, K.C., Bergamasco, R., Cossich, E.S. and Moraes, L.C.K. 2008. Otimização dos tempos de mistura e decantação no processo de coagulação/floculação da água bruta por meio da moringa oleifera lam. Acta Scientiarum - Technology 30(02): 193–198.

Cerutti, W.G., Viégas, J., Barbosa, A.M., Oliveira, R.L., Dias, C.A., Costa, E.S., Nornberg, J.L., Carvalho, G.G.P., Bezerra, L.R. and Silveira, A.M. 2016. Fatty acid profiles of milk and Minas frescal cheese from lactating grazed cows supplemented with peanut cake. J. Dairy Res. 83: 42–49.

Coelho, J.S., Santos, N.D.L., Napoleão, T.H., Gomes, F.S., Ferreira, R.S., Zingali, R.B., Coelho, L.C.B.B., Leite, S.P., Navarro, D.M.A.F. and Paiva, P.M.G. 2009. Effect of *Moringa oleifera* lectin on development and mortality of *Aedes aegypti* larvae. Chemosphere 77: 934–938.

Fakayode, O.A. and Ajav, E.A. 2016. Process optimization of mechanical oil expression from Moringa (Moringa oleifera) seeds. Ind. Crops and Prod. 90: 142–151.

Fernandes, D.M., Souza, R.M.F., Oliveira, A., Morais, S.A.L., Richter, E.M., Muñoz, R.A.A. 2015. *Moringa oleifera*: a potential source for production of biodiesel and antioxidant additives. Fuel 146: 75–80.

Ferreira, R.S., Napoleão, T.H., Santos, A.F.S., Sá, R.A., Carneiro-da-cunha, M.G., Morais, M.M.C., Silva-lucca, R.A., Oliva, M.L.V., Coelho, L.C.B.B. and Paiva, P.M.G. 2011. Coagulant and antibacterial activities of the water-soluble seed lectin from *Moringa oleifera*. Lett. Appl. Microbiol. 53: 186–192.

Foidl, N., Makkar, H.P.S. and Becker, K. 2001. The potential of *Moringa oleifera* for agricultural and industrial uses. pp. 45–76. *In*: Fuglie, L.J. (ed.). The Miracle Tree/The Multiple Attributes of Moringa. CTA, New York.

Gharsallah, K., Resig, L., Msaada, K., Chalh, A. and Soltani, T. 2021. Chemical composition and profile characterization of *moringa oleifera* seed oil. South Afri J. Bot. 137: 475–482.

Ghebremichael, K.A., Gunaratna, K.R., Henriksson, H., Brumer, H. and Dalhammar, G. 2005. A simple purification and activity assay of the coagulant protein from *Moringa oleifera* seed. Water Res. 39: 2338–2344.

Gonzaga-neto, S., Oliveira, R.L., Lima, F.H.S., Medeiros, A.N., Bezerra, L.R., Viégas, J., Júnior, N.G.N. and Neto, M.D.F. 2015. Milk production, intake, digestion, blood parameters, and ingestive behavior of cows supplement with by-products from the biodiesel industry. Trop. Anim. Health Prod. 47: 191–200.

Jacques, A.S., Arnaud, S.S.S., Fréjus, O.O.H. and Jacques, D.T. 2020. Review on biological and immunomodulatory properties of *Moringa oleífera* in animal and human nutrition. J. Pharma and Phito. 12: 1–9.

Jaiswal, D., Rai, P.K., Kumar, A., Mehta, S. and Watal, G. 2009. Effect of *Moringa oleifera* Lam. leaves aqueous extract therapy on hyperglycemic rats. J. Ethnopharmacol. 123: 392–396.

Karadi, R.V., Gadge, N.B., Alagawadi, K.R. and Savadi, R.V. 2006. Effect of *Moringa oleifera* Lam. root-wood on ethylene glycol induced urolithiasis in rats. J. Ethnopharmacol. 105: 306–311.

Khare, C.P. 2007. Indian Medicinal Plants—An Illustrated Dictionary. Springer, Berlin/Heidelberg.

Makkar, H.P.S. and Becker, K. 1996. Nutritional value and antinutritional components of whole and ethanol extracted *Moringa oleifera*. Anim Feed Sci. Technol. 63: 211–228.

Meneghel, A.P., Gonçalves Jr., A.C., Strey, L., Rubio, F., Schwantes, D. and Casarin, J. 2013. Biosorption and removal of chromium from water by using moringa seed cake (*Moringa oleifera* Lam.). Química Nova 36(8): 1104–1110.

Mofijur, M., Masjuki, H.H., Kalam, M.A., Atabani, A.E., Fattah, I.M.R. and Mobarak, H.M. 2014. Comparative evalution of performance and emission characteristics of *Moringa oleifera* and Palm oil biodiesel in a diesel engine. Ind. Crops Prod. 53: 78–84.

Moreira, M.N., Silva, A.M.A., Bezerra, L.R., Carneiro, H., Morais, R.K.O. and Medeiros, F.F. 2016. Effect of replacement of sugarcane by press cake greenhouse gases and volatile fatty acids production *in vitro*. Acta Scientiarum: Anim Sci. 38: 293–300.

Moura, M.C., Pontual, E.V., Gomes, F.S., Napoleão, T.H., Xavier, H.S., Paiva, P.M.G. and Coelho, L.C.B.B. 2011. Preparations of *Moringa oleifera* flowers to treat contaminated water. pp. 269–285. *In*: Daniels, J.A. (ed.). Advances in Environmental Research, v. 21, Nova Sci Publi.. Inc., New York.

Nova, I.C., Moreira, L.R., Torres, D.J.L., Oliveira, K.K.S., Patriota, L.L.S., Coelho, L.C.B.B., Paiva, P.M.G., Napoleão, T.H., Lorena, V.M.B. and Pontual, E.V. 2020. A Trypsin Inhibitor from *Moringa oleifera* flowers modulates the immune response *in vitro* of *Trypanosoma cruzi*-infected human cells. Antib-Basel 9: 515.

Ojiako, E.N. and Okeke, C.C. 2013. Determination of antioxidant of *Moringa oleifera* seed oil and its use in the production of a body cream. Asian Journal of Plant Science and Research 3(3): 1–4.

Okuda, T., Baes, A.U., Nishijima, W. and Okada, M. 2001. Isolation and characterization of coagulant extracted from *Moringa oleifera* seed by salt solution. Water Res. 35: 405–410.

Olagbemide, P.T. and Philip, C.N.A. 2014. Proximate analysis and chemical composition of raw and defatted *Moringa oleifera* kernel. Advan. Life Sci. Technol. 24: 92–99.

Oliveira, A.P.S., Silva, L.L.S., Albuquerque, T.L., Pontual, E.V., Santos, N.D.L., Coelho, L.C.B.B., Navarro, D.M.A.F., Zingali, R.B., Napoleão, T.H. and Paiva, P.M.G. 2016. Biotechnological value of *Moringa oleifera* seed cake as source of insecticidal lectin against *Aedes aegypti*. Proc. Bioche. 51(10): 1683–1690.

Oliveira, C.F.R., Luz, L.A., Paiva, P.M.G., Coelho, L.C.B.B., Marangoni, S. and Macedo, M.L.R. 2011. Evaluation of seed coagulant *Moringa oleifera* lectin (cMoL) as a bioinsecticidal tool with potential for the control of insects. Proc. Bioche. 46: 498–504.

Parrotta, J.A. 2009. *Moringa oleifera* LAM., 1785. *In*: Roloff, A., Weisgerber, H., Lang, U. and STIMM, B. (eds.). Enzyklopädie der Holzgewächse, Handbuch und Atlas der Dendrologie. Enzyklopädie der Holzgewächse – 40. WILEY-VCH Verlag GmbH & Co. KGaA, Weinheim.

Pontual, E.V., Napoleão, T.H., Assis, C.R.D., Bezerra, R.S., Xavier, H.S., Navarro, D.M.A.F., Coelho, L.C.B.B. and Paiva, P.M.G. 2012a. Effect of *Moringa oleifera* flower extract on larval trypsin and acethylcholinesterase activities in *Aedes aegypti*. Archi Ins Bioch. Physio. 79: 135–152.

Pontual, E.V., Carvalho, B.E.A., Bezerra, R.S., Coelho, L.C.B.B., Napoleão, T.H., Paiva, P.M.G. 2012b. Caseinolytic and milk-clotting activities from Moringa oleifera flowers. Food Chem. 135: 1848–1854.

Pontual, E.V., Santos, N.D.L., Moura, M.C., Coelho, L.C.B.B., Navarro, D.M.A.F., Napoleão, T.H. and Paiva, P.M.G. 2014. Trypsin inhibitor from Moringa oleifera flowers interferes with survival and development of *Aedes aegypti* larvae and kills bacteria inhabitant of larvae midgut. Parasitol Res. (1987. Print) 113: 727–733.

Rahman, M.M., Sheikh, M.M.I., Sharmin, S.A., Islam, M.S., Rahman, M.A., Rahman, M.M. and Alam, M.F. 2009. Antibacterial activity of leaf juice and extracts of *Moringa oleifera* Lam. against some human pathogenic bacteria. Chiang Mai Uni J. 8: 219–227.

Ramluckan, K., Moodley, K.G. and Bux, F. 2014. An evaluation of the efficacy of using selected solvents for the extraction of lipids from algal biomass by the soxhlet extraction method. Fuel 116: 103–108.

Rezig, L., Chouaibi, M., Ojeda-Amador, R.M., Gomez-alonso, S., Salvador, M.D., Fregapane, G. and Handi, S. 2018. Curcubita maxima Pumpkin seed oil: from the chemical properties to the different extracting techniques. Not. Bot. Horti. Agro. 46: 663–669.

Rodrigues, M.J. and Barros, L.S.S. 2013. USING *Moringa oleifera* seed extract and solar radiation in the treatment of water intended for human consumption/o extrato de sementes de Moringa oleifera e radiação solar no tratamento de água destinada ao consumo humano. Ars Vet. 29(2): 98–103.

Rolim, L.A.D.M.M., Macedo, M.F.S., Sisenando, H.A., Napoleão, T.H., Felzenswalb, I., Aiub, C.A.F., Coelho, L.C.B.B., Medeiros, S.R.B. and Paiva, P.M.G. 2011. Genotoxicity evaluation of *Moringa oleifera* seed extract and lectin. J. Food Sci. 76: 53–58.

Saini, R.K., Sivanesan, I. and Keum, Y.S. 2016. Phytochemicals of *Moringa oleifera*: a review of their nutritional, therapeutic and industrial significance. 3 Bio. 6(2): 203.

Sánchez-machado, D.I., López-cervantes, J. and Vázquez, N.J.R. 2006. High-performance liquid chromatography method to measure α and γ-tocopherol in leaves, flowers and fresh beans from *Moringa oleifera*. J. Chromat. A. 1105: 111–114.

Santos, A.F.S., Luz L.A., Argolo, A.C.C., Teixeira, J.A., Paiva, P.M.G., Coelho, L.C.B.B. 2009. Isolation of a seed coagulant *Moringa oleifera* lectin. Proc. Bioche. 44: 504–508.

Santos, N.D.L., Moura, K.S., Napoleão, T.H., Santos, G.K.N., Coelho, L.C.B.B., Navarro, D.M.A.F. and Paiva, P.M.G. 2012. Oviposition-stimulant and ovicidal activities of *Moringa oleifera* lectin on *Aedes aegypti*. Plos One 7: 44840.

Santos, N.D.L., Paixão, K.S., Napoleão, T.H., Trindade, P.B., Pinto, M.R., Coelho, L.C.B.B., Eiras, A.E., Navarro, D.M.A.F. and Paiva, P.M.G. 2014. Evaluation of *Moringa oleifera* seed lectin in traps for the capture of *Aedes aegypti* eggs and adults under semi-field conditions. Parasitol. Res. (1987. Print) 113: 1837–1842.

Shan, T.C., Matar, M.A., Makky, E.A. and Ali, E.N. 2016. The use of *Moringa oleifera* seed as a natural coagulant for wastewater treatment and heavy metals removal. App. Water Science 7(3): 1369–1376.

Teixeira, C.M.L.L. and Teixeira, P.C.N. 2017. Evaluation of the flocculation efficiency of chlorella vulgaris mediated by *Moringa oleifera* seed under different forms: flour, seed cake and extracts of flour and cake. Braz J. Chem. Eng. 34(1): 65–74.

Tie, J., Li, P., Xu, Z., Zhou, Y., Li, C. and Zhang, X. 2014. Removal of Congo red from aqueous solution using *Moringa oleifera* seed cake as natural coagulant. Des Water Treat. 54(10): 2817–2824.

Tsunoda, T., Cuong, T.C., Dong, T.D., Yen, N.T., Le, N.H., Phong, T.V. and Minakawa, N. 2014. Winter refuge for *Aedes aegypti* and *Ae. albopictus* mosquitoes in Hanoi during Winter. PloS One 9(4): e95606.

World Health Organization. 2020. Dengue and severe dengue. Fact-sheets.

Zaid, A.Q. and Ghazali, S.B. 2019. Dataset on physicochemical properties of particle-sized *Moringa oleifera* seed cake and its application as bio-coagulants in water treatment application. Chem. Data Collec. 100284.

Chapter 11

The Role of Microorganisms in the Use of Food-Waste

Monalisa de Alencar Lucena,[1] *Igor Frederico S. Ramos,*[1] *Valesca Lima Fernandes,*[2] *Allana Costa Araújo Souza,*[3] *Márcia dos Santos Rizzo,*[4] *Alessandra Braga Ribeiro*[5] and *Marcília Pinheiro da Costa*[1,2,4,*]

Introduction

Bioprospecting is the process of exploring biological resources in search of new products with added social and commercial value. Bioprospecting involves applying advanced technologies to develop innovative resources in various industrial sectors, especially the food sector (Beattie et al., 2011; Müller et al., 2016).

In recent years, food production has grown to meet the rising high demands from population growth (Kaur et al., 2020; Ng et al., 2020). However, many by-products or waste are produced through various food industries worldwide (Sadh et al., 2018). The generation of food waste is relevant, as about a third of all food produced globally is lost in the various stages of the production chain (Zisopoulos et al., 2017). Additionally, these residues can cause severe damage to the environment (Sadh et al., 2018). Therefore, the development of sustainable solutions for food waste management currently represents one of the main challenges faced by society (Girotto et al., 2015). In recent years, the scientific community has focused efforts to

[1] Graduate Program of Science and Engineering of Materials, Federal University of Piauí, Teresina, PI, 64049-550, Brazil.
[2] College of Pharmacy, Federal University of Piauí, Teresina, PI, 64049-550, Brazil.
[3] College of Nutrition, University of Fortaleza (Unifor), Fortaleza, CE, 60811-905, Brazil.
[4] Graduate Program of Pharmaceutical Science, Federal University of Piauí, Teresina, PI, 64049-550, Brazil.
[5] Centre of Biotechnology and Fine Chemistry, Faculty of Biotechnology, Catholic University of Portugal, Porto, 4169-005, Portugal.
* Corresponding author: marciliapc@ufpi.edu.br

optimize food processing and agricultural waste management system, aiming on the bioconversion of these inputs into high-value products (Ng et al., 2020).

In this way, several biotechnological strategies have been designed to maximize the use of food waste and increase the potential income of the entire bioprocessing chain (Girotto et al., 2015). A significant advantage is that these by-products or residues, whose composition depends on their origin, are excellent sustainable raw materials in different bioprocesses to produce bioenergy and other bioproducts of commercial interest (Sadh et al., 2018; Ng et al., 2020). Thus, waste from the food supply chain is considered a resource with significant potential to be used as raw material to produce industrial enzymes, bioactive compounds and fuels due to the amount generated globally, the diversity of chemical constituents and the ability to be used for higher value-added applications (Lin et al., 2013; Ng et al., 2020).

Microorganisms play a crucial role in the search for new derivatives from food inputs. It has also been drawing the attention of food industries as biocatalysts, due to their genetic diversity and metabolic versatility. Currently, several compounds are produced in large-scale processes through microbial biotechnology (Katz and Baltz, 2016; Pessôa et al., 2019).

Sources of food waste generation

Food waste is mainly responsible to produce a considerable amount of biological waste worldwide, which, if not well managed, can result in environmental pollution, directly affecting the public health of the population (Paritosh et al., 2017; Thi et al., 2016b; Kumar et al., 2020). In addition, the increase in food waste generation in recent years reflects a highly industrialized modern food system to suppress the demand caused by the increase in population and changes in their lifestyle (Ju et al., 2017; Kaur et al., 2020).

Food waste can be generated at any stage of food production, including harvesting, processing, manufacturing, surplus not traded and household disposal (Slorach et al., 2019). Some factors interfere in the amount and composition of food waste produced, such as time of year, climate, the standard of living, eating habits and the degree of development of countries, since in developing countries, the most significant loss occurs in the stages immediately after harvest, in contrast, in more industrialized countries most losses occur in the post-consumption stages (Redlingshöfer et al., 2020; Gallardo et al., 2016). According to Xu et al. (2018b), on an average, developed countries generate 100 to 170 kg of food waste per capita per year, equivalent to more than twice the amount produced in developing countries.

The main sources of food waste include the food industries and domestic, commercial and agricultural waste (Xue et al., 2017), described in detail below.

Food processing industries

Data from the study carried out by Baiano (2014) indicated that approximately 26% of food waste comes from the beverage industry and that 21% is generated by the dairy industry, 14.8% by the production and processing of fruits, 12.9% from processing and manufacturing of cereals, 8% for processing and preserving meat products, 3.9% for manufacturing and processing vegetable and animal oils, 4% for processing and preserving fish products, and others (12 to 7%). Generally, these food residues are related to failures in some of the processing or manufacturing steps, such as forecast errors which result in excess stock that, consequently, cannot be marketed within the indicated shelf life; incidents in quality control that make products unsuitable for consumption; in-stock products withdrawn from the market, and also possible damage during manufacture. The concern on the part of the food industries regarding the generation of waste has increased in recent years due to economic, social, ethical and environmental factors related to the management of these materials (Tavill, 2020; Wakefield and Axon, 2020).

Household food waste

The largest production of food waste occurs at the final consumer level, with perishable items such as fruits, vegetables, dairy products and meat being the most likely to be discarded. Among the main reasons to produce food waste in households, the excessive purchase of food, non-consumption on time, which may be related to the lack of attention regarding their shelf lives, as well the lack of awareness of the correct methods to avoid food waste are highlighted. For this reason, it is of great importance to prevent food waste in the final stages of the supply chain, resulting in reduced socioeconomic and environmental impacts (Kavanaugh and Quinlan, 2020; Nazzaro et al., 2018; Patra et al., 2020).

Agro-industrial food waste

Food production processes in the agro-industrial sector are responsible for producing large amounts of food waste, emphasizing organic waste from processed raw materials (Nayak and Bhushan, 2019; Prandi et al., 2019). Leaves, barks, seeds, remains of animal slaughter, pits, pulp, bran and bagasse, among others, stand out as the main residues obtained in the agro-industrial sector (Angiolillo et al., 2015). Generally, these residues are used in animal feed, production of biofuels, biofertilizers, biofibers, among others (Carciochi et al., 2017).

The increase in the availability of food waste from different sources, makes it necessary to dispose it off in a more controlled and effective way with the intention of avoiding greater environmental and socioeconomic impacts. However, the disposal of these wastes through incineration or disposal in landfills, the most commonly used methods, can result in undesirable consequences for the environment, such as the release of toxic gases and leachate production (Slorach et al., 2019; Fausto-Castro et al., 2020). For this reason, interest in the relocation and reuse of this food waste in the most sustainable and economically possible way has increased. Further the conventional uses (animal feed, biofuel production, etc.), these materials have drawn attention as they are cheap sources of important bioactive components, which allow their transformation into additives or new products, increasing their economic value and limiting their availability in the environment (Tsang et al., 2019; Lombardelli et al., 2020).

Microorganisms in the processing of food inputs

Microorganisms have been widely used in industry to produce various chemicals and materials from renewable resources (Lee et al., 2012). Moreover, due to the success of bioprocesses and bioproducts, in recent years, there has been a growing interest from the industry in producing resources of microbial biotechnology (Zhou et al., 2017; Zhu et al., 2018; Pessôa et al., 2019).

Several factors have contributed to the application of microorganisms in bioprocesses, among them: the genetic diversity of microbial cells; their high growth rates; versatility in growing conditions; the metabolic diversity of the strains; the robustness of biotransformation processes; the ease of defining fermentation parameters; the possibility of carrying out genetic improvement of the lines; and large-scale production capacity (Zhu et al., 2018; Pessôa et al., 2019).

Despite all the advantages of microbial bioprocesses, the success of biotransformation depends on the appropriate choice of microorganisms. Choosing the correct microbial strain represents an important step to improve biosynthesis qualitatively and quantitatively to optimize synthesized products (Pessôa et al., 2019). As resources to find the best strain, microorganisms can be isolated from different sources, including fermented foods, adverse environments, contaminated samples or agro-industrial residues. These are important strategies that can provide a more adaptable strain, capable of being used as an efficient biocatalyst, that exhibits resistance to industrial conditions and that guarantees high yields in the biotechnological production of compounds (Pessôa et al., 2019).

Unfortunately, microbial metabolism has not evolved to suit humanity's needs; thus, when microorganisms are isolated from nature, their efficiency in producing any molecule is relatively low (Lee et al., 2012). Therefore, few microorganisms selected from their natural environment can be applied directly to industrial applications due to low productivity, low-stress resistance and accumulation of by-products. In order to overcome these drawbacks, metabolic engineering and synthetic biology strategies have provided new methods of improving the performance of industrial microorganisms (Zhou et al., 2017).

In several biotechnological processes, bacteria and fungi use biomass residues to obtain products with high added value. These agricultural or food residues are mainly rich in carbohydrates, proteins, lipids and traces of inorganic compounds and are considered attractive substrates for bioconversion of bioproducts (Paritosh et al., 2017; Wang et al., 2018; Ng et al., 2020). Table 11.1 presents some microorganisms, their food residues used as substrates and their respective bioproducts.

Microbial biotransformation processes

Currently, anaerobic microorganisms have led to an enormous impact in environmental remediation, in which their metabolic diversity favors the transformation of a variety of organic and inorganic compounds being used for the treatment of urban and industrial waste (Hatti-Kaul and Mattiasson, 2016).

These microorganisms have been widely used in biomanufacturing, which consists of productions that use biological systems (for example, living

Table 11.1. Bioproducts produced from food waste by microbial fermentation.

Microorganisms	Substrates	Bioproducts	References
Aspergillus niger	Wheat bran, soybean meal, cottonseed, orange peels	Protease	De Castro et al., 2015
Aspergillus niger	Grains and brewer's yeast	β-glucosidase	Leite et al., 2019
Aspergillus sp.	Wheat bran	Pullulanase	Naik et al., 2019
Bacillus halodurans	Wheat bran	Fibrinolytic enzyme	Vijayaraghavan et al., 2016
Lactobacillus delbrueckii	Brewery waste	Protease	Mathias et al., 2017
Pichia pastoris	Citrus peel waste	Oligogalacturonides	Yang et al., 2020
Saccharomyces cerevisiae	Pineapple peels	Proteins	Mensah and Twumasi, 2016
Lactic acid bacteria	Fish skins, viscera, and muscles	Bacteriocins	Gomez-Sala et al., 2015
Aspergillus fumigatus	Orange peel waste	Ellagic acid	Sepúlveda et al., 2020

microorganisms, cells, plants, animals, tissues, enzymes or synthetic substances *in vitro*) to produce important value-added products for use in various industry sectors (Zhang et al., 2017).

Agricultural residues often serve as cheap carbon and nitrogen sources for the growth of microorganisms, which can be used for the fermentation of bioactive compounds (Javed et al., 2019). Alternatively, numerous bioproducts, including proteins, enzymes, antioxidants and pigments, can be produced by microbial fermentation using food waste as raw materials (Sadh et al., 2018).

Several microbial biochemical processes are used in biomanufacturing to obtain these products; fermentation, electro-fermentation and anaerobic digestion (Zhang et al., 2017; Ng et al., 2020) are highlighted here. Figure 11.1 represents the steps for obtaining bioproducts by microbial biotechnology.

Fermentation is considered an emerging technology that uses microorganisms (bacteria or fungi) to process food inputs to generate commercially important products (Ng et al., 2020). The formation of fermentation products is directly linked to the specific growth rate of the strain and the yield of the product is related to the mass cell yield, which in turn is determined by the organism's metabolic pathway and the cultivation conditions (Hatti-Kaul and Mattiasson, 2016).

In the biotransformation of food waste into bioproducts, three main types of fermentation can be used: solid-state fermentation and submerged or liquid fermentation (Sadh et al., 2018). The type of fermentation strategy for the bioconversion process is very dependent on the types of food waste used as a substrate (Ng et al., 2020).

Solid-state fermentation is the procedure in which microorganisms develop on solid substrates, its main objective being to maximize the use of nutrients from the substrate by microorganisms (Sadh et al., 2018). Submerged fermentation is used the most for liquefied substrates, and it is mainly used on an industrial scale for the

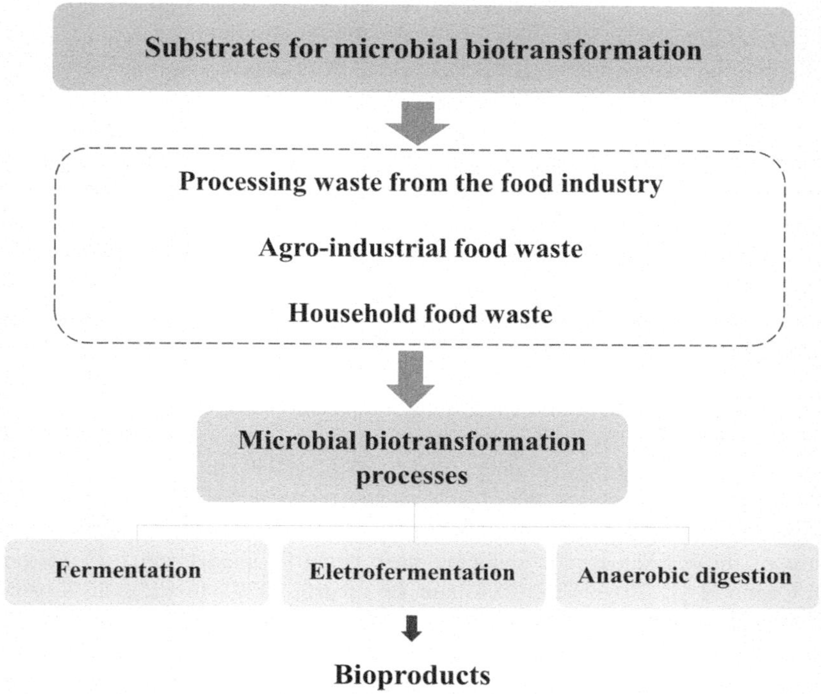

Figure 11.1. Scheme for obtaining bioproducts from food and agro-industrial inputs through microbial processes.

fermentation of food waste due to its low processing cost and high productivity. Furthermore, submerged fermentation is practically implemented to produce various enzymes due to process control and sterilization procedure (Ng et al., 2020).

The production of biogas from food waste can be carried out through different technologies of microbial bioprocesses, such as anaerobic digestion and electro-fermentation. Anaerobic digestion is a biodegradation process that usually involves three stages: hydrolysis, acidogenesis or acid fermentation, acetogenesis and methanogenesis, which happen simultaneously or separately depending on the type of process (single or multiphase) used (Zhang et al., 2014; Kader et al., 2015; Kumar et al., 2018). In anaerobic digestion, food waste works as excellent substrates to produce relatively high yields of biomethane and biogas (Pramanik et al., 2019; Ng et al., 2020). Another innovative fermentation process has also been used for the biotransformation of food waste, electrofermentation. Electrofermentation is a new technique that improves biogas production by incorporating electrodes into microbial fermentation processes (Liu et al., 2019). The addition of electrodes and voltage application has significantly improved bioproduct metabolism rates (Sravan et al., 2018).

The microbial biotransformation process may also involve fermentation mediated by symbiotic microorganisms, such as kefir grains grown in milk or water. Water kefir grains consist of a stable association of several species of microorganisms, which belong to the group of Lactic Acid Bacteria (LAB), acetic acid bacteria (BAA), bifidobacteria, and yeast (Xu et al., 2018a). These microorganisms produce an exopolysaccharide matrix (dextran), which protects the microbial matrix against adverse environmental conditions, such as sudden changes in pH and temperature. Dextran is a biopolymer of microbial origin and can be used in numerous industrial sectors due to its biological and mechanical properties (Ahmed and Ahmad, 2017). In addition to obtaining the biopolymer, water kefir grains are used as a starter culture to produce fermented beverages, characterized by their freshness and low alcohol content, and their acidic and fruity flavor. The beverage is obtained through fermentation, mediated by kefir grains, of substrates containing sucrose and water, under anaerobic conditions and at room temperature for 2 to 4 d (De Roos and De Vuyst, 2018).

The use of these microorganisms in symbiosis is of great importance for several industrial sectors, which are increasingly interested in investing in biotechnological processes that result in the elimination of waste in a more sustainable way and in obtaining energy and new products (García et al., 2019). In this way, food residues obtained in the processing stages of fruits and other vegetables can be used as substrates for the fermentation processes mediated by water kefir grains, thus promoting the reuse of these residues and reducing their negative impacts on the environment (Di Cagno et al., 2016; Vicenssuto and De Castro, 2020). In addition to providing fermented beverages rich in nutrients and bioactive compounds, with excellent sensory characteristics and probiotic properties, important factors for the development of new functional beverages that attract greater attention from consumers as they are considered healthier options (Irkin, 2019; Randazzo et al., 2016).

The study carried out by Azi et al. (2020) shows the use of soy whey, a residue generated during tofu production, as a substrate to produce fermented beverages using three water kefir microbiotas of different origins. Currently, the disposal of soy whey as effluent has generated environmental pollution that can have implications for the health of the population, making its reuse an important strategy to reduce these damages and obtain value-added beverages. This research shows that water kefir grains can be considered a suitable starter culture to obtain new functional drinks with a high content of phenolic acids, aglycone isoflavones and antioxidant activity.

Biological conversion of food waste into commercial products

According to the Food and Agricultural Organization - FAO (2019), it is important to promote the reduction of food losses and waste, solve problems related to production costs, and also increase the efficiency of the food system. Furthermore, contribute in a fundamental way to environmental sustainability. Many alternatives and more attractive solutions have been highlighted for the waste recovery process through the extraction and recovery of compounds that have high added value and great economic and biological potential and can be applied in the production of fuels, in the development of various chemical and biological products, among other applications (Dave et al., 2020).

Various food processing residues, such as fruit peels, are important sources of bioactive compounds, such as water-soluble (polyphenols) and water-insoluble (carotenoids) antioxidants and vitamins flavonoids, fiber and pectin. In addition, some of these residues are rich in carbon and hydrogen sources (carbohydrates, lipids and proteins). In this way, they serve as important sources of raw materials for the bioconversion of high value-added bioproducts, such as enzymes and other bioactive compounds, bioplastics and biofuels, among others obtained by different types of processes. These products obtained from food waste have great potential to be used as additives in the food industry (functional foods, beverages), pharmaceuticals and cosmetics industries (Dave et al., 2020; Joana Gil-Chávez et al., 2013; Galanakis, 2020; Ng et al., 2020). The main components obtained from food waste and their applications are described next.

- ***Enzymes***

Enzymes are considered one of the high-value biological products that have been used for the longest time and can be obtained by various processes. Some types of microorganisms, such as *Saccharomyces cerevisiae*, *Aspergillus* and *Bacillus* sp., are applied in the production of enzymes through the degradation of complex carbohydrates, commonly found in food residues, using sugars released in this process as a source for the metabolism of enzymes (Haddadi et al., 2018; Ravindran and Jaiswal, 2016). Many enzymes used in industrial applications, such as in the production of bioproducts and biorefineries (determining the efficiency and speed of reactions), can be obtained from food waste, with emphasis on oxidative enzymes (laccase, xylanase and cellulase), enzymes capable of hydrolyzing starch (α-amylase and glucoamylase), as well as proteases, pectinases and lipases (Budzianowski, 2017; Vea et al., 2018; Teigiserova et al., 2019). Thus, agro-industrial residues and food residues from other sectors are alternative sources for a more economical production of these enzymes, which can be used in various industrial sectors as biocatalysts in the process of obtaining new bioproducts (Sindhu et al., 2019).

- ***Other bioactive compounds***

Among all the numerous bioactive compounds that can be obtained from food residues, carotenoids stand out for their wide availability in these matrices especially from citric foods. Beyond their excellent biological activities, such an outstanding antioxidant activity, which can be important in reducing cardiovascular diseases and muscle degeneration, in addition to promoting an increase in the immune system. Some residues, such as fruit peels and seeds are rich sources of carotenoids, flavonoids and phenolic compounds, which also stand out for their antioxidant properties (Teigiserova et al., 2019).

Vitamins can also be extracted from food waste, mainly vitamins D3 and E (obtained from plant sources), through biorefineries. After obtaining these vitamins, the waste can still be subjected to other processes to obtain other bioactive compounds or be used in energy production (Budzianowski, 2017). Other bioactive compounds obtained from food residues are sterols, tocopherols, terpenes and polyphenols, which can be used in the food industry to formulate functional foods with enhanced antioxidant properties (Kalogeropoulos et al., 2012).

These compounds have gained prominence because of their biological activities are of great importance to health and are commonly used in the food industry (as an ingredient in manufacturing products rich in nutrients and bioactives). In the pharmaceutical and cosmetics industry, interest in innovative, low-cost and sustainable products has increased, containing natural components such as antioxidants, thickeners and dyes, which are abundant in different types of food waste (Arun et al., 2020; Jiménez-Velázquez et al., 2020).

- ***Bioplastics***

Food waste is a rich source of carbon and can be used as a substrate in producing numerous bioplastics, such as polyhydroxyalkanoates (PHA) and polyhydroxy butyrate (PHB). These byproducts are produced in the cells of microorganisms in the form of granular inclusions of fatty acids and sugars, the reaction is mediated by the action of the synthase (a polymeric enzyme) (Fang et al., 2019; Merrylin et al., 2020). Another application of food waste in the production of bioplastics can be observed in the study by Nistico et al. (2017), in which tomato residues (post-harvest and municipal waste) were used in the production of plastic films combined with poly (vinyl alcohol-co-ethylene). In this way, food waste is considered an appropriate raw material option to produce bioplastics because, in addition to its wide availability, it is biodegradable, which makes it possible to reduce the costs of producing these materials of great applicability in various industrial areas as a more sustainable alternative (Nielsen et al., 2017; Raza et al., 2018).

- ***Proteins, carbohydrates and lipids***

Besides bioactive compounds, food residues are rich in proteins, carbohydrates and lipids, and several processes are used to recover these nutrients from food residues, such as enzymatic and fungal hydrolysis (Pleissner et al., 2013; Awasthi et al., 2018). These components have high added value and can be used in various applications. According to Wang (2013), residues obtained from the production and processing of corn and wheat have high concentrations of proteins, minerals and lipids, serving as a cheaper and sustainable source for obtaining these nutrients. There are still only a few studies on obtaining proteins as a purified fraction from food residues. Generally, protein-rich residues are incorporated into foods to increase their protein value (Ritala et al., 2017). However, some insoluble proteins can be isolated from agri-food residues, mainly from the production of starch and bioethanol, and can be applied to produce bioenergy (Zhang et al., 2014).

In addition to proteins, another compound that can be extracted from food residues, mainly of plant origins such as citrus peels, apple pomace, watermelon peel and passion fruit peel, is pectin, which consists of a polysaccharide found in the cell wall and plant tissues (Caldeira et al., 2020). Chitin, a naturally occurring amino polysaccharide, can also be obtained from food residues. These polysaccharides, due to their structural properties, present numerous possibilities for applications as a coagulant agent in the treatment of effluents, as well as in the production of biomaterials in the food, biomedical, pharmaceutical and cosmetic industries (Liu et al., 2016; Xiong et al., 2019).

Lipids are usually obtained from cooking oil residues, which are subjected to a series of pre-treatments to improve the lipid extraction process, generally used to produce biodiesel (Carmona-Cabello et al., 2018). Other residues can serve as a source for obtaining lipids, as shown in the study by Karmee et al. (2015), in which lipids were obtained through the fungal hydrolysis of bakery residues.

Moreover, sustainable biotechnology plays an important role in the process of valuing food waste as an efficient means of rehandling and reusing these materials, boosting the reduction of waste, and decreasing the impacts on the environment caused by the accumulation of these residues (Nnolim et al., 2020). In this way, biotechnological development significantly contributes to solving several problems related to the accumulation of waste, which may be related to the food production chain, socioeconomic and environmental impacts and health impacts. For this reason, biotechnology is an important tool that allows the development of techniques for the extraction, transformation and production of various biological components obtained from different sources of food waste because of the great potential for the application of these compounds in numerous industrial sectors (Rahman et al., 2020).

The structure on energy production from food waste

Flexibility in accessing energy is considered a major factor in a nation's progress, mainly in production and distribution (Jain and Jain, 2017; Shane et al., 2017; Ye et al., 2016). However, uncontrolled population development opens the way for an increase in energy demand and, consequently, causes an increase in the use of non-renewable energy resources, such as fossil fuels. The economic impact caused, particularly on the environment, remains an unresolved issue (Khalil et al., 2019; Yi et al., 2018). Thus, there is a concern and a necessary demand for an ecologically sustainable solution, which has opened space for advancing research on the production of reusable energy from biomass or biowaste (IPCC, 2018; Panahi et al., 2020).

Within the possibilities, the use of biomass is recognized as one of the most promising alternatives in the replacement of fossil fuels, as it can reduce the impact on the environment since it does not contribute to the increase of carbon dioxide levels in the atmosphere (Lopez et al., 2016). This resource comes mainly in agriculture, the food industry and society, reaching 1.6 billion tons generated/year in the world (Nizami et al., 2017; Farmanbordar et al., 2018; FAO, 2018). This number, which is growing, creates a problem of handling and disposal. However, when applied to energy production, it contributes to sustainable socio-economic development (Ma et al., 2018; Rehan et al., 2018; Tan et al., 2018).

Europe, followed by Japan, are the main models to have achieved biomass for primary energy production (WEC, 2016). However, most countries, such as Africa and India, still face difficulties implementing this type of technology due to limited domestic food production and inadequate waste disposal, lack of public policies to support bioenergy production, and less technological know-how in the design and construction of facilities/digestors (Thi et al., 2016a; Dahunsi et al., 2020; Solarin, 2020), making it necessary to export biomass and, in some cases, biofuel.

The export of energy from biomass can generate income, helping to reduce the socioeconomic inequality that still exists. With this objective in mind, many countries have become pioneers in the global distribution of biofuels (Karthikeyan et al., 2018). Brazil is considered the largest producer and major exporter of bioenergy (bioethanol) recovered from sugarcane biomass, firewood and coal, representing 25.8% of the national energy supplement and at least 15% of annual exports (Welfle, 2017; MME, 2019).

However, Brazil is still taking its first steps and depends largely on foreign technology when it comes to biomass from food waste. Contributing to the worsening of the Brazilian situation is the lack of firm legislation regarding management and disposal, as well as the low capacity to treat these residues and the diversified food consumption pattern that requires specific technologies for energy production, increasing implementation costs (De Clercq et al., 2017; Dhanya et al., 2020).

The pattern of consumption, eating habits and the composition of food waste vary from region to region, depending on the time of year, cultural habits, climate and economic level of the region in question (Parfitt et al., 2010; Karthikeyan et al., 2018). Depending on the proximate composition of the waste, the production of bioenergy can vary, and with it, the technologies and processes used (Wang et al., 2012; Negri et al., 2020). Bioenergy production techniques are divided into biological (anaerobic digestion, esterification, fermentation) and physicochemical (pyrolysis, gasification, incineration and landfills) (Cucchiella et al., 2017; Sun et al., 2018; Ramos et al., 2018; Wang et al., 2020).

Among the available processes, the present highlight is anaerobic digestion, which produces energy under anaerobic conditions (Zhang et al., 2014). A study carried out verified that if bioenergy from 1 ton of waste were all through this technique, about 1915 GWh of energy would be generated, while this same mass would generate approximately around 847 KWh through sanitary landfills (Lou et al., 2013). The Biomass to Bioenergy (BtB) technology market is worth approximately US$25.32 billion worldwide and is expected to increase to US$40 billion by 2023 (WEC, 2016). Associated with the economic potential of these technologies, social and sustainable development is also added through the generation of jobs, reduction of food waste production and emission of harmful gases into the atmosphere (IPCC, 2018; Caldeira et al., 2020).

Conclusions

Currently, there is a growing concern with managing large amounts of food waste generated around the world. This serious global problem has led industries to enable the microbial transformation of food waste into bioproducts of great commercial value. Despite the obstacles related to the costs of selective collection and industrial facilities for the bioconversion of this food waste, it is possible to consider the success of this new venture not only because of the innovative bioproducts generated, but also for the development of biotechnological industries, the jobs generated and the preservation of the environment.

References

Ahmed, Z. and Ahmad, A. 2017. Chapter 8 - Biopolymer Produced by the Lactic Acid Bacteria: Production and Practical Application. Microb. Prod. Food Ing Add, Academic Press, 2017: 217–257. DOI: 10.1016/B978-0-12-811520-6.00008-8.

Angiolillo, L., Del Nobile, M.A. and Conte, A. 2015. The extraction of bioactive compounds from food residues using microwaves. Curr. Opin. Food Sci. 5: 93–98. DOI: 10.1016/j.cofs.2015.10.001.

Arun, K.B., Madhavan, A., Sindhu, R., Binod, P., Pandey, A., Reshmy, R. et al. 2020. Remodelingagro-industrial and food wastes into value-added bioactives and biopolymers. Ind. Crops Prod. 154: 112621. DOI: 10.1016/j.indcrop.2020.112621.

Awasthi, M.K., Selvam, A., Chan, M.T. and Wong, J.W.C. 2018. Bio-degradation of oily food waste employing thermophilic bacterial strains. Bioresour. Technol. 248: 141–147. DOI: 10.1016/j.biortech.2017.06.115.

Azi, F., Tu, C., Rasheed, H.A. and Dong, M. 2020. Comparative study of the phenolics, antioxidant and metagenomic composition of novel soy whey-based beverages produced using three different water kefir microbiota. Int. J. Food Science Technology 55(4): 1689–1697. DOI: 10.1111/ijfs.14439.

Baiano, A. 2014. Recovery of biomolecules from food wastes—a review. Molecules 19(9): 14821–14842. DOI: 10.3390/molecules190914821.

Beattie, A.J., Hay, M., Magnusson, B., De Nys, R. James Smeathers, J. and Vincent, L.F.V. 2011. Ecology and bioprospecting. Aust. Ecol. 36(3): 341–356. DOI: 10.1111/j.1442-9993.2010.02170.x.

Budzianowski, W.M. 2017. High-value low-volume bioproducts coupled to bioenergies with potential to enhance business development of sustainable biorefineries. Renew. Sust. Energ. Rev. 70: 793–804. DOI: 10.1016/j.rser.2016.11.260.

Caldeira, C., Vlysidis, A., Fiore, G., De Laurentiis, V., Vignali, G. and Sala, S. 2020. Sustainability of food waste biorefinery: a review on valorisation pathways, techno-economic constraints, and environmental assessment. Bioresour. Technol. 312: 123575. DOI: 10.1016/j.biortech.2020.123575.

Carciochi, R.A., d'Alessandro, L.G., Vauchel, P., Rodriguez, M.M., Nolasco, S.M. and Dimitrov, K. 2017. Valorization of agrifood by-products by extracting valuable bioactive compounds using green processes. pp. 191–228. *In*: Ingredients Extraction by Physicochemical Methods in Food. Academic Press. DOI: 10.1016/B978-0-12-811521-3.00004-1.

Carmona-Cabello, M., Leiva-Candia, D., Castro-Cantarero, J.L., Pinzi, S. and Dorado, M.P. 2018. Valorization of food waste from restaurants by transesterification of the lipid fraction. Fuel 215: 492–498. DOI: 10.1016/j.fuel.2017.11.096.

Cucchiella, F., D'adamo, I. and Gastaldi, M. 2017. Sustainable waste management: Waste to energy plant as an alternative to landfill. Energy Convers Manag. 131: 18–31. DOI: 10.1016/j.rser.2014.02.0151364-0321.

Dahunsi, S.O., Fagbiele, O.O. and Yusuf, E.O. 2020. Bioenergy technologies adoption in Africa: A review of past and current status. J. Clean. Prod. 121683. DOI: 10.1016/j.jclepro.2020.121683.

Dave, A., Bhatt, M. and Kumar, M.A. 2020. Approaches for recovering bio-based products from municipal and industrial wastes. pp. 199–216. *In*: Current Developments in Biotechnology and Bioengineering. Elsevier. DOI: 10.1016/B978-0-444-64321-6.00010-0.

De Castro, R.J.S., Ohara, A., Nishide, T.G., Bagagli, M.P., Dias, F.F.G. and Sato, H.H. 2015. A versatile system based on substrate formulation using agroindustrial wastes for protease production by *Aspergillus niger* under solid state fermentation. Biocatal. Agric Biotechnol. 4(4): 678–684. DOI: 10.1016/j.bcab.2015.08.010.

De Clercq, D., Wen, Z., Gottfried, O., Schmidt, F. and Fei, F. 2017. A review of global strategies promoting the conversion of food waste to bioenergy via anaerobic digestion. Renew. Sust. Energ. Rev. 79: 204–221. DOI: 10.1016/j.rser.2017.05.047.

De Roos, J. and De Vuyst, L. 2018. Acetic acid bacteria in fermented foods and beverages. Curr. Opin. Biotechnol. 49: 115–119. DOI: 10.1016/j.copbio.2017.08.007.

Dhanya, B.S., Mishra, A., Chandel, A.K. and Verma, M.L. 2020. Development of sustainable approaches for converting the organic waste to bioenergy. Sci. Total Environ. 723: 138109. DOI: 10.1016/j.scitotenv.2020.138109.

Di Cagno, R., Filannino, P. and Gobbetti, M. 2016. Fermented Foods: Fermented Vegetables and Other Products. Encyclopedia of Food and Health. 3rd Edition. Publisher: Academic Press, Oxford. p. 668–674. DOI: 10.1016/B978-0-12-384947-2.00284-1.

Fang, F., Xu, R.-Z., Huang, Y.-Q., Wang, S.-N., Zhang, L.-L., Dong, J.-Y. et al. 2019. Production of polyhydroxyalkanoates and enrichment of associated microbes in bioreactors fed with rice winery wastewater at various organic loading rates. Bioresour. Technol. 292: 121978. DOI: 10.1016/j.scitotenv.2020.138109.

FAO. 2018. World fertilizer trends and outlook to 2018. Data available at. http://www.fao. org/3/a-i4324e.pdf.

Farmanbordar, S., Karimi, K. and Amiri, H. 2018. Municipal solid waste as a suitable substrate for butanol production as an advanced biofuel. Energy Convers Manag. 157: 396–408. DOI: 10.1016/j.enconman.2017.12.020.

Fausto-Castro, L., Rivas-García, P., Gómez-Nafte, J.A., Rico-Martínez, R., Rico-Ramírez, V., Gomez-Gonzalez, R. et al. 2020. Selection of food waste with low moisture and high protein content from Mexican restaurants as a supplement to swine feed. J. Clean. Prod. 256: 120137. DOI: 10.1016/j.jclepro.2020.120137.

Food and Agriculture Organization (FAO). 2019. The State of Food and Agriculture 2019. Moving forward on food loss and waste reduction. Rome. Licence: CC BY-NC-SA 3.0 IGO.

Galanakis, C. 2020. Food waste valorization opportunities for different food industries. pp. 341–422. *In*: The Interaction of Food Industry and Environment. Academic Press. DOI: 10.1016/B978-0-12-816449-5.00011-4.

Gallardo, A., Edo-Alcón, N., Carlos, M. and Renau, M. 2016. The determination of waste generation and composition as an essential tool to improve the waste management plan of a university. Waste Manag. 53: 3–11. DOI: 10.1016/j.wasman.2016.04.013.

García, C., Rendueles, M. and Díaz, M. 2019. Liquid-phase food fermentations with microbial consortia involving lactic acid bacteria: A review. Int. Food Res. J. 119: 207–220. DOI: 10.1016/j.foodres.2019.01.043.

Girotto, F., Alibardi, L. and Cossu, R. 2015. Food waste generation and industrial uses: a review. Waste Manag. 45: 32–41. DOI: 10.1016/j.wasman.2015.06.008.

Gómez-Sala, B., Muñoz-Atienza, E., Sánchez, J., Basanta, A., Herranz, C., Hernández, P.E. et al. 2015. Bacteriocin production by lactic acid bacteria isolated from fish, seafood and fish products. Eur. Food Res. Technol. 24(3): 341–356. DOI: 10.1007/s00217-015-2465-3.

Haddadi, M.H., Aiyelabegan, H.T. and Negahdari, B. 2018. Advanced biotechnology in biorefinery: a new insight into municipal waste management to the production of high-value products. Int. J. Environ. Sci. Technol. 15(3): 675–686. DOI: 10.1007/s13762-017-1424-x.

Hatti-Kaul, R. and Mattiasson, B. 2016. Anaerobes in industrial-and environmental biotechnology. Adv. Biochem. Eng. Biotechnol. 156: 1–33. DOI: 10.1007/10_2016_10.

Hatti-Kaul, R. and Mattiasson, B. 2018. Anaerobes in industrial-and environmental biotechnology. *In*: IPCC. Trends in Energy Use and Associated Greenhouse Gas Emissions. Http:// WwwIpccCh 2018.

IPCC. 2018. Global warming of 1.5°C, an IPCC special report on the impacts of global warming of 1.5°C above pre-industrial levels and related global greenhouse gas emission pathways, in the context of strengthening the global response to the threat of climate change, sustainable development, and efforts to eradicate poverty. Summary for policy makers, Available online at: Available online at: https://unfccc.int/topics/science/workstreams/cooperation-with-the-ipcc/ipcc-special-report-on-global-warming-of-15-degc.

Irkin, R. 2019. Natural fermented beverages. pp. 399–425. *In*: Natural Beverages. Academic Press, v. 13. DOI: 10.1016/B978-0-12-816689-5.00014-6.

Jain, S. and Jain, P.K. 2017. The rise of renewable energy implementation in South Africa. Energy Procedia 143: 721–726. DOI: 10.1016/j.egypro.2017.12.752.

Javed, U., Ansari, A., Aman, A. and Qader, S.A.U. 2019. Fermentation and saccharification of agro-industrial wastes: A cost-effective approach for dual use of plant biomass wastes for xylose production. Biocatal. Agric Biotechnol. 21: 101341. DOI: 10.1016/j.bcab.2019.101341.

Jiménez-Velázquez, R., Valle-Guadarrama, S., Alia-Tejacal, I., Salinas-Moreno, Y., García-Cruz, L., Pérez-López, A. et al. 2020. Separation of bioactive compounds from epicarp of 'Hass' avocado fruit through aqueous two-phase systems. Food Bioprod. Process 2020. DOI: 10.1016/j.fbp.2020.07.004.

Joana Gil-Chávez, G., Villa, J.A., Ayala-Zavala, J.F., Basilio Heredia, J., Sepulveda, D., Yahia, E.M. et al. 2013. Technologies for extraction and production of bioactive compounds to be used as nutraceuticals and food ingredients: an overview. Compr. Rev. Food Sci. Food Saf. 12(1): 5–23. DOI: 10.1111/1541-4337.12005.

Ju, M., Osako, M. and Harashina, S. 2017. Food loss rate in food supply chain using material flow analysis. Waste Manag. 61: 443–454. DOI: 10.1016/j.wasman.2017.01.021.

Kader, F., Baky, A.H., Khan, M.N.H. and Chowdhury, H.A. 2015. Production of biogas by anaerobic digestion of food waste and process simulation. Am. J. Mech. Eng. 3(3): 79–83. DOI: 10.1115/ES2014-6756.

Kalogeropoulos, N., Chiou, A., Pyriochou, V., Peristeraki, A., Vaios T. and Karathanos, V.T. 2012. Bioactive phytochemicals in industrial tomatoes and their processing byproducts. LWT-Food Sci. Technol. 49(2): 213–216. DOI: 10.1016/j.lwt.2011.12.036.

Karmee, S.K., Linardi, D., Lee, J. and Lin, C.S.K. 2015. Conversion of lipid from food waste to biodiesel. Waste Manag. 41: 169–173. DOI: 10.1016/j.wasman.2015.03.025.

Karthikeyan, O.P., Trably, E., Mehariya, S., Bernet, N., Wong, J.W.C. and Carrere, H. 2018. Pretreatment of food waste for methane and hydrogen recovery: a review. Bioresour. Technol. 249: 1025–1039. DOI: 10.1016/j.biortech.2017.09.105.

Katz, L. and Baltz, R.H. 2016. Natural product discovery: past, present, and future. J. Ind. Microbiol. Biotechnol. 43: 155–176. DOI: 10.1007/s10295-015-1723-5.

Kaur, J., Rani, G. and Yogalakshmi, K.N. 2020. Problems and issues of food waste-based biorefineries. pp. 343–357. *In*: Food Waste to Valuable Resources. Academic Press. DOI: 10.1016/B978-0-12-818353-3.00016-X.

Kavanaugh, M. and Quinlan, J.J. 2020. Consumer knowledge and behaviors regarding food date labels and food waste. Food Control 115: 107285. DOI: 10.1016/j.foodcont.2020.107285.

Khalil, M., Berawi, M.A., Heryanto, R. and Akhmad Rizalie, A. 2019. Waste to energy technology: The potential of sustainable biogas production from animal waste in Indonesia. Renew. Sust. Energ. Rev. 105: 323–331. DOI: 10.1016/j.rser.2019.02.011.

Kumar, A., Gudiukaite, R., Gricajeva, A., Sadauskas, M., Malunavicius, V., Kamyab, H. et al. 2020. Microbial lipolytic enzymes–promising energy-efficient biocatalysts in bioremediation. Energy 192: 116674. DOI: 10.1016/j.energy.2019.116674.

Kumar, P., Chandrasekhar, K., Kumari, A., Sathiyamoorthi, E. and Kim, B.S. 2018. Electro-fermentation in aid of bioenergy and biopolymers. Energies 11(2): 343. DOI: 10.3390/en11020343.

Lee, J.W., Na, D., Park, J.M., Lee, J., Choi, S. and Lee, S.Y. 2012. Systems metabolic engineering of microorganisms for natural and non-natural chemicals. Nat. Chem. Biol. 8(6): 536. DOI: 10.1038/nchembio.970.

Leite, P., Silva, C., Salgado, J.M. and Belo, I. 2019. Simultaneous production of lignocellulolytic enzymes and extraction of antioxidant compounds by solid-state fermentation of agro-industrial wastes. Ind. Crops Prod. 37: 315–322. DOI: 10.1016/j.indcrop.2019.04.044.

Lin, C.S.K., Pfaltzgraff, L.A., Herrero-Davila, L., Mubofu, E.B., Abderrahim, S., Clark, J.H. et al. 2013. Food waste as a valuable resource for the production of chemicals, materials and fuels. Current situation and global perspective. Energy Environ. Sci. 6(2): 426–464. DOI: 10.1039/c2ee23440h.

Liu, Z., Liao, W. and Liu, Y. 2016. A sustainable biorefinery to convert agricultural residues into value-added chemicals. Biotechnol Biofuels 9(1): 197. DOI: 10.1186/s13068-016-0609-8.

Liu, S., Deng, Z., Li, H. and Feng, K. 2019. Contribution of electrodes and electric current to process stability and methane production during the electro-fermentation of food waste. Bioresour. Technol. 288: 121536. DOI: 10.1016/j.biortech.2019.121536.

Lombardelli, C., Liburdi, K., Benucci, I. and Esti, M. 2020. Tailored and synergistic enzyme-assisted extraction of carotenoid-containing chromoplasts from tomatoes. Food Bioprod. Process 121: 43–53. DOI: 10.1016/j.fbp.2020.01.014.

Lopez, G., Alvarez, J., Amutio, M., Arregi, A., Bilbao, J. and Olazar, M. 2016. Assessment of steam gasification kinetics of the char from lignocellulosic biomass in a conical spouted bed reactor. Energy 107: 493–501. DOI: 10.1016/j.energy.2016.04.040.

Lou, X.F., Nair, J. and Ho, G. 2013. Potential for energy generation from anaerobic digestion of food waste in Australia. Waste Manag. Res. 31(3): 283–294. DOI: 10.1177/0734242x12474334.

Ma, C., Liu, J., Ye, M., Zou, L., Qian, G. and Li, Y.Y. 2018. Towards utmost bioenergy conversion efficiency of food waste: Pretreatment, co-digestion, and reactor type. Renew. Sust. Energ. Rev. 90: 700–709. DOI: 10.1016/j.rser.2018.03.110.

Mathias, T.R.S., Aguiar, P.F., Silva, J.B.A., Mello, P.P.M. and Sérvulo, E.F.C. 2017. Brewery waste reuse for protease production by lactic acid fermentation. Food Technol. Biotechnol. 55(2): 218–224. DOI: 10.17113/ftb.55.02.17.4378.

Mensah, J.K.M. and Twumasi, P. 2016. Use of pineapple waste for single cell protein (SCP) production and the effect of substrate concentration on the yield. J. Food Process Eng. 40(3): 1–9. DOI: 10.1111/jfpe.12478.

Merrylin, J., Preethi, M., Saratale, G.D. and Banu, J.R. 2020. Production of biopolymers and feed protein from food wastes. Food Waste to Valuable Resources. p. 143–162. DOI: 10.1016/b978-0-12-818353-3.00007-9.

MME, Synthesis Report - Base Year 2018, Brazilian Energy Balanc, 2019, pp. 1e67. http://www.epe.gov.br/pt/publicacoes-dados-abertos/publicacoes/ balanco-energetico-nacional-2019.

Müller, C.A., Obermeier, M.M. and Berg, G. 2016. Bioprospecting plant-associated microbiomes. J. Biotechnol. 235: 171–180. DOI: 10.1016/j.jbiotec.2016.03.033.

Naik, B., Goyal, S.K., Tripathi, A.D. and Kumar, V. 2019. Screening of agro-industrial waste and physical factors for the optimum production of pullulanase in solid-state fermentation from endophytic *Aspergillus* sp. Biocatal. Agric Biotechnol. 22: 678–684. DOI: 10.1016/j.bcab.2019.101423.

Nayak, A. and Bhushan, B. 2019. An overview of the recent trends on the waste valorization techniques for food wastes. J. Environ. Manage 233: 352–370. DOI: 10.1016/j.jenvman.2018.12.041.

Nazzaro, F., Fratianni, F., Ombra, M.N., d' Acierno, A. and Coppola, R. 2018. Recovery of biomolecules of high benefit from food waste. Curr. Opin. Food Sci. 22: 43–54. DOI: 10.1016/j.cofs.2018.01.012.

Negri, C., Ricci, M., Zilio, M., D'Imporzano, G., Qiao, W., Dong, R. et al. 2020. Anaerobic digestion of food waste for bio-energy production in China and Southeast Asia: A review. Renew. Sust. Energ. Rev. 133: 110138. DOI: 10.1016/j.rser.2020.110138.

Ng, H.S., Kee, P.E., Yim, H.S., Chen, P.-T., Wei, Y.-H. and Lan, J.C.-W. 2020. Recent advances on the sustainable approaches for conversion and reutilization of food wastes to valuable bioproducts. Bioresour. Technol. 302: 122889. DOI: 10.1016/j.biortech.2020.122889.

Nielsen, C., Rahman, A., Rehman, A.U., Walsh, M.K. and Miller, C.D. 2017. Food waste conversion to microbial polyhydroxyalkanoates. Microbial Biotechnology 10(6): 1338–1352. DOI: 10.1111/1751-7915.12776.

Nisticò, R., Evon, P., Labonne, L., Vaca-Medina, G., Montoneri, E., Vaca-Garcia, C. et al. 2017. Post-harvest tomato plants and urban food wastes for manufacturing plastic films. J. Clean. Prod. 167: 68–74. DOI: 10.1016/j.jclepro.2017.08.160.

Nizami, A.S., Rehan, M., Waqas, M., Naqvi, M., Ouda, O.K.M., Shahzad, K. et al. 2017. Waste biorefineries: enabling circular economies in developing countries. Bioresour. Technol. 241: 1101–1117. DOI: 10.1016/j.biortech.2017.05.097.

Nnolim, N.E., Okoh, A.I. and Nwodo, U.U. 2020. Proteolytic bacteria isolated from agro-waste dumpsites produced keratinolytic enzymes. Biotechnol. Rep. e00483. DOI: 10.1016/j.btre.2020.e00483.

Panahi, H.K.S., Dehhaghi, M., Aghbashlo, M., Karimi, K. and Tabatabaei, M. 2020. Conversion of residues from agro-food industry into bioethanol in Iran: An under-valued biofuel additive to phase out MTBE in gasoline. Renewable Energy 145: 699–710. DOI: 10.1016/j.renene.2019.06.081.

Parfitt, J., Barthel, M. and Macnaughton, S. 2010. Food waste within food supply chains: quantification and potential for change to 2050. Philosophical Transactions of the Royal Society B: Biological Sciences 365(1554): 3065–3081. DOI: 10.1098/rstb.2010.0126.

Paritosh, K., Kushwaha, S.K., Yadav, M., Pareek, N., Chawade, A. and Vivekanand, V. 2017. Food waste to energy: an overview of sustainable approaches for food waste management and nutrient recycling. BioMed Research International 2017: 1–19. DOI: 10.1155/2017/2370927.

Patra, D., Leisnham, P.T., Tanui, C.K. and Pradhan, A.K. 2020. Evaluation of global research trends in the area of food waste due to date labeling using a scientometrics approach. Food Control 107307. DOI: 10.1016/j.foodcont.2020.107307.

Pessôa, M.G., Vespermann, K.A.C., Paulino, B.N., Barcelos, M.C.S., Pastore, G.M. and Molina, G. 2019. Newly isolated microorganisms with potential application in biotechnology. Bioresour. Technol. 37(2): 319–339. DOI: 10.1016/j.biotechadv.2019.01.007.

Pleissner, D., Lam, W.C., Sun, Z. and Lin, C.S.K. 2013. Food waste as nutrient source in heterotrophic microalgae cultivation. Bioresour. Technol. 137: 139–146. DOI: 10.1016/j.biortech.2013.03.088.

Pramanik, S.K., Suja, F.B., Zain, S.M. and Pramanik, B.K. 2019. The anaerobic digestion process of biogas production from food waste: Prospects and constraints. Bioresource Technology Reports 8: 100310. DOI: 10.1016/j.biteb.2019.100310.

Prandi, B., Faccini, A., Lambertini, F., Bencivenni, M., Jorba, M., Droogenbroek, B.V. et al. 2019. Food wastes from agrifood industry as possible sources of proteins: A detailed molecular view on the composition of the nitrogen fraction, amino acid profile and racemisation degree of 39 food waste streams. Food Chemistry 286: 567–575. DOI: 10.1016/j.foodchem.2019.01.166.

Rahman, M., Billah, M.M., Hack-Polay, D. and Alam, A. 2020. The use of biotechnologies in textile processing and environmental sustainability: An emerging market context. Technological Forecasting and Social Change 159: 120204. DOI: 10.1016/j.techfore.2020.120204.

Ramos, A., Monteiro, E., Silva, V. and Rouboa, A. 2018. Co-gasification and recent developments on waste-to-energy conversion: A review. Renewable and Sustainable Energy Reviews 81: 380–398. DOI: 10.1016/j.rser.2017.07.025.

Randazzo, W., Corona, O., Guarcello, R., Francesca, N., Germanà, M.A., Erten, H. et al. 2016. Development of new non-dairy beverages from Mediterranean fruit juices fermented with water kefir microorganisms. Food Microbiol. 54: 40–51. DOI: 10.1016/j.fm.2015.10.018.

Ravindran, R. and Jaiswal, A.K. 2016. Exploitation of food industry waste for high-value products. Trends Biotechnol. 34(1): 58–69. DOI: 10.1016/j.tibtech.2015.10.008.

Raza, Z.A., Abid, S. and Banat, I.M. 2018. Polyhydroxyalkanoates: Characteristics, production, recent developments and applications. Int. Biodeterior. Biodegradation 126: 45–56. DOI: 10.1016/j.ibiod.2017.10.001.

Redlingshöfer, B., Barles, S. and Weisz, H. 2020. Are waste hierarchies effective in reducing environmental impacts from food waste? A systematic review for OECD countries. Resour. Conserv. Recycl. 156: 104723. DOI: 10.1016/j.resconrec.2020.104723.

Rehan, M., Gardy, J., Demirbas, A., Rashid, U., Budzianowski, W.M., Pant, D. et al. 2018. Waste to biodiesel: A preliminary assessment for Saudi Arabia. Bioresour. Technol. 250: 17–25. DOI: 10.1016/j.biortech.2017.11.024.

Ritala, A., Häkkinen, S.T., Toivari, M. and Wiebe, M.G. 2017. Single cell protein—state-of-the-art, industrial landscape and patents 2001–2016. Front. Microbiol. 8: 2009. DOI: 10.3389/fmicb.2017.02009.

Sadh, P.K., Kumar, S., Chawla, P. and Duhan, J.S. 2018. Fermentation: A boon for production of bioactive compounds by processing of food industries wastes (By-Products). Molecules 23: 2560. DOI: 10.3390/molecules23102560.

Sepúlveda, L., Laredo-Alcalá, E., Buenrostro-Figueroa, J.J., Ascacio-Valdés, J.A., Genisheva, Z., Aguilar, C. et al. 2020. Ellagic acid production using polyphenols from orange peel waste by submerged fermentation. Electron J. Biotechnol. 43: 1–7. DOI: 10.1016/j.ejbt.2019.11.002.

Shane, A., Gheewala, S.H. and Kafwembe, Y. 2017. Urban commercial biogas power plant model for Zambian towns. Renewable Energy 103: 1–14. DOI: 10.1016/j.renene.2016.11.017.

Sindhu, R., Gnansounou, E., Rebello, S., Binod, P., Varjani, S., Thakur, I.S. et al. 2019. Conversion of food and kitchen waste to value-added products. J. Environ. Manage 241: 619–630. DOI: 10.1016/j.jenvman.2019.02.053.

Slorach, P.C., Jeswani, H.K., Cuéllar-Franca, R. and Azapagic, A. 2019. Environmental sustainability of anaerobic digestion of household food waste. J. Environ. Manage 236: 798–814. DOI: 10.1016/j.jenvman.2019.02.001.

Solarin, S.A. 2020. An environmental impact assessment of fossil fuel subsidies in emerging and developing economies. Environ. Impact Assess Rev. 85: 106443. DOI: 10.1016/j.eiar.2020.106443.

Sravan, J.S., Butti, S.K., Sarkar, O., Krishna, K.V. and Mohan, S.V. 2018. Electrofermentation of food waste–Regulating acidogenesis towards enhanced volatile fatty acids production. Chem. Eng. J. 334: 1709–1718. DOI: 10.1016/j.cej.2017.11.005.

Sun, L., Fujii, M., Tasaki, T., Dong, H. and Ohnishi, S. 2018. Improving waste to energy rate by promoting an integrated municipal solid-waste management system. Resour. Conserv. Recycl. 136: 289–296. DOI: 10.1016/j.resconrec.2018.05.005.

Tan, X.B., Lam, M.K., Uemura, Y., Lim, J.W., Wong, C.Y., Ramli, A. et al. 2018. Semi-continuous cultivation of *Chlorella vulgaris* using chicken compost as nutrients source: Growth optimization study and fatty acid composition analysis. Energy Convers Manag. 164: 363–373. DOI: 10.1016/j.enconman.2018.03.020.

Tavill, G. 2020. Industry challenges and approaches to food waste. Physiology & Behavior 223: 112993. DOI: 10.1016/j.physbeh.2020.112993.

Teigiserova, D.A., Hamelin, L. and Thomsen, M. 2019. Review of high-value food waste and food residues biorefineries with focus on unavoidable wastes from processing. Resour. Conserv. Recycl. 149: 413–426. DOI: 10.1016/j.resconrec.2019.05.003.

Thi, N.B.D., Lin, C.-Y. and Kumar, G. 2016a. Waste-to-wealth for valorization of food waste to hydrogen and methane towards creating a sustainable ideal source of bioenergy. J. Clean. Prod. 122: 29–41. DOI: 10.1016/j.jclepro.2016.02.034.

Thi, N.B.D., Lin, C.-Y. and Kumar, G. 2016b. Electricity generation comparison of food waste-based bioenergy with wind and solar powers: A mini review. Sustain. Environ. Res. 26(5): 197–202. DOI: 10.1016/j.serj.2016.06.001.

Tsang, Y.F., Kumar, V., Samadar, P., Yang, Y., Lee, J., Ok, Y.S. et al. 2019. Production of bioplastic through food waste valorization. Environ. Int. 127: 625–644. DOI: 10.1016/j.envint.2019.03.076.

Vea, E.B., Romeo, D. and Thomsen, M. 2018. Biowaste valorisation in a future circular bioeconomy. Procedia Cirp 69: 591–596. DOI: 10.1016/j.procir.2017.11.062.

Vicenssuto, G.M. and De Castro, R.J.S. 2020. Development of a novel probiotic milk product with enhanced antioxidant properties using mango peel as a fermentation substrate. Biocatal. Agric Biotechnol. 101564. DOI: 10.1016/j.bcab.2020.101564.

Vijayaraghavan, P., Vincent, S.G.P., Arasu, M.V. and Al-Dhabi, N.A. 2016. Bioconversion of agro-industrial wastes for the production of fibrinolytic enzyme from *Bacillus halodurans* IND18: Purification and biochemical characterization. Electron J. Biotechnol. 20: 1–8. DOI: 10.1016/j.ejbt.2016.01.002.

Wakefield, A. and Axon, S. 2020. "I'm a bit of a waster": Identifying the enablers of, and barriers to, sustainable food waste practices. J. Clean. Prod. 122803. DOI: 10.1016/j.jclepro.2020.122803.

Wang, J., Cheng, G., You, Y., Xiao, B., Liu, S., He, P. et al. 2012. Hydrogen-rich gas production by steam gasification of municipal solid waste (MSW) using NiO supported on modified dolomite. Int. J. Hydrog. Energy 37(8): 6503–6510. DOI: 10.1016/j.ijhydene.2012.01.070.

Wang, L.J. 2013. Production of bioenergy and bioproducts from food processing wastes: a review. Trans ASABE 56(1): 217–230. DOI: 10.13031/2013.42572.

Wang, P., Wang, H., Qiu, Y., Ren, L. and Jiang, B. 2018. Microbial characteristics in anaerobic digestion process of food waste for methane production—A review. Bioresour. Technol. 248: 29–36. DOI: 10.1016/j.biortech.2017.06.152.

Wang, S., Yu, S., Lu, Q., Liao, Y., Li, H., Sun, L. et al. 2020. Development of an alkaline/acid pre-treatment and anaerobic digestion (APAD) process for methane generation from waste activated sludge. Sci. Total Environ. 708: 134564. DOI: 10.1016/j.scitotenv.2019.134564.

WEC. Waste to energy; 2016. <https://www.worldenergy.org/wp-content/uploads/2017/03/WEResources_Waste_to_Energy_2016.pdf>.

Welfle, A. 2017. Balancing growing global bioenergy resource demands-Brazil's biomass potential and the availability of resource for trade. Biomass Bioenergy 105: 83–95. DOI: 10.1016/j.biombioe.2017.06.011.

Xiong, X., Yu, I.K.M., Tsang, D.C.W., Bolan, N.S., Ok, Y.S., Igalavithana, A.D. et al. 2019. Value-added chemicals from food supply chain wastes: State-of-the-art review and future prospects. Chem. Eng. J. 375: 121983. DOI: 10.1016/j.cej.2019.121983.

Xu, D., Fels, L., Wefers, D., Behr, J., Jakob, F. and Vogel, R.F. 2018a. *Lactobacillus hordei* dextrans induce *Saccharomyces cerevisiae* aggregation and network formation on hydrophilic surfaces. Int. J. Biol. Macromol. (115): 236–242. DOI: 10.1016/j.ijbiomac.2018.04.068.

Xu, F., Li, Y., Ge, X., Yang, L., Li, Y. 2018b. Anaerobic digestion of food waste—Challenges and opportunities. Bioresour. Technol. 247: 1047–1058. DOI: 10.1016/j.biortech.2017.09.020.

Xue, L., Liu, G., Parfitt, J., Liu, X., Herpen, E.V., Stenmarck, Å. et al. 2017. Missing food, missing data? A critical review of global food losses and food waste data. Environ. Sci. Technol. 51(12): 6618–6633. DOI: 10.1021/acs.est.7b00401.

Yang, G., Tan, H., Li, S., Zhang, M., Che, J., Li, K. et al. 2020. Application of engineered yeast strain fermentation for oligogalacturonides production from pectin-rich waste biomass. Bioresour. Technol. 300: 122645. DOI: 10.1016/j.biortech.2019.122645.

Ye, W., Gao, Y., Ding, H., Liu, M., Liu, S., Han, X. et al. 2016. Kinetics of transesterification of palm oil under conventional heating and microwave irradiation, using CaO as heterogeneous catalyst. Fuel 180: 574–579. DOI: 10.1016/j.fuel.2016.04.084.

Yi, H., Jiang, M., Huang, D., Zeng, G., Lai, C., Qin, L. et al. 2018. Advanced photocatalytic Fenton-like process over biomimetic hemin-Bi2WO6 with enhanced pH. J. Taiwan Inst. Chem. Eng. 93: 184–192. DOI: 10.1016/j.jtice.2018.06.037.

Zisopoulos, F.K., Rossier-Miranda, F.J., van der Goot, A.J. and Boom, R.M. 2017. The use of exergetic indicators in the food industry—A review. Crit. Rev. Food Sci. Nutr. 57(1): 197–211. DOI: doi:10.1080/10408398.2014.975335.

Zhang, C., Su, H., Baeyens, J. and Tan, T. 2014. Reviewing the anaerobic digestion of food waste for biogas production. Renew. Sust. Energ. Rev. 38: 383–392. DOI: 10.1016/j.rser.2014.05.038.

Zhang, C., Sanders, J.P.M. and Bruins, M.E. 2014. Critical parameters in cost-effective alkaline extraction for high protein yield from leaves. Biomass Bioenergy 67: 466–472. DOI: 10.1016/j.biombioe.2014.05.020.

Zhang, Y.H.P., Sun, J. and Ma, Y. 2017. Biomanufacturing: history and perspective. J. Ind. Microbiol. Biotechnol. 44: 773–784. DOI: 10.1007/s10295-016-1863-2.

Zhou, S., Du, G., Kang, Z., Li, J., Chen. J., Li, H. et al. 2017. The application of powerful promoters to enhance gene expression in industrial microorganisms. World J. Microbiol. Biotechnol. 33(2): 23. DOI:10.1007/s11274-016-2184-3.

Zhu, Z., Zhang, J., Ji, X., Fang, Z., Wu, Z., Chen, J. et al. 2018. Evolutionary engineering of industrial microorganisms—strategies and applications. Appl. Microbiol. Biotechnol. 102(11): 4615–4627. DOI:10.1007/s00253-018-8937-1.

Chapter 12

General Toxicological Aspects of Toxins from Conventional and Non-Conventional Foods

Antonia Amanda Cardoso de Almeida,[1] *Ranyelison Silva Machado,*[2] *Aldenora Maria Ximenes Rodrigues,*[3] *Railson Pereira Souza*[4] and *Paulo Michel Pinheiro Ferreira*[1,4,*]

Introduction

The expression "food waste" refers to organic waste generated from the preparation of human food, from our kitchen home or other types of arrangements. They are abundantly generated over the world in proportions that can reach up to 65% of total organics, and have great potential to be used as raw material for the production of value-added products, for example, fuels and chemicals (Viana et al., 2006; Elkhalifa et al., 2019).

Regarding the chemical composition of food waste, it varies in different regions of the world, depending on the production and composition (Ward et al., 2008). According to Zhang et al. (2014), the generation of food waste is associated with population growth, as it is linked to all stages of the food chain. Thus, the greater the demand for food, the greater the generation of food waste (Zhang et al., 2014). Around 14% of food produced is lost between harvest and retail, while an estimated 17% of total global food production is wasted. Food that is lost and wasted accounts for 38% of total energy usage in the global food system (United Nations, 2021), and the value of this wasted food is worth over US$ 1 trillion.

[1] Laboratory of Experimental Cancerology (LabCancer), Department of Biophysics and Physiology, Federal University of Piauí, Teresina, Brazil.
[2] Department of Biochemistry, Federal University of São Paulo, São Paulo, Brazil.
[3] Research Ethics Committee, Federal University of São Paulo, São Paulo, Brazil.
[4] Postgraduation Program in Pharmacology, Federal University of Piauí, Teresina, Brazil.
* Corresponding author: pmpf@ufpi.edu.br

Food waste usually consists of cereals, meats, pasta, sausages, eggs, fruits and vegetables. In general, the residues are composed of carbohydrate polymers (starch, cellulose and hemicellulose), lignin, proteins, lipids, organic acids and minor quantities of inorganic compounds (Kiran et al., 2014; Gueri et al., 2018; Braguglia et al., 2018).

Products of animal source

Meat is a very complex *ex vivo* system under a continuous process of changes in its components of high and low molecular weight that can be useful to support the demands of the human organism for metabolic energy and other relevant functions. The majority of the compounds has quality and safety to the customer. However, by-products contain harmful substances, including toxins and carcinogens (Püssa, 2013).

Animal residues are obtained from meat processing of sheep, beef, swine, poultry and fishes and/or their by-products, like bones, trimmings, organs and glands, lard, feathers, hair and skins, blood, horns and hooves, husks, scales and carcass residues in retail trade (Fig. 12.1) (Barros, 2007).

Some of these residues can be modified into value-added by-products. Industrial tallow, processed animal meal for animal feed, fish skin-based gelatins and the use of eggshells can be cited. However, it is known that most of these by-products are not palatable, are unsuitable for human consumption, and there are not enough studies confirming the absence of toxicity (Barros, 2007; Jayathilakan et al., 2012; Ruff et al., 2012; Raja et al., 2020).

In relation to human and animal health, biogenic amines are compounds that, when in high concentrations, can cause toxic events. Miles and Wilson (2000) analyzed the effects of the amines cadaverine, histamine, putrescine, spermidine, spermine, tyramine, tryptamine and phenylethylamine at different concentrations (0 to 1500 ppm) in chicken receiving pre-established doses of these compounds and the animals' performance presented no changes.

Figure 12.1. Main types of food waste of animal origin.

Polyamines (putrescine, spermidine and spermine) are present in different levels in foods, and studies indicate they can cause toxicosis when ingested by animals (Bellaver, 2009). Investigations with commercial diets for chickens showed high amounts of biogenic amines, especially phenylethylamine, putrescine, cadaverine, histamine and tyramine (Keirs and Bennett, 1993; Poole, 1994; Tamim and Doerr, 2003), resulting in reduced performance, intestinal disorders and birds with intense symptoms of gizzard and proventriculus erosion (Keirs and Bennett, 1993; Poole, 1994; Barnes et al., 2001), reduced growth rate and presence of excessive mucus in the upper intestine (Tamim and Doerr, 2003). Spermine, for example, is classified as toxic when inserted in the diet at around 0.2% (Sousa Dias and Smith, 1995).

Tamim and Doerr (2003) revealed the presence of amines such as phenylethylamine, putrescine, cadaverine, histamine and tyramine in several animal diets such as bird, meat and fish by-product flours. In birds, they found average levels of amines as follows: 12.5 mg/kg of phenylethylamine, 92 mg/kg of putrescine, 154 mg/kg of cadaverine, 36 mg/kg of histamine and 50 mg/kg of tyramine. In humans, the ingestion of foods with high concentrations of biogenic amines can induce digestive problems, allergies and headaches, but the toxic action depends on the individual's sensitivity (Tofalo et al., 2016).

Tyramine and β-phenylethylamine are involved in the development of hypertensive crises, characterized by elevated blood pressure, heart failure or, in more severe cases, cerebral hemorrhage (Til et al., 1997; Mohan et al., 2009; Kalac, 2009) (Fig. 12.2). Histamine, on the other hand, can cause food poisoning, with allergic manifestations such as difficulty in breathing, intense itching, rashes, fever and hypertension (Hernandez-Jover et al., 1997; Yongmei et al., 2009; Naila et al., 2009).

Regarding fish farming, it is known that environmental problems are recurrent since a high load of fish by-product residues (head, liver, bones, skin, viscera) is generated from the filleting processes, canning, smoking, salting, among others (Visentainer et al., 2000; Oetterer, 2002; Souza et al., 2005; Sockalingam and Abdullah, 2015; Merina et al., 2017). Another way of using waste is to obtain

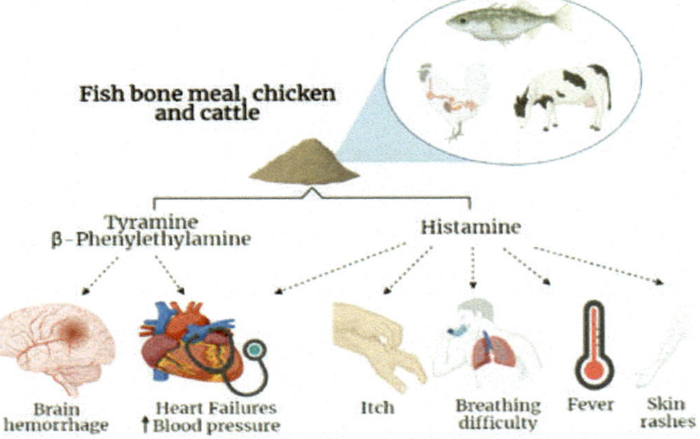

Figure 12.2. Toxic effects of biogenic amines in animal waste flours.

gelatin from fish skins, bones and scales (Huang et al., 2016), whose characteristics have been very similar to conventional gelatins (Al-Saidi et al., 2012; Gelatin Handbook, 2012).

Raja et al. (2020) investigated the anticancer properties of skin crude extracts from the marine catfish *Tachysurus dussumieri* on human colon cancer cell line (HT-29) by the MTT assay. Additional examinations revealed DNA fragmentation analyzed by flow cytometry and showed that the skin extracts cause chromatin condensation, apoptotic cell death and promote cell cycle arrest of colon cancer cells in G0/G1 phase.

An alternative by-product that can be used as a nutriment, as well as for cosmetics, fertilizers and bioceramic bases industry, is the eggshell (Murakami, 2006). The eggshell has three layers: (1) the cuticle (the outermost), (2) a spongy limestone layer, and (3) an inner lamellar (or nipple) layer. Its chemical constitution presents (by weight): 94% calcium carbonate, 1% magnesium carbonate, 1% calcium phosphate and 4% organic matter (Stadelman, 2000).

The Natural Eggshell Membrane (NEM®) is an innovative ingredient rich in glycosaminoglycans and proteins for the maintenance of joints and connective tissues. The toxicological safety of NEM® was analyzed for cytotoxicity, genotoxicity, acute oral toxicity and 90-day repeated-dose oral toxicity. NEM® did not exhibit cytotoxicity at a dose of 100 μg, showed no genotoxic effects and a single dose of up to 2,000 mg/kg body weight did not showed signs of acute oral toxicity in rats, suggesting it may be safe for human ingestion (Ruff et al., 2012).

Products of vegetal source

Cereals

Brazil is one of the largest cereal producers and has intensified its grain production, using genetic improvement, implementation of technology in the countryside and soil management to improve agricultural production. Rice and corn are the two main grains in Brazil according to the estimates of the Systematic Survey of Agricultural Production (Levantamento Sistemático da Produção Agrícola, LSPA) (IBGE, 2020).

Rice, the second most-produced cereal, is widely cultivated, occupying about 11% of the world's arable land. It is an excellent source of energy because of its high concentration of starch (with low lipid content) and provides proteins, vitamins and minerals (Walter et al., 2008; Prasad et al., 2017). Since it is generally cultivated in flooded soils, the excess water causes greater mobilization of arsenic (especially the inorganic one). Consequently, an increased content of arsenic in plants and their grains has been detected. As3+, the most toxic species found in food, has high water solubility, which increases its mobility in the soil, being efficiently absorbed by the roots, reaching the grains and therefore, introduced in the diet (Souza et al., 2015). It is worth noting that the long-term intake of high doses of arsenic has been associated with toxic effects such as skin lesions, cancers (skin, lung, bladder and kidney), cardiovascular diseases (hypertension, arrhythmias and vascular dysfunction), glucose metabolism imbalance and diabetes (Martinez et al., 2011; Molin et al., 2015). The toxicity of rice has been directly linked to the presence of high levels of arsenic (EFSA, 2010) and the *European Food Safety Authority* demonstrated that children fed with a rice-based infant formula are exposed to a higher intake of inorganic arsenic than other consumers, since children consume more food per body weight when compared to adults.

Corn, the third most important food crop, after rice and wheat, is cultivated mainly in the United States, China and Brazil, which produced 563 of the 717 million tons/year (Pretes et al., 2019). The toxicological effect of the aqueous extract of corn silk (*Stigma maydis*), frequently used in folk medicine, was evaluated in Wistar rats. The study exhibited a significant increase in serum levels of white blood cells, platelets, lymphocytes and high-density lipoprotein cholesterol, while the concentrations of total cholesterol, low-density lipoprotein cholesterol and atherogenic index value were significantly reduced. These findings are suggestive of non-hepatotoxic potential of the extract, suggesting a safe use (Saheed et al., 2015).

For the first time, Guerra et al. (2019) demonstrated the effects of corn oil on the prostate after oral administration, indicating its interference in homeostasis and revealing its potential as an endocrine disruptor. Treatment with corn oil increased the expression of positive cells for estrogen receptors (ERα and ERβ), and the difference in the expression of these hormone receptors led to prostate morphological changes.

Another corn-derived product is corn starch fiber, used as dietary fiber, and its toxicological potential was analyzed at doses of 1,000, 3,000 or 10,000 mg/kg/d for 90 d. The animals did not show relevant toxicological or biological signs, revealing that corn starch fiber is an edible non-toxic material, even at high levels (Crincoli et al., 2016).

Edible carboxymethyl cellulose (CMC) is prepared from corn agricultural residues. Its subchronic oral toxicity was evaluated in Swiss albino mice at daily oral doses ranging from 5 to 20 mg/g body weight/day. The results showed no treatment-related changes in terms of hematological and biochemical parameters. Weekly body weight showed no significant differences between control and CMC-treated mice, as well as no microscopical abnormalities were detected in the liver, heart, lungs and kidneys (Mondal and Yeasmin, 2016).

Fruits and vegetables

Fruits and vegetables are essential for a healthy life. Vegetables are herbaceous plants. The green parts used are called vegetables; when its fruit or seed is used, it is called a legume and the root parts are the tubers or rhizomes (Brasil, 2020). They are rich sources of vitamins, fibers, minerals, polyphenolic compounds, glucosinolates and other bioactive compounds (Delaquis and Mazza, 1998; Farnham, 2003; Ferreira et al., 2019). Experimental *in vitro* and *in vivo* tests have shown the protective effects and therapeutic attributes of carotenoids, coumarins, flavonoids, glucosinolates and phenolic compounds terpenes (Pereira-Freire et al., 2018; Silva et al., 2020; Ferreira et al., 2021).

Non-conventional parts of fruits and vegetables are generally discarded, but they can be an alternative source of nutrients because of their high nutritional value, especially for people from poor regions as a way to give better opportunities and improve their quality of life (Pereira-Freire et al., 2018). It is also known that some leaves, which are not usually included in daily diets, are considered excellent sources of fibers to prevent cardiovascular and gastrointestinal diseases (Espíndola, 1987; Pereira-Freire et al., 2019). The fruits of *Mauritia flexuosa* a palm tree widely distributed in South America can be cited as a good example, especially in the Amazon region and Brazilian Cerrado, where it has demonstrated high economic potential for biotechnology development (Pereira-Freire et al., 2016). The antioxidant analysis of *M. flexuosa* fruits and its by-products showed promising chemopreventive potentialities, and the peels demonstrated higher quantities of bioactive compounds and phenolic substances before and after *in vitro* bioaccessibility investigation. Since the processing of *M. flexuosa* fruits generates a large quantity of agricultural residues such as peels, endocarps and seeds, most of them are commonly discarded or are used as feed for ruminant animals only, especially after the production of sweets and oil extraction. Consequently, it is extremely important to explore the nutritional characteristics of these by-products for human/livestock foods and to install biofriendly techniques and sustainable biotechnology handling of natural resources (Pereira-Freire et al., 2018).

The production of fruits and vegetables is concentrated in the Brazilian South and Southeast regions (75% of the total). Following the processing of fruits and vegetables, waste generated include leaves, stalks, bark and seeds, which are usually discarded as common waste or for animal feed (Gondim et al., 2005; Reetz, 2014; Pereira-Freire et al., 2018, 2019). Some studies using unconventional parts of fruits and vegetables, such as peels, rinds and by-products of tubers, roots and fruits are used for the production of flour, cakes and cookies, from these by-products (Fernándes-López et al., 2005; Guimarães et al., 2010; Aquino et al., 2016). Additionally, the conventional parts of raw or processed fruits and vegetables, as well as their residues are similar sources of several bioactive compounds as lectins, phenolic compounds, terpenoids and essential oils (Delaquis and Mazza, 1998; Farnham, 2003; Pereira-Freire et al., 2018, 2019), which can reduce the action of gram-negative and positive bacteria (Lee et al., 2003; Mahfuzul Hoque et al., 2007; Siqueira et al., 2014).

Agro-industrial residues generate natural antimicrobials that control the growth of *Listeria monocytogenes* (Listeriaceae), a pathogenic bacterium that has become a major challenge for the meat industry that has drawn the attention of national and international health surveillance agencies. It was possible to observe the inhibition halos of *L. monocytogenes* (Corrêa, 2011; Martin, 2011). Arbos et al. (2013) also investigated the antimicrobial action of peels and kernels from mango (*Mangifera indica*, Anacardiaceae), whose antibiotic activity was attributed to the phenols.

The antimicrobial action of unconventional fruit and vegetable residues is mainly due to the presence of bioactive compounds such as polyphenols and glucosinolates that disrupt the bacterial outer membranes, generate complexes with the cell wall, interact with the genetic material, inactivate enzymes and interrupt the flow of protons, electrons and the active transport (Martin, 2011). These compounds are also chemical defense mechanisms against carcinogens (Morant et al., 2008), the chemoprevention, which involves natural or synthetic pharmacological agents to suppress, interrupt or reverse carcinogenesis' early stages or alter cancer growth and aggressiveness, implicating disruption of one or more steps (initiation, promotion and progression) of the disease. Such agents include drugs, vitamins or *in natura* substances used to attempt to reduce the risk of cancers or delay the development or recurrence of them (Melo-Cavalcante et al., 2019; Ferreira et al., 2019).

Other compounds commonly found in these foods are isothiocyanates, whose function is to act as hepatic enzyme inducers of phases 1 and 2 during the detoxification of xenobiotics (Morant et al., 2008; Ferreira et al., 2019), which clearly improves the cellular performance of biochemical defense processes (Volden et al., 2008). Isothiocyanates can be divided into six phytocompounds. Such a classification takes into consideration their biological activities based on molecules found in our diets, such as allyl, benzyl, phenethyl, erucin, iberin and sulforaphane (Yao et al., 2011). Sulforaphane (4-methyl-sulfinyl-butyl-glucosinate), the most studied of the isothiocyanates, is derived from glucoraphanin (4-methyl-sulfinyl-butyl-glucosinate), being one of the most widely studied and commonly found isothiocyanates in vegetables from *Brassica* spp., especially in broccoli (0.8–21.7 mmol/g of dry matter), cabbage and cauliflower, members of the Brassicaceae family (Ferreira et al., 2019).

The interaction of the isothiocyanate central carbon of sulforaphane with cysteine residues of Kelch-like ECH-associated protein-1 (Keap-1) results in the stabilization of (erythroid-derived 2)-like 2 (Nrf2) and higher expression of the phase II metabolism enzyme responsible for antioxidant activity. Nrf2 switches antioxidant proteins' expression, detoxification enzymes and xenobiotic transporters as a way of protection against oxidative damages initiated by chemical lesion, hypoxia and inflammation. Such antioxidant proteins act as a cytoprotective antioxidant defense system due to the intensifying of transcription genes associated with intracellular redox status restoration and prevention of EROs-induced damages. Nrf2 and antioxidant proteins have important functions in the control of phase II enzymes' expression (Bryan et al., 2013; Ferreira et al., 2019). Interestingly, sulforaphane and cruciferous vegetables show relatively atoxic oxidative stress inhibition, delay or blockade *in vivo* carcinogenesis, cause biochemical and epigenetic changes in different stages and grades of prostate cancers, and frequently activate cell death by intrinsic ways of apoptosis. Faster absorption, higher bioavailability and greater plasma levels of sulforaphane were found after consumption of raw broccoli in comparison with processed samples (Ferreira et al., 2019).

Flavonoids are also bioactive glycosides or aglycones usually found in fruits, leaves, seeds and bark. They have low molecular weight and antioxidant activity, which depends on their chemical structure and can be determined by factors such as reactivity being a hydrogen and electron donor, stability of the flavonyl radical, reactivity against other antioxidants and interaction with membranes (Barreiros et al., 2006; Alexandre et al., 2020; Silva et al., 2020).

Chel-Guerrero et al. (2018) verified the toxicity index of extracts from vegetable peels and fruits. Extracts from *Annona squamosa* (Annonaceae) and *Chrysophyllum cainito* (Sapotaceae) exhibited moderate toxicity, while other fruits analyzed in their study showed low toxicity (Chel-Guerrero et al., 2018). These and other plant studies contain flavonoids, anthraquinones and triterpenoids as determined by colorimetric methods. These discoveries indicate tropical fruit peels as underexploited residues and can be the source of bioactive compounds, and our findings provide new information about their pharmacologic potential so that they can be used as raw material for the development of new drugs aimed at treating a variety of ailments (Chel-Guerrero et al., 2018; Pereira-Freire et al., 2018).

Bioassays with *Artemia salina* (Artemiidae) to assess the toxicity of Brazilian cerrado fruit seeds and grains. Extracts from the seeds of araticum (*Annona crassiflora*, Annonaceae), mangaba (*Hancornia speciose*, Apocynaceae), cagaita (*Eugenia dysenterica*, Myrtaceae), jatobá (*Hymenaea stigonocarpa*, Fabaceae) and tucumã (*Astrocaryum aculeatum*, Arecaceae) were considered toxic to *A. salina*, and some of the dose-response curves were very similar to those obtained with pure potassium cyanide standards, while the samples of baru (*Dipteryx alata*), cajá-manga (*Spondias Cytherea*, Anacardiaceae), siriguela (*Spondias purpurea*, Anacardiaceae), tarumã (*Vitex montevidensis*, Lamiaceae) and veludo (*Guettarda viburnoides*, Rubiaceae) were not toxic on *Artemia* nauplii (Fonseca et al., 2013).

Amedu et al. (2016) investigated the toxicological properties of the methanol extract from passion fruit seeds (*Passiflora edulis* var. *flavicarpa*, Passifloraceae). The acute toxicity study revealed that the extract was safe up to 5,000 mg/kg but subchronic analysis revealed a significant increase ($p < 0.05$) in body weight at 3,000 mg/kg and neutrophilia in all exposed groups ($p < 0.05$), while extra hematological parameters showed no significant differences. On the other hand, liver function biomarkers presented mild to moderate changes [alanine aminotransferase (ALT), aspartate aminotransferase (AST), alkaline phosphatase (ALP), and glutamyl transpeptidase (GGT)] ($p < 0.05$) with a slight distortion of hepatic architecture at the highest dose (3,000 mg/kg). No pathological changes in kidneys were observed but such investigations suggest that the extract should be used with caution (Amedu et al., 2016).

Leguminous plants-based products

Food contamination is an important public health problem and, consequently, a cause of great concern for regulations at the industrial and consumer level. Technical knowledge is necessary to understand the scientific and social complexity of toxicological aspects related to food, which requires a descriptive study that takes into account the classes of toxins and toxic substances, such as food adulterants, pesticides, natural toxins, food additives, packaging toxins, industrial pollutants and toxic metabolites generated during the food process. Then, using scientific criteria, it is able to infer the influence of environmental contamination along the food chain, and correlate the contaminants and epidemiological aspects that affect food safety (Morales et al., 2014).

There are a large number of toxins (which act in the short term) associated with food, as well as anti-nutritional factors (which act in the long term), and they have the ability to cause irreversible physiological and/or anatomical changes. After many years of phytochemical, pharmaceutical and clinical studies, several chemical structures widely distributed in the plant kingdom were discovered and are associated with the classical toxicity signs and symptoms (García et al., 2005).

Legumes from the Fabaceae family, for example, has played a leading role as a food source, due to their high protein value, amino acid balance and acceptable content, fibers and minerals and variable concentrations of micronutrients, such as Mn, Cu, Fe, Co, Ni, S, Se and Cd, most of them performing primary roles in the enzymatic metabolism as cofactors (Savón et al., 2005; Garcia et al., 2005). Legumes are the best-adapted plant species. Their adaptability reflects a large amount and diversity of chemical compounds from secondary metabolism which are produced by vegetal cells as a defense against herbivores, phytopathogens and tissue diseases (Valdés and Balbín, 2000). Their components are extremely diverse and change according to the species, regrowth age and environmental conditions (García and Ojeda, 2004). Table 12.1 shows the distribution of the most common secondary metabolites present in leguminous specimens.

Among the most studied compounds—because of their diversity and distribution—are polyphenols and non-protein nitrogen compounds, members with the greatest toxic potential (García and Ojeda, 2004). Polyphenols are compounds by tannins (Mimosoid subfamily) and coumarins (Faboid subfamily), both with structures widely disseminated in these plants, while the second group consists of alkaloids and non-protein amino acids.

Since 1992 reports detail common cases describing toxicological conditions in birds, pigs, horses and rabbits fed with a diet based on taniferous plants (Mueller-Harvey and Mc Allan, 1992). Moreover, many secondary metabolites simultaneously cause nutritional disorders because they constitute Anti-Nutritional Factors (ANFs), which, in turn, induce severe toxic conditions after oral ingestion. Figure 12.3 shows the structure of a condensed tannin present in leguminous specimens and the probable harmful mechanism of action.

Coumarins have very stable conjugated flat structures (which give them a long half-life in the body), if concentrations in the edible biomass are significant. Some coumarins can also cause phototoxicity in the presence of ultraviolet light due to

Table 12.1. Most common secondary metabolites in leguminous specimens.

Secondary metabolites	Presence
Simple phenols	High
Taninns	High
Flavonoids	High
Cumarins	Low
Cyanidins	High
Non-protein amino acids	Low
Triterpenoids	High
Alkaloids	High
Saponins	High
Cyanogens	Low
Cardenolides	Low

Structure

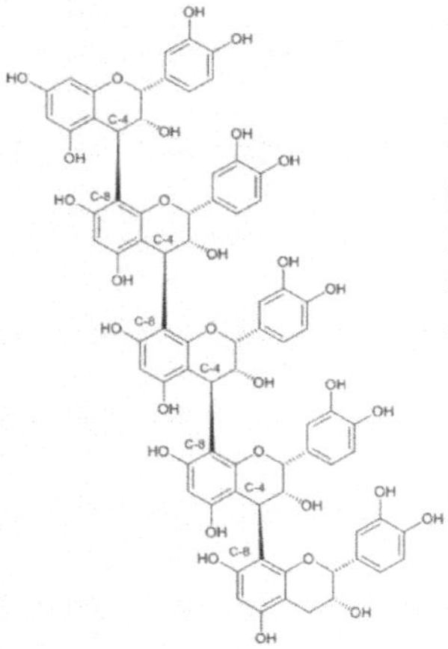

Mechanism of action

Interference on the enzymatic oxidation of the hydroxyl groups (-OH) due to the structural complexity of the polyphenolic units. They can cause irreparable membrane damages, obstruction in the peripheral nervous system and imbalance of the enzyme activity.

Figure 12.3. Structure of the "linear" condensed tannin present in leguminous plants with toxic potential and its mechanism of action.

Structure

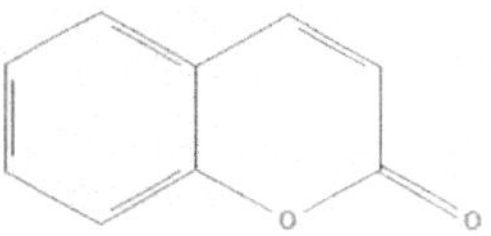

Mechanism of action

They interfere in the blood clotting process, through direct action on factors VII, IC and X, and in the conversion of prothrombin to thrombin.

Figure 12.4. Structure of a coumarin and its mechanism of action.

allergic reactions. Figure 12.4 presents the classical structure of the pyran pattern of coumarin with severe toxic properties.

Allergenicity is an important and typical aspect of coumarin toxicity but is not well established in terms of acute toxicity. It arises especially when members of the Amarcadiaceae (urushiol), Asteraceae (thiophenes, sesquiterpenlactone), Hypericaceae (hypericin) and Apiaceae (furanocoumarins) are used. Almost 30 years ago, Lampe and McCann (1985) had already registered the allergenic capacity of many legumes and they also cited their mutagenic action.

And when describing DNA damage, many leguminous plants like celery, turnip, etc., contain furocoumarins which, when activated by light, produce powerful carcinogens (Vega, 1986). The quinones, widely distributed in nature (e.g., rhubarb), can act as electron donors or acceptors to generate semiquinones, which in turn directly react with DNA or indice the creation of superoxides, affecting the lipid peroxidation (De La Rosa Acosta, 2005) (Table 12.2).

Table 12.2. Main toxic compounds present in legumes.

Source	Type of substance
Toxic substances of natural origin	Lectins Phytates Oxalates Cyanogenic glycosides Thioglycosides Oligosaccharides Enzyme inhibitors Saponins Vice and Convicine
Toxic of intentional origin (additives)	Preservatives Dyes Flavor enhancers Antioxidants
Toxic of accidental origin	Pesticides - Organochlorine - Carbamates - Nicotinic - Pyrethrins

Additionally, lectins are ubiquitously distributed glycoproteins characterized for their carbohydrate-binding properties, the most investigated plant molecules. They belong to the family Leguminosae and most of them have been purified from mature seeds. Legume lectins have broad specificity in binding with carbohydrate moieties and glycoconjugates. About 2 to 10% of the protein content in leguminous seeds is represented by lectins, a type of glycoproteins with the ability to agglutinate erythrocytes in a way similar to the antibodies. Some of these were found to be toxic or antinutritional for man and animals. So, the ability of lectins to selectively bind to different types of blood cells has also been proposed as a means to predict the oral toxicity of these proteins. It has been suggested that the hemagglutinating ability could form a basis for *in vitro* screening of potentially toxic lectins (Vasconcleos and Oliveira, 2004).

Lectins have been identified in animals, plants and microorganisms, and the wide distribution of lectins in all tissues of plants and their ubiquitous presence in the plant kingdom suggest important roles for these proteins. One possible physiological function that has emerged is the defensive role of these carbohydrate-binding proteins against phytopathogenic microorganisms, phytophagous insects and plant-eating animals (Vasconcleos and Oliveira, 2004; Katoch and Tripathi, 2021). Based on their carbohydrate-binding specificity lectins have been classified as (i) Mannose/glucose binding; (ii) Galactose/*N*-acetyl-D-galactosamine binding; (iii) *N*-acetyl-D-glucosamine binding; (iv) L-Fucose binding lectin; (v) Sialic acid binding (Katoch and Tripathi, 2021).

The ingestion of lectins causes nutritional adverse effects because they bind to intestinal tissue cells, inhibit the transport and absorption of nutrients and cause the death of enterocytes (Carrillo and Martínez, 2012). Symptoms in humans due to lectin ingestion occurs mainly at the gastrointestinal level and nausea, bloating, vomiting and diarrhea characterize the oral acute toxicity of lectins on humans exposed to them. In experimental animals fed on diets containing plant lectins the evident symptoms are loss of appetite, decreased body weight and eventually death (Vasconcleos and Oliveira, 2004; Dolan et al., 2010). Many lectins either directly or indirectly cause profound morphological and physiological modifications in the small intestine. Such alterations characteristically lead to increased shedding of brush border membranes, accelerated cell loss and shortened, sparse and irregular enterocyte microvilli (Vasconcleos and Oliveira, 2004). In addition to the disruptive effects on cell membranes, lectins have been shown to inhibit various intestinal and brush border enzymes. Sucrose, maltase, alkaline phosphatase, leucine aminopeptidase and g-glutamyltransferase, all suffered a significant decrease in activities by inclusion of some lectins in a basal diet given to rats (Vasconcleos and Oliveira, 2004).

Since many of these plants are eaten raw, these plant lectins are considered non-toxic for humans and mammals in general. However, some legume lectins, e.g., ConA and PHA have been known for exerting toxic effects in mammals (Vasconcelos and Oliveira, 2004). This indicates the need of conducting toxicity studies concerning the safety issues associated with genetically engineered plants with lectin genes.

The most known lectins extracted from leguminous plants were obtained from the genus *Vicia* and *Erythrina* (Fabaceae), *Physalis vulgaris* (Solanaceae), *Pisum sativum* (Fabaceae), *Lens culinaris* (Fabaceae) and *Lathyrus odoratus* (Fabaceae) (Revoredo, 2017).

López-Sánchez et al. (2010) conducted a study that evaluated the effect of semi-pure fractions of bean lectins on different types of human cancer cells and their *in vivo* toxic effects. While the semi-pure fraction showed a dependent concentration cytotoxic action on human cancer cells, the lectin-concentrated fraction showed low oral toxicity, both at acute and subchronic exposure and results suggested that a dose of 50 mg/kg dose could be considered to initiate cancer studies.

Phytates are also usually found in a class in leguminous plants, mainly as the salt of mono and divalent cations (K^+, Mg^{2+} and Ca^{2+}), which accumulate in seeds during the ripening period. Phytate is considered the main storage pattern of phosphate and inositol in seeds and grains (Fekadu and Ratta, 2014), and the presence of these substances affects the bioavailability of minerals.

Oxalates, another class of mineral sequestering agents, are present more in spinach, rhubarb, beet leaves, tea and cocoa and at lower levels in lettuce, celery, squash, cauliflower, carrots, potatoes, peas, beans and turnip (Benevides et al., 2011). These substances result from the combination of vegetable sodium oxalate with calcium and magnesium in the body and are toxic because calcium ions generate calcium oxalate (CaC_2O_4), a very poorly soluble salt (Fekadu and Ratta, 2014). Diets rich in oxalic acid cause tissue irritation by the production of small insoluble crystals, especially in the stomach and kidneys. In addition, oxalates can cause hypocalcemia due to the ability to bind the blood calcium. There are registers of acute poisoning with oxalates in humans, causing corrosive gastroenteritis, convulsions, shock, decreased plasma calcium, kidney damage, breathing difficulties, abdominal pain, nausea, vomiting, diarrhea, coma and death (if the coronary system collapses). This type of intoxication is unlikely, as it would involve excessive ingestion of oxalate-enriched foods (5 kg of rhubarb, 500 g of spinach leaves or 2.5 kg of tomato, for example) (Dolan et al., 2010).

The cyanogenic glycosides are amino-acid-derived constituents produced as secondary metabolites found in at least 2000 plant species, whose hydrolysis releases hydrogen cyanide, a process called cyanogenesis. In the presence of water, the non-toxic cyanogenic glycosides are hydrolyzed by the enzyme producing cyanohydrins which quickly decompose to the toxic hydrogen cyanide. In this way, cyanogenic plants are protected against predators. Cyanogenic glycosides, cyanohydrins and hydrogen cyanide are collectively known as cyanogens (The Government of the Hong Kong Special Administrative Region, 2007).

Some leguminous plants are rich in amygdalin (almonds), linamarin (cassava, lima beans), lotaustralina (cassava, lima beans) (Ballhorn, 2011). The toxicity of cyanogenic glycosides occurs during the production of cyanide after eating foods that contain these natural toxins. Plant species that produce these cyanogenic glycosides usually also have a corresponding hydrolytic enzyme (β-glucosidase) (Dolan et al., 2010).

Cyanide inhibits cytochrome oxidase and the oxygen-saturated hemoglobin fails to release its oxygen at the tissues since this phenomenon depends on the enzyme cytochrome oxidase is inhibited by cyanide. Cyanide prevents oxygen utilization leading to cytotoxic anoxia. Additionally, increases in blood glucose and lactic acid levels and a decrease in the ATP/ADP ratio are observed, indicating a shift from aerobic to anaerobic metabolism (Nyirenda, 2020)

The acute lethal dose of hydrogen cyanide for humans is reported to be 0.5–3.5 mg/kg body weight. Approximately 50 to 200 mg of free cyanide constitutes a lethal dose for an adult man. In humans, the clinical signs of acute cyanide intoxication can include: rapid respiration, drop in blood pressure, rapid pulse, dizziness, headache, stomach pains, vomiting, diarrhea, mental confusion, stupor, cyanosis with twitching and convulsions followed by terminal coma. If untreated death is rarely delayed more than 60 min (WHO, 1993; The Government of the Hong Kong Special Administrative Region, 2007; Nyirenda, 2020).

Health manifestations such as malnutrition, congenital malformations, neurological disorders and myelopathy have been attributed to chronic cyanide toxicity. Adverse effects on the central nervous system are the most prevalent because of the high metabolic demand for oxygen in neurons and its control of respiratory function. Neurological findings among the affected individuals include symmetrical hyperreflexia of the upper limbs, symmetrical spastic paraparesis of the lower limbs, spastic dysarthria, diminished visual acuity, peripheral neuropathy, cerebellar signs and deafness (Tylleskär et al., 1992; Nyirenda, 2020).

Leguminous plants also contain goitrogenic substances (thioglycosides), of which species of *Brassica* (rapeseed, cabbage, turnip), peanuts, sesame, among others, are richer in such molecules, as goitrin, a goitrogenic compound derived from glucosinolates that inhibit iodine organification (Nyirenda, 2020). Individuals suffering from goiter and cretinism due to iodine deficiency may then experience an exacerbation of the condition following continuous dietary cyanide exposure. The endocrine effect may be due to the formation of thiocyanate, a lesser toxic metabolite of cyanide. Thiocyanate blocks iodine uptake and competes with iodide ion (I^-) as a substrate for the thyroid peroxidase, thereby decreasing the iodination of tyrosine to form iodotyrosine by the thyroid gland and inhibiting the formation and release of thyroid hormones.

Additionally, the uptake of iodide from the circulation through the thyroid gland is stimulated by the Thyroid-Stimulating Hormone (TSH) and blocked by thiocyanate, perchlorate and high concentrations of iodide. After iodide uptake by thyroid follicular cells, the oxidation of iodide to iodine can be antagonized by goitrogenic agents such as thiouracil or thiourea. Moreover, the formation of Triiodothyronine (T3) and Tetraiodothyronine (T4) can be blocked by the presence of sulfa thioureas and paraminobenzoic acid (PABA) (Aires, 2018).

Flatulence promoters (e.g., oligosaccharides) are molecules that can be added to a food or produced *in situ*, usually found in beans, chickpeas, peas and lentils. These leguminous contain oligosaccharides that may be not properly digested due to a digestive enzymatic deficiency. Such oligosaccharides are taken to the large intestine where digestion by the microbiota occurs, producing intestinal gases, diarrhea, abdominal pain, cramps, etc. Both cooking and fermentation or germination reduce the total content of these oligosaccharides and reduce flatulence (Dolan et al., 2010).

Faba bean is also an important plant for human consumption, containing a valuable nutritional composition, particularly rich in high-quality protein and providing a balanced diet of lysine-rich protein, carbohydrates, fiber and phytochemicals, but like other legumes, it contains compounds showing an "anti-nutritional" effects (Rizzello et al., 2016). Such beans are rich in two glucosidic aminopyrimidine derivatives—vicine and convicine—which, on hydrolysis of the β-glucosidic bond between glucose and the hydroxyl group at C-5 on the pyrimidine ring, generate the aglycones divicine (2,6-diamino-4,5-dihydroxypyrimidine) and isouramil (6-amino-2,4,5-trihydroxypyrimidine), respectively (Carrillo and Martínez, 2012; Rizzello et al., 2016) (Fig. 12.5).

Vicine and convicine are substances found exclusively in *Vicia* genus and whose concentration in the seeds decreases as they mature, and divicine and isouramil are responsible for favism, an inherited metabolic disorder consisting of a deficiency of glucose-6-phosphate dehydrogenase (G6PD) in erythrocytes (Carrillo and Martínez, 2012), a life-threatening hemolytic crisis that can result from the ingestion of faba beans by susceptible individuals. Since G6PD regulates the production of NADPH in the red blood cell by the hexose monophosphate shunt, individuals affected by G6PD deficiency are unable to regenerate reduced glutathione (GSH) and become vulnerable to the oxidative stress. As a result, G6PD deficiency accelerates normal senescence of erythrocytes, the best natural model of oxidative damage (Arese et al., 2012).

Soybean, on the other hand, contains phytoestrogens. A study evaluated the pubertal toxicity of a diet rich in soy milk during the prepubertal period in male rats, and detected endocrine disruption through the reduction of testosterone levels and the number of Sertoli cells, and a significant increase of degenerated Sertoli and Leydig cells, which indicate phytoestrogens affect the sexual development if consumed during the prepubertal period (Nardi et al., 2017). In a cross-sectional study involving 99 men, Chavarro et al. (2008) showed that dietary intake of soy and isoflavones was inversely related to sperm concentration. Men from higher exposure

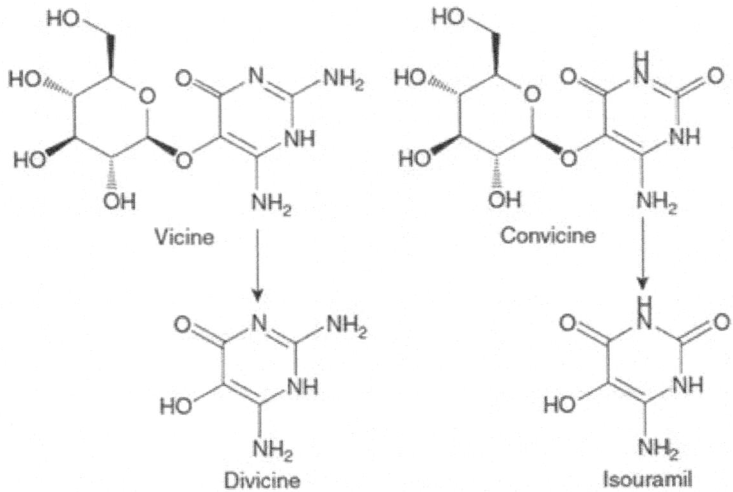

Figure 12.5. Molecular structures of glucosidic aminopyrimidine derivatives found in faba beans.

intake of soy-based foods showed about 41 million sperm cells/ml less than men who did not consume soy-based foods. Research with 19,972 women revealed that the greatest risk of developing benign tumors of the smooth muscle (leiomyoma or myoma uterine) is associated with soy-based formulas consumed during childhood (D' Aloisio et al., 2010).

And finally, mycotoxins, widely present in peanuts, are produced by fungi and with harmful consequences for human and animal health. Fungi of the *Aspergillus*, *Penicillium* and *Fusarium* are producers of several mycotoxins present in distinct stages of the food chain. Among the various toxic effects, hepatic, renal, circulatory, gastrointestinal and nervous systems are the most affected. The most studied toxic mycotoxins, which have maximum residue limits established in several countries and frequently found in grains, are those generated during their cultivation [T-2 toxin, fumonisins (FB1 and FB2), deoxynivalenol and zearalenone] and those from harvest to storage [aflatoxins (AFB1, AFB2, AFG1 and AFG2) and ochratoxin A] (Scussel et al., 2016) (Fig. 12.6).

Levels of aflatoxins are regulated in animal feedstuffs and human foods because of their toxicity worldwide. Due to their prevalence, they are responsible for a major outbreak of acute hepatitis around the world due to the intake of contaminated food, being an important problem for public health. They are one of the highly toxic secondary metabolites derived from polyketides produced by fungal species such as *Aspergillus flavus, A. parasiticus* and *A. nomius* (Iqbal et al., 2014; Kumar et al., 2017) and the level of toxicity associated with aflatoxin varies with the types of toxin as follows: AFTs-B_1 > AFTs-G_1 > AFTs-B_2 > AFTs-G_2 (Jaimez et al., 2000). Once formed, aflatoxins are relatively stable and resistant to heat and therefore, are not eliminated by heat processing (Campagnollo et al., 2016).

They are potent hepatotoxins, mutagens, carcinogens, immunosuppressants, teratogens, have adverse epigenetic effects and are linked with reproductive

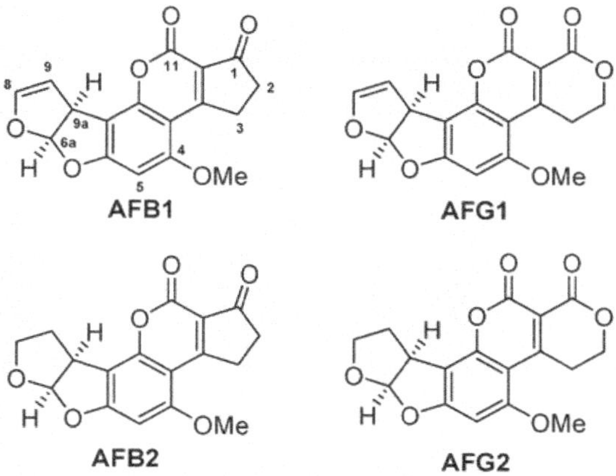

Figure 12.6. Molecular structures of aflatoxins (AF).

dysfunctions and stunted growth. Aflatoxin toxicity is associated with hepatocellular damage and necrosis, cholestasis, hepatomas, acute hepatitis, periportal fibrosis, hemorrhage, jaundice, fatty liver changes, cirrhosis in malnourished children, the disruption of basic metabolic processes and protein synthesis and Kwashiorkor. Early symptoms of hepatotoxicity of the liver caused by aflatoxins involve fever, malaise and anorexia followed by abdominal pain, vomiting and hepatitis (Kumar et al., 2017; Min et al., 2021). Chronic toxicity by aflatoxins comprises immunosuppressive and carcinogenic effects (Qian et al., 2014).

AFTs-B1 is activated by cytochrome p450 monooxygenase system enzymes, which are converted to AFTs-B1-8, 9-epoxide, which is responsible for carcinogenic effects. The hepatocarcinogenicity of aflatoxins is mainly due to lipid peroxidation and oxidative damage to DNA. Once formed, the epoxide can react further to form DNA adducts (aflatoxin-N7-guanine) and induce mutations and cancer. A big number of studies suggest a high correlation between aflatoxin exposure and point mutations at a specific location, the third base of codon 249, of the p53 tumor suppressor gene. This mutation, a transversion of guanine to thymidine, is present in some Chinese and African patients with hepatocarcinoma. On the other hand, though classified as a human carcinogen, there is evidence that aflatoxin requires the presence of cofactors to induce hepatic tumors, as simultaneous chronic infection caused by Hepatitis B or C virus (HBV or HCV) (Haschek et al., 2002; Kumar et al., 2017).
The toxic effects of vegetal compounds are summarized in Fig. 12.7.

Conclusion

Food waste is an inefficient market, the kind of which does not persist in other industries but most waste products and related intoxications can be followed from an epidemiological point of view. Therefore, it is critical to ensure food safety and quality through the establishment and strengthening of national food control systems to protect public health and promote business and food production, considering the severe control and environmental toxicological aspects to predict the origin,

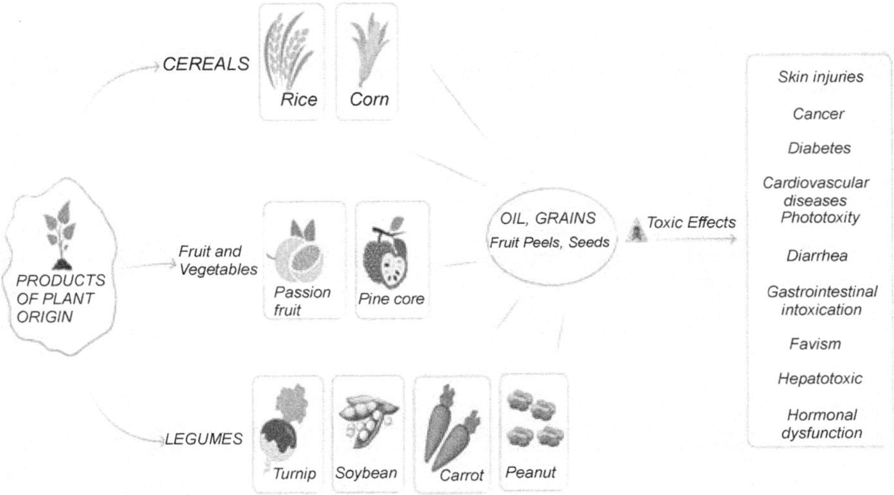

Figure 12.7. Main toxic effects of products of plant origin.

mechanisms of formation and action, the concentration of a toxin in contaminated foods and the frequency of contaminations.

References

Aires, M.M. 2018. Physiology. 4° edition Rio de Janeiro: Guanabara Koogan. 1352.

Alexandre, L.S., Oliveira, M.S., Dittz, D., Sousa, R.W.R., Ferreira, P.M.P., Pessoa, C., Varotti, F.P., Sanchez, B.A.M., Banfi, F.F., Chaves, M.H. and Vieira Júnior, G.M. 2020. Flavonoids, cytotoxic, and antimalarial activities of Dipteryx lacunifera. Rev. Bras. Farmacogn. 30: 544–550.

Al-Saidi, G., Al-Alawi, A., Rahman, M. and Guizani, N. 2012. Fourier transform infrared (FTIR) spectroscopic study of extracted gelatin from shaari (*Lithrinus microdon*) skin: Effects of extraction conditions. Int. Food Res. J. 19(3): 1167–1173.

Amedu, A.M., Wasagu, R.S.U., Lawal, M., Sabir, A.A., Sahabi, S.M. and Zaharadeen, A. 2016. Acute and subchronic toxicity study of methanol seed extract of passion fruit (*Passiflora edulis* var. *flavicarpa*) in albino rats. Food Science and Quality Management 26: 33405–36350-1.

Aquino, J.S., Vasconcelos, M.H., Pessoa, D.C.N.D., Barbosa, J.K., Prado, J.P.S., Magnani, M. and Stamford, T.L.M. 2016. Intake of cookies made with buriti oil (*Mauritia flexuosa*) improves vitamin A status and lipid profiles in young rats. Food Funct. 7(10): 4442–50.

Arbos, K.A., Stevani, P.C. and Castanha, R.F. 2013. Atividade antimicrobiana, antioxidante e teor de compostos fenólicos em casca e amêndoa de frutos de manga. Revista Ceres. 60(2): 161–165.

Arese, P., Gallo, V., Pantaleo, A. and Turrini, F. 2012. Life and death of glucose-6-phosphate dehydrogenase (G6PD) deficient erythrocytes–role of redox stress and band 3 modifications. Transfus. Medi. Hemoth. 39: 328–334.

Ballhorn, D.J. 2011. Cyanogenic Glycosides in Nuts and Seeds overlay panel. Nuts and Seeds in Health and Disease Prevention. Chapter 14: 129–136.

Barnes, D.M., Kirby, Y.K. and Oliver, K.G. 2001. Effects of biogenic amines on growth and incidence of proventricular lesions in broiler chickens. Poultry Sci. 80: 906–911.

Barros, F.D. 2007. Reciclagem de resíduos de origem animal: um estudo qualitativo entre processos contínuos e descontínuos e a geração de odores fugitivos. 136f. Dissertação (Mestrado em Engenharia de Processos Químicos e Bioquímicos). Escola de Engenharia Mauá do Centro Universitário do Instituto Mauá de Tecnologia. São Caetano do Sul: IMT-CEUN, 2007.

Bellaver, C. 2009. Qualidade no processamento em fábricas de farinhas e gorduras animais. EMBRAPA: Encontro técnico Unifrango.

Benevides, C., Souza, M.V., Souza, R.D.B. and Lopes, M.V. 2011. Fatores antinutricionais em alimentos: Revisão. Segurança Alimentar e Nutricional. 18(2): 67–79.

Barreiros, A.L.B.S., David, J.M. and David, J.P. 2006. Estresse oxidativo: relação entre geração de espécies reativas e defesas do organismo. Quím. Nova. 29(1): 113–123.

Braguglia, C.M., Gallipoli, A., Gianico, A. and Pagliaccia, P. 2018. Anaerobic bioconversion of food waste into energy: A critical review. Bioresour. Technol. 248: 37–56.

Brasil. 2020. Ministério da Saúde. Agência Nacional de Vigilância Sanitária. Resolução CNNPA n. 12 de 1978. Disponível em <http://portal.anvisa.gov.br/wps/wcm/connect/e57b7380474588a39266d63fbc4c6735/RESOLUCAO_12_ 1978.pdf?MOD=AJPERES>. Acesso em 01 de nov de 2020.

Bryan, H.K., Olayanju, A., Goldring, C.E. and Park, B.K. 2013. The Nrf2 cell defence pathway, Keap1-dependent and -independent mechanisms of regulation. Biochem. Pharmacol. 85: 705–17.

Campagnollo, F.B., Ganev, K.C., Khaneghah, A.M., Portela, J.B., Cruz, A.G., Granato, D., Corassine, C.H., Oliveira, C.A.F. and Sant'Anaa, A.S. 2016. The occurrence and effect of unit operations for dairy products processing on the fate of aflatoxin M1: a review. Food Control. 68: 310–329.

Carrillo, M.C.C. and Martínez, E. 2012. Toxicologia de los alimentos. Espanha: McGraw-Hill, 467p.

Chavarro, J.E., Toth, T.L., Sadio, S.M. and Hauser, R. 2008. Soy food and isoflavone intake in relation to semen quality parameters among men from an infertility clinic. Hum. Reprod. 23: 2584–2590.

Chel-Guerrero, L.D., Sauri-Duch, E., Fragoso-Serrano, M.C., Pérez-Flores, L.J., Gómez-Olivares, J.L., Salinas-Arreortua, N., Sierra-Palacios, E.D.C. and Mendoza-Espinoza, J.A. 2018. Phytochemical profile, toxicity, and pharmacological potential of peels from four species of tropical fruits. J. Med. Food. 21(7): 734–743.

Conab. 2020. Companhia Nacional de Abastecimento. Acompanhamento da Safra Brasileira. v. 7 - Safra 2019/20 - Sétimo levantamento, Brasília: CONAB.

Corrêa, C.B. 2011. Potencial antimicrobiano de resíduos agroindustriais sobre Listeria monocytogenes. 85p. Dissertação (Mestrado em Ciências) – Escola Superior de Agricultura "Luiz de Queiroz"- Universidade de São Paulo, São Paulo-SP.

Crincoli, A.M., Nikiforov, A.I., Rihner, M.O., Lambert, E.A., Greeley, M.A., Godsey, J., Eapen, A.K. and Ligt, J.L.G. 2016. A 90-day oral (dietary) toxicity and mass balance study of corn starch fiber in Sprague Dawley rats. Food Chem. Toxicol. 97: 57–69.

D'aloisio, A.A., Baird, D.D., Deroo, L.A. and Sandler, D.P. 2010. Association of intrauterine and early-life exposures with diagnosis of uterine leiomyomata by 35 years of age in the sister study. Environ. Health Perspect. 118: 375–381.

De La Rosa Acosta, M.Á. 2005. La luz em Biología: Aplicaciones de interés industrial y agrícola. 18 ed. Espanha: Editorial Universidad de Sevilla. 108p.

Delaquis, P. and Mazza, G. 1998. Functional vegetable products. Technomic. Functional Foods 1(1): 193–201.

Dolan, L.C., Matulka, R.A. and Burdock, G.A. 2010. Naturally occurring food toxins. Toxins. 2(9): 2289–2332.

Elkhalifa, S., Al-Ansari, T., Mackey, H.R. and Mckay, G. 2019. Food waste to biochars through pyrolysis: A review. Resour. Conserv. Recycl. 144: 310–320.

Espíndola, F.S. 1987. Fracionamento dos vegetais verdes e obtenção de concentrados protéicos de folhas (CPF) para suplementação de alimentos e ração animal, com aproveitamento dos subprodutos. 140p. Trabalho de Conclusão de Curso (Graduação em Agronomia) – Universidade Federal de Uberlândia, Uberlândia.

European Food Safety Authority. 2010. Scientific Opinion on Arsenic in Food. EFSA Panel on Contaminants in the Food Chain (CONTAM).

Farnham, M.W. 2003. Vegetable cruciferous - status report. Disponível em: <lillp://www\v.arsrin.uov.npc|s/cqsreporls/crucirerl 201.htm>. Acesso em 01 de nov de 2020.

Fekadu, H. and Ratta, N. 2014. Antinutritional factors in plant foods: potential health benefits and adverse effects. Int. J. Food Sci. Nutr. 3(4): 284–289.

Ferreira, P.M.P., Lopes, L.A.R., Carnib, L.P.A., Sousa, P.V.L., Lugo, L.M.N., Nunes, N.M.F., Silva, J.N., Araujo, L.S. and Frota, K.M.G. 2019. Cruciferous vegetables as antioxidative, chemopreventive and antineoplasic functional foods: Preclinical and clinical evidences of sulforaphane against prostate cancers. Curr. Pharm. Des. 25: 4779–4793.

Ferreira, P.M.P., Drumond, R.R., Silva, J.N., Sousa, I.J.O., Alencar, M.V.O.B., Mata, A M.O.F., Moncao, N.B.N., Cito, A.M.G.L., Carvalho, A.F.F.U., Farias, D.F., Costa, P.M., Nunes, A.M.V., Sousa, J.M.C. and Melo-Cavalcante, A.A.C. 2021. Chemotherapeutic and safety profile of a fraction from *Mimosa caesalpiniifolia* stem bark. J. Oncol. 2021: 1–12.

Fernández-López, J., Zhi, N., Aleson-Carbonell, L., Pérez-Alvarez, J.A. and Kuri, V. 2005. Antioxidant and antibacterial activities of natural extracts: application in beef meatballs. Meat Sci. 69(3): 371–380.

Fonseca, R.C., Souza, N.A., Correa, T.C.L., Garcia, L.F., Reis, L.G.V. and Rodriguez, A.G. 2013. Assessment of toxic potential of Cerrado fruit seeds using *Artemia salina* bioassay. J. Food Sci. Technol. 33(2): 251–256.

García, D.E. and Ojeda, F. 2004. Evaluación de los principales factores que influyen en la composición fitoquímica de Morus alba (Linn.). II. Polifenoles totales. Pastos y Forrajes. 27: 59.

García, D.E., Medina, M.G., Soca, M. and Montejo, I.L. 2005. Toxicidad de las leguminosas forrajeras en la alimentación de los animales monogástricos. Pastos y Forrajes. 28(4): 279–289.

Gelatin Manufacturers Institute of America. 2012. Gelatin Handbook. Membros da GMIA.

Gondim, J.A.M., Moura, M.D.F.V., Dantas, A.S., Medeiros, R.L.S. and Santos, K.M. 2005. Composição centesimal e de minerais em cascas de frutas. J. Food Sci. Technol. 25(4): 825–827.

Guerra, L.H.A., Tamarindo, G.H., Campos, S.G.P., Taboga, S.R. and Vilamaior, P.S.L. 2019. Do mineral and corn oil serve as potential endocrine disruptors in the gerbil prostate? Reprod. Toxicol. 90: 141–149.

Gueri, M.V.D., Souza, S.N.M.S., Kuczman, O., Schirmer, W.N., Buratto, W.G., Ribeiro, C.B. and Besinella, G.B. 2018. Digestão anaeróbia de resíduos alimentares utilizando ensaios bmp. Biofix Sci. J. 3(1): 08–16.

Guimarães, R.R., Freitas, M.C.J. and Silva, V.L.M. 2010. Bolos simples elaborados com farinha da entrecasca de melancia (*Citrullus vulgaris*, sobral): avaliação química, física e sensorial. Ciênc. Tecnol. Aliment. 30(2): 354–363.

Haschek, H.W.M., Rousseaux, C.G. and Wallig. M.A. 2002. Book of Toxicologic Pathology. Academic Press.

Hernandez-Jover, T., Izquierdo-Pulido, M., Veciana-Nogués, T., Marine-Font, A. and Vidal-Carou, C. 1997. Biogenic amine and polyamine contents in meat and meat products. J. Agric. Food Chem. 45(6): 2098–2102.

Huang, C.Y., Kuo, J., Wu, S. and Tsai, H. 2016. Isolation and characterization of fish scale collagen from tilapia (*Oreochromis* sp.) by a novel extrusion-hydro-extraction process. Food Chem. 190: 997–1006.

IBGE 2020 - IBGE. 2019. Para 2020, IBGE prevê recorde de 240,9 milhões de toneladas na safra de grãos. Disponível em: <https://agenciadenoticias.ibge.gov.br/agencia-sala-de-imprensa/2013-agencia-de-noticias/releases/26303-para-2020-ibge-preve-recorde-de-240-9-milhoes-de-toneladas-na-safra-de-graos>. Acesso em: 01 de dez de 2020.

Iqbal, S.Z., Mustafa, H.G., Asi, M.R. and Jinap, S. 2014. Variation in vitamin E level and aflatoxins contamination in different rice varieties. J. Cereal Sci. 60: 352–355.

Jaimez, J., Fente, C.A., Vazquez, B.I., Franco, C.M., Cepeda, A., Mahuzier, G. and Prognon, P. 2000. Application of the assay of aflatoxins by liquid chromatography with fluorescence detection in food analysis. J. Chromatogr. A 882: 1–10.

Jayathilakan, K. Sultana, K., Radhakrishna, K. and Bawa, A.S. 2012. Utilization of byproducts and waste materials from meat, poultry and fish processing industries: a review. J. Food Sci. Technol. 49(3): 278–293.

Kalac, P. 2009. Recent advances in the research on biological roles of dietary polyamines in man. J. Appl. Biomed. 7(2): 65–74.

Katoch, R. and Tripathi, A. 2021. Research advances and prospects of legume lectins. J. Biosci. 46(4): 104.

Keirs, R.W. and Bennett, L. 1993. Broiler performance loss associated with biogenic amines. Proc. Maryland Nutr. Conf. Univ. Maryland, College Park, MD 31–34.

Kiran, E.U., Trzcinski, A.P., Ng, W.J. and Liu, Y. 2014. Bioconversion of food waste to energy: A review. Fuel. 134: 389–399.

Kumar, P., Mahato, D.K., Kamle, M., Mohanta, T.K. and Kang, S.G. 2017. Aflatoxins: A global concern for food safety, human health and their management. Front Microbiol. 17(7): 2170.

Lampe, K.F. and Mc Cann, M.A. 1985. AMA Handbook of Poisonous and Injurious Plant. Chicago: American Medical Association. 500 p.

Lee, Y.L., Cesario, T., Wang, Y., Shanbrom, E. and Thrupp, L. 2003. Antibacterial activity of vegetables and juices. Nutrition. 19(1): 994–996.

López-Sánchez, C., López-Martínez, F.J., Castañeda-Cuevas, A.L., Yllescas-Gasca, L., Ferriz-Martínez, R.A., Torres-Arteaga, I.C. and García-Gasca, T. 2010. Evaluación de la Toxicidad *in Vitro* e *in Vivo* de Lectinas de Frijol Tépari. CIENCIA@UAQ 3(1): 3–13.

Mahfuzul Hoque, M.D., Bari, M.L., Inatsu, Y., Juneja, V.K. and Kawamoto, S. 2007. Antibacterial activity of guava (Psidium guajava L.) and neem (Azadirachta indica A. Juss.) extracts against foodborne pathogens and spoilage bacteria. Foodborne Pathog. Dis. 4(1): 481–488.

Martinez, V.D., Vucic, E.A., Becker-Santos, D.D., Gil, L. and Lam, W.L. 2011. Arsenic exposure and the induction of human cancers. J. Toxicol. 2011: 431287.

Martin, J.G.P. 2011. Atividade antimicrobiana de produtos naturais: erva-mate e resíduos agroindustriais. 100p. Dissertação (Mestrado em Ciências) – Escola Superior de Agricultura "Luiz de Queiroz" – Universidade de São Paulo, São Paulo-SP.

Merina, P., Suguna, P.R., Karpuram, P., Vijaylakshmi, J.V. and Renuka, M. 2017. Extraction and characterization of gelatin: a funcional biopolymer. Int. J. Pharm. Pharm. Sci. 9(9): 239–242.

Melo-Cavalcante, A.A.C., Sousa, L.R., Alencar, M.V.O.B., Santos, J.V.O., Mata, A.M.O.F., Paz, M.F.C.J., Ricardo, M.C., Nunes, N.M.F., Ahmed, M.I., Mendes, A.N., Goncalves, J.C.R., Silva, F.C.C., Ferreira, P.M.P. and Sousa, J.M.C. 2019. Retinol palmitate and ascorbic acid: Role in oncological prevention and therapy. Biomed. Pharmacother. 109: 1394–1405.

Miles, R.D. and Wilson, H.R. 2000. Biogenic amines: I. Influence of feeding various dietary concentrations of eight biogenic amines individually or in combination to broilers. Poultry Sci. 79: suppl. 125.

Min, L., Fink-Gremmels, J., Li, D., Tong, X., Tang, J., Nan, X., Yu, Z., Chen, W. and Wang, G. 2021. An overview of aflatoxin B1 biotransformation and aflatoxin M1 secretion in lactating dairy cows. Animal nutrition (Zhongguo xu mu shou yi xue hui) 7(1): 42–48.

Mohan, C.O., Ravishankar, C.N., Srinivasa Gopal, T.K., Ashok Kumar, K. and Lalitha, K.V. 2009. Biogenic amines formation in seer fish (*Scomberomorus commerson*) steaks packed with O_2 scavenger during chilled storage. Food Res. Int. 42(3): 411–416.

Molin, M., Ulven, S.M., Meltzer, H.M. and Alexander, J. 2015. Arsenic in the human food chain, biotransformation and toxicology—review focusing on seafood arsenic. J. Trace Elem. Med. Biol. 31(249).

Mondal, I. and Yeasmin, S. 2016. Toxicity study of food-grade carboxymethyl cellulose synthesized from maize husk in Swiss albino mice. Int. J. Biol. Macromol. 92: 965–971.

Morales, J.J., Ligardo, Y.M. and Sierra, C.S. 2014. Tóxicos en el ambiente y la seguridad alimentaria. Cap&Cua: Ciencia, Tecnología y Cultura 11: 16–23.

Morant, A.V., Jorgensen, K., Jorgensen, C., Paquette, S.M., Sánchez-Perez, R., Moller, B.L. and Bak, S. 2008. beta-Glucosidases as detonators of plant chemical defense. Phytochemistry 69(9): 1795–813.

Mueller-Harvey, I. and Mc Allan, A.B. 1992. Their biochemistry and nutritional properties. *In*: Advances in Plant Cell Biochemistry and Biotechnology I. Morrison London IM: JAI Press Ltd. 151p.

Murakami, F.S. 2006. Estudo termoanalítico entre carbonato de cálcio industrial e carbonato de cálcio obtido da casca do ovo. V Congresso Brasileiro de Análise Térmica e Calorimetria – V CBRATEC.

Naila, A., Flint, S., Fletcher, G., Bremer, P. and Meerdink, G. 2010. Control of biogenic amines in food-existing and emerging approaches. J. Food Sci. 75(7): 139–150.

Nardi, J., Moras, P.B., Koeppe, C., Dallegrave, E., Leal, B. and Rossato-Grando, L.G. 2017. Prepubertal subchronic exposure to soy milk and glyphosate leads to endocrine disruption. Food Chem. Toxicol. 100: 247–252.

Nyirenda, K.K. 2020. Toxicity potential of cyanogenic glycosides in edible plants. *In*: Pınar Erkekoglu and Tomohisa Ogawa (eds.). Medical Toxicology. IntechOpen.

Oetterer, M. 2002. Industrialização do pescado cultivado. Guaíba, RS: Livraria e editora Agropecuária.

Pereira-Freire, J.A., Barros, K.B.N.T., Lima, L.K.F., Martins, J.M., Araujo, Y.C., Oliveira, G.L.S., Aquino, J.S. and Ferreira, P.M.P. 2016. Phytochemistry profile, nutritional properties and pharmacological activities of *Mauritia flexuosa*. J. Food Sci. 81: R2611–R2622.

Pereira-Freire, J.A., Oliveira, G.L.S., Lima, L.K.F., Ramos, C.L.S., Medeiros, S.R.A., Lima, A.C.S., Teixeira, S.A., Oliveira, G.A.L., Nunes, N.M.F., Amorim, V.R., Lopes, L.S., Rolim, L.A., Costa-Junior, J.S. and Ferreira, P.M.P. 2018. *In vitro* and *ex vivo* chemopreventive action of *Mauritia flexuosa* products. Evid.-Based Complementary Altern. Med. 2018: 1–12.

Pereira-Freire, J.A., Amorim, V.R., Ribeiro, F.M.C., Medeiros, S.R.A., Silva, J.N. and Ferreira, P.M.P. 2019. Compostos bioativos e potencial nutracêutico do fruto de buriti (*Mauritia flexuosa* L.) na terapia coadjuvante em portadores de dislipidemia. In: Nayara Araújo Cardoso; Renan Rhonalty Rocha; Maria Vitória Laurindo. (Org.). As Ciências Biológicas e da Saúde na Contemporaneidade 2. 1ed. Ponta Grossa: Atena Editora. 2: 103–117.

Poole, D. 1994. Biogenic amines: an update. Proc. Western Poult. Dis. Conf. Univ. California, Davis, CA, 40–42.

Prasad, R., Shivay, Y.S. and Kumar, D. 2017. Current status, challenges, and opportunities in rice production. pp. 1–32. *In*: Rice Production Worldwide. Springer, Cham.

Prestes, I.D., Rocha, L.O., Nunes, K.V.M. and Silva, N.C.C. 2019. Fungi and mycotoxins in corn grains and their consequences. Sci. Agric. 10(4): 559–570.

Püssa, T. 2013. Toxicological issues associated with production and processing of meat. Meat Sci. 95: 844–853.

Qian, G., Tang, L., Guo, X., Wang, F., Massey, M.E., Su, J. et al. 2014. Aflatoxin B1 modulates the expression of phenotypic markers and cytokines by splenic lymphocytes of male F344 rats. J. Appl. Toxicol. 34: 241–249.

Raja, K., Martin, L.C., Bose, L., Sahayanathan, G.J., Padmanaban, D. and Chinnasamy, A. 2020. Anti-proliferative and apoptotic effects of by-product (skin extract) from marine catfish *Tachysurus dussumieri*. Biocatal. Agric. Biotechnol. 29(101816).

Reetz, E.R. 2014. Anuário brasileiro de hortaliças 2014. Santa Cruz do Sul: Editora Gazeta Santa Cruz. 188p.

Revoredo, P.G.R. 2017. Químicos naturales em alimentos vegetales que atentan contra la inocuidade alimentaria. 62p. Tese (Doutorado em Engenharia agroindustrial), Universidad Nacional de Trujillo, Peru.

Rizzello, C.G., Losito, I., Facchini, L., Katina, K., Palmisano, F., Gobbetti, M. and Coda, R. 2016. Degradation of vicine, convicine and their aglycones during fermentation of faba bean flour. Sci. Rep. 6: 32452.

Ruff, K.J., Endres, J.R.E., Clewell, A.E., Szabo, J.R. and Shauss, A.G. 2012. Schauss, Safety evaluation of a natural eggshell membrane-derived product. Food Chem. Toxicol. 50(3–4): 604–611.

Savón, L., Odilla, G., Ojeda, F. and Scull, I. 2005. Harinas de follajes tropicales: una alternativa potencial para la alimentación de espécies monogástricas. Pastos y Forrajes. 28(1): 69–79.

Saheed, S., Oladipipo, A.E., Abdulazeez, A.A., Olarewaju, A., Ismaila, N.O., Emmanuel, I.A., Fatimah, Q.D. and Aisha, A.Y. 2015. Toxicological evaluations of Stigma maydis (corn silk) aqueous extract on hematological and lipid parameters in Wistar rats. Toxicol. Rep. 2: 638–644.

Scussel, V.M., Beber, M. and Souza, K.K. 2016. Problemas de micotoxinas nos grãos e os novos limites toleráveis na cadeia alimentar. 2016. Disponível em: http://eventos.abrapos.org.br/anais/paperfile/16_20160821_20-54-53_439.pdf Acesso em 22 fev 2021.

Silva, J.N., Moncao, N.B.N., Farias, R.R.S., Cito, A.M.G.L., Chaves, M.H., Araujo, M.R.S., Lima, D.J.B., Pessoa, C., Lima, A., Araujo, E.C.C., Militão, G.C.G., Costa, M.P., Capasso, R. and Ferreira, P.M.P. 2020. Toxicological, chemopreventive, and cytotoxic potentialities of rare vegetal species and supporting findings for the Brazilian Unified Health System (SUS). J. Toxicol. Environ. Health Part A. 83: 525–545.

Siqueira, E.P., Andrade, A.A., Souza-Fagundes, E.M., Ramos, J.P., Kohlhoff, M., Nunes, Y.R.F. et al. 2014. *In vitro* antibacterial action on methicillin-susceptible (MSSA) and methicillin-resistant (MRSA) *Staphylococcus aureus* and antitumor potential of *Mauritia flexuosa* L. f. J. Med. Plants Res. 8(48): 1408–1417.

Sockalingam, K. and Abdullah, H.Z. 2015. Extraction and characterization of gelatin biopolymer from black tilapia (*Oreochromis mossambicus*) scales. *In*: AIP Conference Proceedings. AIP Publishing. p. 020053.

Sousa Dias, M.G. and Smith, T.K. 1995. Toxicity and growth-promoting potencial of spermine when fed to chicks. J. Animal Sci. 73: 2375–2381.

Sousa, J.M., Carneiro, M.F.H., Paulelli, A.C.C., Grotto, D., Magalhães Junior, A.M., Barbosa Junior, F. and Batista, B.L. 2015. Arsenic and rice: toxicity, metabolism, and food safety. Quím. Nova. 38(1): 118–127.

Souza, N.E., Matsushita, M., Franco, M.R.B., Prado, I.N. and Visentainer, J.V. 2005. Composição química, perfil de ácidos graxos e quantificação dos ácidos A-linolênico, eicosapentaenóico e docosahexaenóico em vísceras de tilápias (*Oreochromis niloticus*). Acta Sci-Technol. 27(1): 7376.

Stadelman, W.J. 2000. Eggs and egg products. Encyclopedia of Food Science and Technology, p. 593–599. John Wiley & Sons, New York.

Tamim, N.M. and Doerr, J.A. 2003. Effect of putrefaction of poultry carcasses prior to rendering on biogenic amine production. Poultry Sci. 12: 456–460.

The Government of the Hong Kong Special Administrative Region. 2007. Natural Toxins in Food Plants. Risk Assessment Studies. Report No. 27. Centre for Food Safety: Food and Environmental Hygiene Department.

Til, H.P., Falke, H.E., Prinsen, M.K. and Willems, M.I. 1997. Acute and subacute toxicity of tyramine, spermidine, spermine, putrescine and cadaverine in rats. Food Chem. Toxicol. 35(3-4): 337–348.

Tofalo, R., Perpetuini, G., Schirone, M. and Suzzi, G. 2016. Biogenic Amines: toxicology and health effect. *In*: Caballero, B., Finglas, P.M. and Toldrá, F. (eds.). Encyclopedia of Food and Health. Elsevier.

Tylleskär, T., Rosling, H., Banea, M., Bikangi, N., Cooke, R.D. and Poulter, N.H. 1992. Cassava cyanogens and konzo, an upper motor neuron disease found in Africa. Lancet. 339(8787): 208–211.

United Nations. 2021. Stop Food Loss and waste, for the people, for the planet. Available at: https://www.un.org/en/observances/end-food-waste-day. Accessed on: 6 jul. 2022.

Valdés, R. and Balbín, M.I. 2000. Curso de fisiología y bioquímica vegetal. La Habana: UNAH. p. 89.

Vasconcelos, I.M. and Oliveira, J.T.A. 2004. Antinutritional properties of plant lectins. Toxicon. 44: 385–403.

Vega, P.V. 1986. Toxicología de alimentos. Londres: Palgrave Centro panamericano de ecologia humana y salud.

Viana, E., Schultz, H.E., Albuquerque, R. and Noronha, A.B. 2006. Resíduos alimentares do lixo domiciliar: Estudo do uso na alimentação de frangos de corte. Rev. Bras. de Eng. Agricola e Ambient. 10(1): 203–211.

Visentainer, J.V. et al. 2000. Quantificação de ácidos graxos ômega-3 (LNA, EPA E DHA), caracterização físico-químico e composição em ácidos graxos em cabeças de tilápias jovens. pp. 5–221. *In*: Congresso Brasileiro de Ciência e Tecnologia de Alimentos, 2000, Fortaleza: SBCTA/UFC.

Volden, J., Borge, G.I., Bengtsson, G.B., Hansen, M., Thygesen, I.E. and Wicklund, T. 2008. Effect of thermal treatment on glucosinolates and antioxidant-related parameters in red cabbage (*Brassica oleracea* L. ssp. *capitata f. rubra*). Food Chem. 109(3): 595–605.

Walter, M., Marchezan, E. and Avila, L.A. 2008. Rice: composition and nutritional characteristics. Ciênc. rural. 3(4): 1184–1192.

Ward, A.J., Hobbs, O.J., Holliman, P.J. and Jones, D.L. 2008. Optimisation of the anaerobic digestion of agricultural resources. Bioresour. Technol. 99(17): 7928–7940.

World Health Organization (WHO). 1993. Cyanogenic Glycosides. Toxicological evaluation of certain food additives and naturally occurring toxicants. WHO Food Additive Series 30. Geneva: WHO.

Yao, H., Xu, W., Shi, X. et al. 2021. Dietary flavonoids as cancer prevention agents. J. Environ. Sci. Health C Environ. Carcinogenesis Ecotoxicol. Rev. 29: 1–31.

Yongmei, L., Chen, X., Mei, J., Xin, Lv., Rahman, N., Mingsheng, D. and Yan, G. 2009. Biogenic amines in Chinese soy sauce. Food Control. 20(6): 593–597.

Zhang, C., Su, H., Baeyens, J. and Tan, T. 2014. Reviewing the anaerobic digestion of food waste for biogas production. Renew. Sust. Energ. Rev. 38: 383–392.

Chapter 13

Annonaceae Family

Characteristics, Properties and Applications

Valéria Lima Silva Veras,[1] *Pedro Vitor Oliveira Silva Furtado,*[2] *Pelrry da Silva Costa,*[2] *Mahendra Rai,*[3] *Chistiane Mendes Feitosa*[4,*] and *João Marcelo de Castro e Sousa*[5]

Introduction

Annonaceae is the largest family in the order Magnoliales, with 112 genera and 2,440 species. There are about 900 species in the neotropical region and about 1200 species in the tropical regions of Asia and Australia, but the most common genera occur in the Americas (about 40 genera), (about 35 genera) dominate in South America and Africa is the continent with the least number of species, about 450 species.

Members of this family appear in the form of trees and rarely as shrubs or lianas. (Silva, 2019). The morphological characteristics important for the identification of the species of this family are dystic alternate leaves, simple, without stipules and the entire margin. The flowers are isolated or gathered in inflorescences, hemicyclic, hermaphroditic, diclamyds with differentiated perianth in the calyx and corolla, in general, they are trymere and fleshy; numerous stamens, arranged in a spiral fashion; ovary súperous with numerous apocarpic carpels with one to many ovules (Joly, 2002).

[1] Center of Agricultural Sciences, Postgraduate Program in Biotechnology - RENORBIO, Federal University of Piauí, 64048-550, Teresina, PI, Brazil.

[2] Department of Chemistry, Federal University of Piauí, 64049-550, Teresina, PI, Brazil.

[3] Department of Biotechnology. SGB Amravati University, Amravati-444 602, Maharashtra State, India.

[4] Federal University of Piaui, Department of Chemistry and Pharmacy, 64049-550, Teresina, PI, Brazil.

[5] Department of Biochemistry and Pharmacology, Federal University of Piauí, 64049-550, Teresina, PI, Brazil.

* Corresponding author: chistiane@ufpi.edu.br

Figure 13.1. Tree and fruit of *Annona squamosa.*

Most species of Annonaceae are almost completely confined to the tropics and grow in low-altitude areas. However, these plants are easily adaptable to different soil and climate conditions and thus can be commercially cultivated in different regions of the world (Aiolfi and Basso, 2016).

The main economic value of this family is the fruits. The genus *Annona* is well known for producing edible fruits. The main varieties grown in tropical regions are *Annona squamosa*, popularly known as ata, count's fruit or pinha (Fig. 13.1); *A. muricata*, better known as soursoup, guanabana; and *A. reticulata*, known as a countess (Lavagnini et al., 2014). Its fruits are consumed fresh or processed in the form of juices, ice cream, jellies, compotes, liqueurs and sweets (Pimenta et al., 2013). Other genera with edible but little-known fruits are the genera *Artabotrys*, *Asimina*, *Duguetia*, *Polyalthia* and *Uvaria* (Murilo and Restrepo, 2000).

Some species of the genera *Cananga* and *Artabotrys* are used in the cosmetics industry for the manufacture of perfumes (Kessler, 1993). In other species, the essential oils from the seeds are used for the manufacture of edible oils and soap, and species of the genera *Bocageopsis*, *Guatteria*, *Oxandra* and *Xylopia* are important in lumber, due to the wood of their trunks and the presence of long and resistant fibers in the bark of the stem, known as enviras (Murilo and Restrepo, 2000). In addition to a large number of species that are used in traditional medicine. In view of the above, it can be seen that these plants have a relevant economic role, especially in the following industrial segments: food, cosmetics, wood and pharmaceuticals.

The objective of this chapter is to bring a general presentation about the Annonaceae family, focusing on its chemical composition, nutritional aspects, pharmacological properties and biotechnological applications (Fig. 13.2).

Chemical composition

There are many species of Annonaceae that release odors, due to the essential oils (EOs) present in them. In the natural environment, these oils perform important actions for plants, such as attracting or repelling insects and even promoting allelopathic communication of the plants themselves (Leite et al., 2021). Moreover, they have

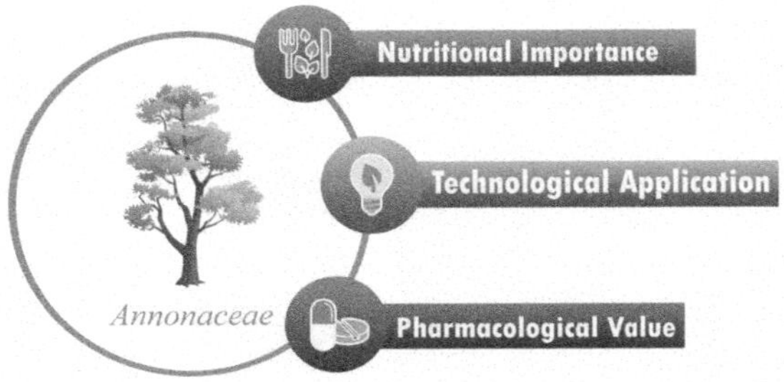

Figure 13.2. Distinctive applications of Annonaceae Family.

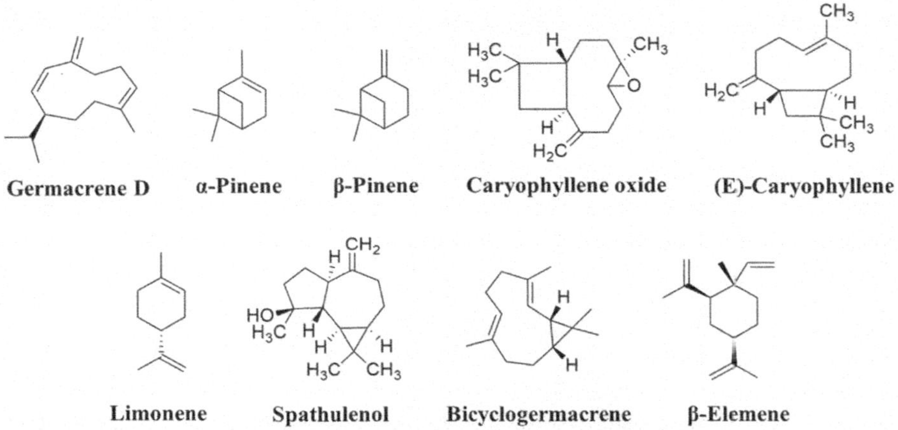

Figure 13.3. Main constituents of the Annonaceae family in Brazil.

antibacterial, antiviral, anti-inflammatory and antifungal properties, among others (Shaaban et al., 2012).

The main compounds present in Annonaceae EOs are monoterpene hydrocarbons in fruits and seeds; sesquiterpene hydrocarbons are found in the leaves; and oxygenated sesquiterpenes are found in the bark and roots (Fournier and Leboeuf, 1999). In this way, it is understood that the chemical composition of these EOs varies widely, and the chemical species present in greater quantities are usually mono and sesquiterpenes.

Making a literary prospection about the essential oils of Annonaceae present in Brazil, it was noted that the most abundant chemical compounds were germacrene D, α-pinene, β-pinene, caryophyllene oxide, (E)-caryophyllene, limonene, spatulenol, bicyclogermacrene and β-elemene, Fig. 13.3.

Araújo et al. (2015), conducted a study in order to evaluate the chemical constituents and antioxidant activity of EOs in *Annona vepretorum* leaves. Twenty-one constituents were found for the oil, with 17 identified (a total of 93% of the crude oil). The main compounds found were Spatulenol (43.7%), caryophyllene oxide (8.1%), limonene (20.5%) and α-pinene (5.5%), the first two of which are known as chemotaxonomic markers of this genus. The antioxidant activity presented by the EOs was low.

The chemical constitution of the EO of *Guatteria pogonopus* leaves was studied by Fontes et al. (2013), along with its antitumor activity. The total amount of compounds identified and quantified was 29. The predominant chemical species were γ-patchoulene (13.55%), (E)-caryophyllene (11.36%), β-pinene (10.37%), germacrene D (6.72%), bicyclogermacrene (5.97%), α-pinene (5.33%) and germacrene B (4.69%). Moreover, the antitumor activity was significant both *in vitro* and *in vivo*.

Costa et al. (2016) studied the chemical composition of EOs from fresh *Xylopia laevigata* fruits and evaluated their cytotoxic activity. The predominant compounds were monoterpenes (95% of the whole oil). The majority of compounds were limonene (56.2%), α-pinene (28.0%) and β-pinene (5.5%). Furthermore, the cytotoxic activity was not effective.

A study conducted by Moura et al. (2016) sought to verify the chemical composition, antitumor activity and toxicity of the EO of fresh *Xylopia langsdorffiana* fruits. The α-pinene (34.57%) and limonene (31.75%) were the most abundant identified constituents. The essential oil showed activity against leukemia and ovarian tumors and showed no induction for genotoxicity.

Gontijo et al. (2019) evaluated the antiplasmodial activity and the chemical constitution of *Xylopia sericea* leaves from both the EO and the dichloromethane extract. The main compounds identified were α-Pinene, β-pinene, o-cymene and D-limonene. The antiplasmodial activity was confirmed.

Brito et al. (2018) determined the photochemical constitution, antitumor activity and toxicity of the essential oil of *Annona leptopetala* leaves. The majority of chemical components were α-Limonene, linalool, α-terpineol, (E)-caryophyllene, bicyclogermacrene, spathulenol and guaiol. Furthermore, the antitumor activity of the oil was confirmed, in addition to moderate toxicity.

The antioxidant activity of *Annona salzmannii* and *A. pickelii* essential oils, as well as their chemical compositions, were studied by Costa et al. (2011). The compounds identified in the highest amount were in the EO of *A. salzmannii* were bicyclogermacrene (20.3%), (E)-caryophyllene (19.9%), δ-cadinene (15.3%), α-copaene (10.0%) and allo-aromadendrene (5.7%), while for *A. pickelii* were bicyclogermacrene (45.4%), (E)-caryophyllene (14.6%) and α-copaene (10.6%). In addition, the antioxidant activity of the EOs was quite significant.

Meira et al. (2015) studied the trypanocidal and antimalarial activities of the EOs obtained from the leaves of *Annona vepretorum* (AVOE) and *Annona squamosa* (ASOE). The EOs of *A. squamosa* had as major components (E)-caryophyllene (27.4%), germacrene D (17.1%) and bicyclogermacrene (10.8%). The EOs of *A. vepretorum*, on the other hand, had bicyclogermacrene (39.0%), spatulenol (14.0%) and α-felandrene (11.5%) as major components. Moreover, the observed trypanocidal and antimalarial activities were high.

Alcantara et al. (2017) conducted research to identify the phytochemical constitution and activity against bacteria of the EOs of *Bocageopsis multiflora* (Mart.) R.E. Fr., *Ephedranthus amazonicus* R.E. Fr., *Guatteria blepharophylla* Mart.

and *Xylopia aromatica* (Lam.) Mart. For B. multiflora the major constituents of the EO were caryophyllene (55.7%), spatulenol (20.3%) and P-bisabolene (11.9%), a sequiterpene. *X. aromatica* showed 21.5% spatulenol and the monoterpenes trans-pinocarveol (10.2%) and dihydrocarveol (11.6%) as the majority species. In the case of *E. amazonicus*, the most abundant compounds were spatulenol (16.9%) and humulene epoxide (16.3%). *G. blepharophylla* had the sesquiterpene caryophyllene oxide (55.7%) as the majority. Furthermore, the essential oils of *X. aromatica* and *G. blepharophylla* showed great antibacterial activity against bacteria of the genus *Streptococcus*.

Costa et al. (2020) evaluated the chemical composition and antileukemic activity of *G. megalophylla* leaf essential oils. The main constituents were spatulenol, γ-muurolene, bicyclogermacrene, β-elemene and δ-elemene. *In vitro* and *in vivo* tests showed that EOs have potential antileukemic activity.

Cascaes et al. (2021) studied the composition and bioactivity of *Annona exsucca* DC. essential oil and identified 58 compounds, where the most abundant chemical species types were monoterpenes and sesquiterpenes, respectively 43.36 and 31.29%. Furthermore, computational assays showed that some of these compounds can present significant antibacterial activity.

According to the above, it was noted that the main compounds identified in the analyzed works corroborate the information previously described by Fournier et al. (1999). Moreover, there are numerous possible applications for them, which will be addressed later.

Phytochemical investigations show that species of the Annonaceae family are rich in alkaloids. According to Lúcio et al. (2015), Annonaceae alkaloids can be classified into 14 types, namely: simple isoquinolines, isoquinolones and phenethylammonium (Type I); benzyltetrahydroisoquinolines, benzylisoquinolines-Seco and benzylisoquinolines (Type II, subtypes 1–7); bisbenzylisoquinolines (Type III); bisbenzyltetrahydroisoquinolines (Type IV, subtypes 1–22); protoberberines (Type V, subtypes 1–5); tetrahydroprotoberberines (Type VI, subtype 1); proaporphines (Type VII); Senso lato aporphinoids (Type VIII, subtypes 1–8); dehydroaporphines (Type IX, subtypes 1–5); 4- or 7-substituted aporphines (Type X, subtypes 1–3); oxoaporphines (Type XI, subtypes 1–14); phenanthrenes (Type XII, subtype 1); miscellaneous isoquinoline alkaloids (Type XIII) and non-isoquinoline alkaloids (Type XIV).

A class of compounds found almost exclusively in species of the Annonaceae family are the acetogenins (Wanke et al., 2015). They are derived from long-chain fatty acids (32–34 carbons) in the polyketide pathway. They are generally characterized by a combination of fatty acids with a 2-propanol unit at C-2 that forms an α, β-unsaturated methyl-substituted γ-lactone (Bermejo et al., 2005). They can be found in all parts of annonaceous plants, but are especially abundant in seeds. Figure 13.4 shows the general structure of an acetogenin.

Annonaceae also produce a wide variety of compounds belonging to different phytochemical groups, such as terpenes, flavonoids, lignans and some aromatic compounds (Lucio et al. 2015).

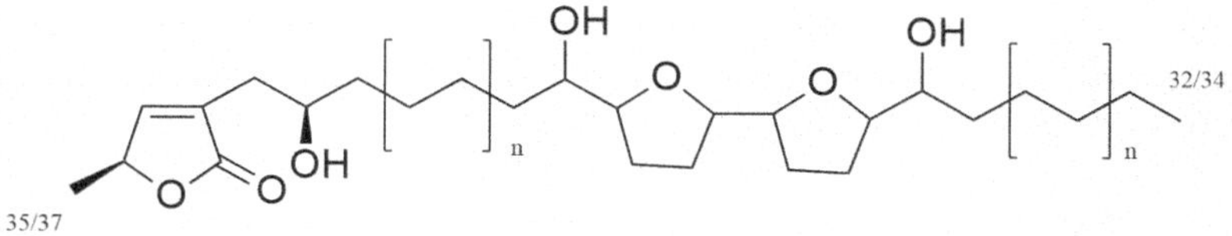

Figure 13.4. General structure of an acetogenin.

Nutritional aspects of the Annonaceae Family

Annonaceae is known for its food applications and for the consumption of its fruits, especially those of the genus *Annona*. Besides the unique aroma and flavor, the fruits of the Annonaceae are important sources of nutrients such as carbohydrates, lipids, proteins, fibers, vitamins and minerals, and many other compounds that give them properties not only nutritional but also pharmacological, such as phenolic compounds, terpenes, alkaloids and acetogenins. In the nutritional aspect, the applications of some fruits (Fig. 13.5) will be highlighted and discussed.

Annona mucosa, popularly known as biribá, is a tropical fruit tree found in Latin America, particularly in the Amazon region. Its fruit has a white, translucent, juicy flesh with a slightly acidic to sweet flavor and unique aroma (Lim, 2012). The biriba pulp has low acidity (0.34 g acid/100 g), pH of 5.80 and moisture content of 92.08%, which makes it interesting for the production of jams, jellies and yogurts. It has a high content of micronutrients such as copper, magnesium, vitamin C, manganese, iron and zinc, where 100 g of fruit pulp represents 52.00; 25.54; 16.67; 16.50; 8.50, and 6.87% of the recommended daily intake, respectively; in addition to considerable phenolic content, which confers antioxidant activity (Becker et al., 2018). The lipid fraction of *A. mucosa* fruits, on the other hand, is mainly composed of polyunsaturated fatty acids (101.12 mg/100 g pulp), followed by saturated and monounsaturated fatty acids (56.21 and 23.27 mg/100 g pulp). Linoleic (57.36), palmitic (45.77), α-linolenic (42.91), and oleic (21.11 mg/100 g) acids were the main fatty acids found (Berto et al., 2015). Thus, the consumption of biribá is a good

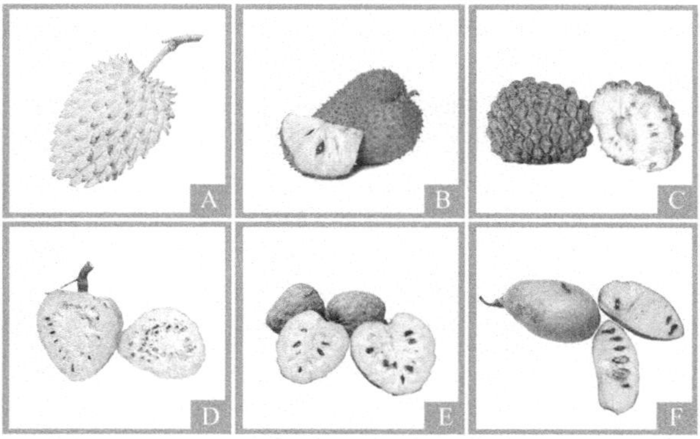

Figure 13.5. Some fruits of species of the Annonaceae family. A. *Annona mucosa* (biribá); B. *A. muricata* (soursop); C. *A. crassiflora* (araticum); D. *A. reticulata* (countess); E. *A. cherimola* (cherimoya); F. *Asimina triloba* (paw paw).

alternative to prevent and combat diseases such as scurvy, anemia and neutropenia, which are the result of vitamin C and mineral deficiencies, and the presence of essential fatty acids contributes to prevent cardiovascular diseases.

In the area of food technology, Annona trees also stand out. Senadeera et al. (2018) developed a probiotic yogurt from the pulp of the soursop (*Annona muricata*), ata (*Annona squamosa*) and countess (*Annona reticulata*) fruits and evaluated the chemical, microbiological, sensory and antioxidant characteristics over 28 days, on a weekly basis. This study showed that the addition of fruit pulp significantly increased the antioxidant activity of yogurts, reduced the pH, resulting in higher acidity, and favored the growth and viability of the bacterium *Bifidobacterium animalis* subsp. *lactis* BB-12, in yogurts during storage. In this study, the yogurt containing soursop fruit pulp recorded the highest scores for flavor, aroma, color and overall acceptance, showing that soursop can be used to make yogurt with good antioxidant properties, sensory attributes and probiotic counts during storage (Senadeera et al., 2018).

Another Annonaceae with potential applications in food technology is *Annona crassiflora*, known as "araticum", "araticum-do-cerrado", "marolo", "bruto", "cabeça-de-negro", "cascudo", "panã", "pinha-do cerrado" and "pasmada" (Arruda et al., 2015; Arruda et al., 2018). The pulp of the fruit of *A. crassiflora* has an energy value of 113.65 kcal/100 g, with 18.65% carbohydrates, 3.78% lipids and 1.27% protein. Furthermore, the fruit is an excellent source of vitamin A, fiber, copper, manganese, potassium, magnesium, zinc, vitamin C and folates (de Cardoso et al., 2013). The oil from the seeds, on the other hand, has a high concentration of monounsaturated fatty acids, which are capable of reducing bad cholesterol levels and increasing good cholesterol levels, in addition to the presence of omega family polyunsaturated fatty acids and functional phytochemicals, including phytosterols, tocopherols and carotenoids (Araújo et al., 2018; Luzia and Jorge, 2013).

In addition to consumption *in natura*, processing the pulp and making flour from the araticum pulp has already led to research that has developed food products such as nectar (Morzelle et al., 2011), yogurt (Della Lucia et al., 2012), ice cream (Morzelle et al., 2012), bread (Villela et al., 2013), fermented milk drink (da Silva et al., 2015), jam (Botrel et al., 2016), compote (Damiani et al., 2017), juice (Schiassi et al., 2018) and cereal bar (Silva et al., 2018). These products are an interesting option for the food industry since they showed good sensory acceptance and consumer purchase intention. In addition, studies have shown that incorporating the fruit pulp of *A. crassiflora* in food products improves their nutritional and functional characteristics. Fruit juice, for example, showed high antioxidant activity (Schiassi et al., 2018), the ice cream had an increase in dietary fiber (Morzelle et al., 2012), fermented dairy drinks had an increase in protein, lipid and mineral contents (de Lima et al., 2016) and cereal bars made with flour from the pulp of araticum resulted in a product with substantially improved amounts of fiber, vitamin C, carotenoids and minerals such as calcium and magnesium (da Silva et al., 2014; Silva et al., 2018).

The cherimoya (*Annona cherimola*) is an exotic subtropical fruit, native to the Andes Mountains, has a sweet taste and pleasant aroma and is a source of nutrients such as vitamins, carbohydrates, minerals, fiber and other bioactive compounds (Barreca et al., 2011). Isas et al. (2020) evaluated the growth capacity of starter cultures of Lactic Acid Bacteria (LAB) of fruit origin during fermentation of cherimoya juice, with the objectives of extending its shelf life and improving its biochemical and functional characteristics, since these bacteria are widely used as natural food bio-preservatives and are able to increase the nutritional properties of foods. This research demonstrated that LABs, isolated from fruits grown in Argentina, are able to grow and ferment cherimoya juice and survive during a cold storage period. The result was a juice with reduced sugar content, potential antiplatelet activity, high antioxidant properties, long shelf life and free of artificial preservatives (Isas et al., 2020).

Although the most well-known fruits are those of the genus *Annona*, other genera also produce edible fruits, such as *Asimina*. The species *Asimina triloba*, known as "pawpaw" or "poor man's banana" (Kral, 1960; Layne, 1996), for example, has fruits that are considered a good source of carbohydrates (19.0%), protein (1.2%), lipids (1.2%), fiber (2.6%), minerals (0.6%) and vitamins (0.02%) as well as polyphenols (Brindza et al., 2019). Brannan et al. (2012) suggested that the fruit pulp has the potential to be added to various food products to provide greater nutritional benefits or flavor enhancement.

Pharmacological properties of the Annonaceae

The plants of the Annonaceae family besides having economic and nutritional relevance, are also known for their pharmacological properties revealed through their traditional use against many ailments and diseases. Some of these properties have been proven, such as antibacterial (Omar et al., 1992); antitumor (Omar et al., 1992; Li et al., 1997) antimalarial (Nkunya et al., 1993; Boyom et al., 2003), leishmanicidal (Mahiou et al., 2000; Costa et al., 2006), anticonvulsant (González-Trujano et al., 2001), antiviral (Paredes et al., 2001), trypanocidal (Akendengue et al., 2002) and antiprotozoal (Osorio et al., 2007), among others.

The pharmacological activities reported in this family are attributed to secondary metabolites such as alkaloids, flavonoids, acetogenins, diterpenes, triterpenes, sesquiterpenes, steroids and other subcategories such as lignans, monoterpenes and xanthones (Bruginski, 2016). Alkaloids and acetogenins are the main metabolites of Annonaceae. Alkaloids are especially responsible for their pharmacological properties, due to their free electron pair on the nitrogen atom, which gives them the ability to interact with drug targets such as proteins, enzymes and receptors (Kittakoop et al., 2013).

Some Annonaceae alkaloids have already had pharmacological properties studied. The bisbenzylisoquinoline alkaloids, for example, have antiparasitic activity, especially against Leishmania sp. (Fournet et al., 1993; Munshi et al., 1972) *Trypanosoma cruzi* (Rojas De Arias et al., 1994) and *Plasmodium* sp. (Likhitwitayawuid et al., 1993; Valentin et al., 1997; Angerhofer et al., 1999). The alkaloid liriodenine, which is a chemotaxonomic marker of the family, has a bactericidal action against *Rhodococcus equi*, a bacterium that can cause pyogranulomatous broncho-pneumonia, ulcerative enteritis and mesenteric lymphadenitis in horses, and showed better results than the reference drug against this bacterium (Costa et al., 2009).

Another property exhibited by many species of Annonaceae is the anticholinesterase action, i.e., they are able to inhibit the enzyme acetylcholinesterase

Table 13.1. Some species of the Annonaceae family with anticholinesterase action.

Species	Extract/compound evaluated	References
Annona glabra	Alkaloids pseudocolumbamine, palmatine and pseudopalmatine Alkaloids (-)-anolobine and (-)-roemeroline	Tsai and Lee, 2010 Lee et al., 2015
Annona crassiflora	Alkaloids stephalagine and xylopine, Methanolic extract of the leaves Leaf and seed methanolic extract	Hidalgo, 2017 Melo, 2017 Formagio et al., 2015
Annona coriacea	Leaf and seed methanolic extract	Formagio et al., 2015
Annona hypoglauca	Seed Oil	Santos et al., 2015
Annona squamosa	Methanolic extract of the leaves	Bangou et al., 2011
Duguetia furfuracea	Methanolic extract of the leaves and seeds	Formagio et al., 2015
Guatteria blepharophylla	Alkaloid isomoschatoline	Feitosa et al., 2015
Guatteria friesiana	Guatteriopsiscine alkaloid	Feitosa et al., 2015
Xylopia leavigata	Diterpenes	Feitosa et al., 2015

causing the neurotransmitter acetylcholine to remain longer in the nerve synapses, having clinical applicability in diseases that result in reduced levels of acetylcholine in the brain, as is the case, for example, of Alzheimer's disease. Table 13.1 shows some Annonaceae species with anticholinesterase activity.

Acetogenins are compounds consisting of a series of polyketides found almost exclusively in Annonaceae (Wanke et al., 2015). Pharmacological studies have shown that acetogenins possess antineoplastic, antiparasitic, cytotoxic, immunosuppressive and neurotoxic effects (Alali et al., 1999; Zafra-Polo et al., 1998). According to the review by Liaw and colleagues (2016) cytotoxic and antitumor properties were the most explored in studies on the pharmacological aspects of acetogenins from Annonaceae. Another secondary metabolite with biological activities is terpenes. The essential oil from the leaves of the species *Guatteria friesiana* is composed mostly of the sesquiterpenes α, β and γ-eudesmol which have cytotoxic activity on human hepatocellular carcinoma HepG2 cells (Bomfim et al., 2013). This same cytotoxic action was exhibited by the alkaloids 4,5-dehydro-9-methoxyguatterfriesine (Branches et al., 2019) and guadiscidine (Costa et al., 2021) from *G. friesiana*.

Ibrahim and colleagues (2019) studied the effects of hexanic, ethyl acetate and ethanolic extracts of *Annona senegalensis* leaves on the inhibition of α-amylase and α-glucosidase, enzymes related to diabetes, and found that the extracts inhibited the action of these enzymes, showing that the species has potential as a treatment in the development of antidiabetic medication. This plant is popularly used for the treatment of diabetes and some diseases caused by helminths (Eyog-Matig et al., 2006).

The genus *Xylopia* is composed of about 150 species, many of which are known for their uses in folk medicine against gastrointestinal disorders and the species *Xylopia staudtii*, also belonging to the Annonaceae family, a study by Nguiam et al. (2021) showed activity against *Shigella flexnerii*, a species of bacteria responsible for about 2/3 of cases of dysentery caused by *Shigella* spp. in developing countries. In this study, the hydroethanolic extract of the barks was tested in mice with Shigellosis and the results showed that the *X. staudtii* extract had a potent bactericidal action by rupturing the plasma membrane of the microorganism and could thus be a candidate for a standardized phytomedication in the treatment of dysentery caused by *Shigella* spp. The ethanolic extract of the leaves of another species, *Xylopia sericeae* has anti-inflammatory, antioxidant and anti-*Helicobacter pylori* actions (Gomes et al., 2022).

Other pharmacological properties of Annonaceae are antifungal (Silva et al., 2001; Lima et al., 2011), anti-inflammatory (Coelho et al., 2006; Rocha et al., 2015), anxiolytic (Rejon-Orantes et al., 2011), larvicidal (Rodrigues et al., 2006), insecticidal (Coelho et al., 2007), antioxidant (Lima et al., 2010; Khallouki et al., 2011) and antipyretic (Jamkhande and Wattamwar, 2015) and it is observed that the studies are more focused on checking the cytotoxic, antibacterial and antiparasitic activities of Annonaceae.

Biotechnological aspects and future perspectives

As discussed so far, due to their chemical constituents, there are many aspects explored in Annonaceae. In addition to the nutritional and pharmacological potential, this family also stands out in the biotechnological field, through biopesticide formulations for agricultural application (Ribeiro et al., 2020), green synthesis of nanoparticles and nanosuspensions with pharmaceutical applications (Chabatulla et al., 2021; Han et al., 2021).

The development of bioinsecticides is important for organic agriculture, which is constantly growing in the world and needs ecological alternatives for pest management. Approximately 57.8 million hectares are cultivated with organic agriculture, integrating a world market of US$ 87.9 billion (Willer and Lernoud, 2019). Synthesis of metal oxide nanoparticles (NPs) via the green approach is more favorable than chemical synthesis, as it requires excessive use of organic solvents and toxic chemicals, so green synthesis is a valuable alternative that leads to the production of non-toxic metal oxide NPs (El Shafey, 2020). Microencapsulation or nanosuspensions of phytochemicals are important when it is necessary to increase the solubility of these substances so that they can exert their action on living organisms.

The Annonaceae family produces edible fruits whose seeds have a large amount of acetogenins, which have insecticidal action and are inhibitors of the mitochondrial complex I, affecting the action of the enzyme NADH-ubiquinone oxidoreductase (Krinski et al., 2014). Based on this, Ribeiro et al. (2020) isolated and purified acetogenins from the seeds of Annona mucosa, for application as a bioinsecticide in the management of the corn weevil, a pest that affects many corn crops worldwide. In this study, the authors found that acetogenins protected corn kernels against weevils. In another study, Maciel et al. (2019) microencapsulated the extract of *A. squamosa* seeds to evaluate its toxicity against Tetranychus urticae, a mite that causes losses in the production of fruits, vegetables and ornamental plants and showed that microencapsulation was efficient in mite mortality and could be used for their control.

In addition to their insecticidal action, acetogenins are also known for their antitumor and cytotoxic properties, however, their low solubility limits their clinical application. Thus, Han et al. (2021) prepared nanosuspensions (NSps) of acetogenins with the amphiphilic pharmaceutical adjuvant P188 and found that NSps significantly improved the solubility of acetogenins and showed stronger cytotoxicity against 4T1 (murine mammary carcinoma) and HeLa (cervical carcinoma) cell lines. Compared to the acetogenins-only solution and exhibited good antitumor efficacy and safety at 0.4 mg/kg (iv) *in vivo* (Han et al., 2021).

In recent years, the use of phytocompounds has received a lot of attention for the green synthesis of nanoparticles (NPs) from metal oxides. Zinc oxide NPs, for example, have been widely used in bioelectronics, biosensors, bioimaging applications, drug delivery and tissue engineering due to their non-toxic nature (Giner-Casares et al., 2016). In this context, Selvam et al. (2022) prepared, for the first time, low-cost and environmentally friendly zinc oxide nanoparticles using the extract of Annona reticulata leaves as a reducing agent. This study showed that the synthesized nanoparticles exhibited high cytotoxicity against human lung cancer cell lines (A549) (Selvan et al., 2022).

Another plant of the Annonaceae family was also used for the green synthesis of metallic nanoparticles, i.e., *Annona muricata.* Chabattula et al. (2021) fabricated zinc oxide nanoparticles using the aqueous extract of *A. muricata* leaves, which showed strong anticancer activity on human T lymphoid cells (MOLT4) and human lung cancer cells (A549). Badmus et al. (2020) also used the aqueous extract of *A. muricata* leaves, but for the green synthesis of silver nanoparticles, which showed strong antioxidant activity *in vitro*, inhibition of lipid peroxidation, antidiabetic and antimicrobial activities *in vitro*, as well as cytotoxicity in HaCaT cells (human keratinocytes) (Badmus et al., 2020).

In this way, the use of Annonaceae for various purposes can be highlighted and in the biotechnological field, the potential for their use as bioactive matrices places the species of this family as agents for future applications in the chemical, pharmaceutical, agricultural, materials industries, etc.

Conclusion

As can be seen in this chapter, some substances, such as flavonoids, alkaloids and acetogenins, are extracted from seeds, bark, roots and leaves to provide antibacterial, insect repellent, antioxidant and anticancer properties. In this way, alkaloids and acetogenins are commonly used in the preparation of pesticides, which are widely applied in agriculture for pest control. In addition, Annonaceae fruits have high nutritional value being rich in macronutrients and micronutrients.

Among the main constituents presented by the essential oils of the family, are the presence of germacrene D, α-pinene, β-pinene, caryophyllene oxide, (E)-caryophyllene, limonene, spathulenol, bicyclogermacrene and β-elemene stands out. In addition, their herbal properties such as antioxidant, anti-inflammatory and anticancer make them a potential raw material for the formulation of natural medicines.

Thus, according to the above, the Annonaceae family, in addition to being broad, is capable of providing several sectors of interest, being an exponent for the development of science and technology and ratifying its importance as a natural product.

References

Aiolfi, A.H. and Basso, C. 2016. Preparações elaboradas com o aproveitamento integral dos alimentos. Disciplinarum Scientia Saúde. 14(1): 109–114.

Akendengue, B., Roblot, F., Loiseau, P.M., Bories, C., Ngoumilama, E., Laurens, A. and Hocquemiller, R. 2002. Klaivanolide, an antiprotozoal lactone from *Uvaria klaineana*. Phytochemistry 59(8): 885–888.

Alali, F.Q., Liu, X.X. and McLaughlin, J.L. 1999. Annonaceous acetogenins: recente progress. J. Nat. Prod. 62: 504–540. DOI:10.1021/np980406d.

Alcântara, J.M., De Lucena, J.M.V.M., Facanali, R., Marques, M.O.M. and Da Paz Lima, M. 2017. Chemical composition and bactericidal activity of the essential oils of four species of annonaceae growing in Brazilian amazon. Nat. Prod. Commun. 12: 619–622.

Angerhofer, C.K., Guinaudeau, H., Wongpanich, V., Pezzuto, J.M. and Cordell, G.A. 1999. Antiplasmodial and cytotoxic activity of natural bisbenzylisoquinoline alkaloids. Journal of Natural Products 62: 59–66.

Araújo, C.S., Oliveira, A.P., Lima, R.N., Alves, P.B., Diniz, T.C. and Almeida, J.R.G.S. 2015. Chemical constituents and antioxidant activity of the essential oil from leaves of Annona vepretorum Mart. (Annonaceae). Pharmacogn. Mag. 11(43): 615–618.

Araújo, R.G., Rodriguez-Jasso, R.M., Ruiz, H.A., Pintado, M.M.E. and Aguilar, C.N. 2018. Avocado by-products: Nutritional and functional properties. Trends in Food Science & Technology 80: 51–60. DOI:10.1016/J.TIFS.2018.07.027.

Arruda, H.S. and de Almeida, M.E.F. 2015. Frutos do cerrado: Panorama, resgate cultural e aproveitamento culinário. Saarbrücken: Novas Edições Acadêmicas. Retrieved from https://www.morebooks.de/store/pt/book/frutos-do-cerrado/isbn/978-3-639-83794-0.

Arruda, H.S., Pereira, G.A., de Morais, D.R., Eberlin, M.N. and Pastore, G.M. 2018. Determination of free, esterified, glycosylated and insoluble-bound phenolics composition in the edible part of araticum fruit (*Annona crassiflora* Mart.) and its byproducts by HPLC-ESI-MS/MS. Food Chemistry 245: 738–749. DOI:10.1016/j.foodchem.2017.11.120.

Badmus, J.A., Oyemomi, S.A., Adedosu, O.T., Yekeen, T.A., Azeez, M.A., Adebayo, E.A., Lateef, A., Badeggi, U.M., Botha, S., Hussein, A.A. and Marnewick, J.L. 2020. Photo-assisted bio-fabrication of silver nanoparticles using *Annona muricata* leaf extract: exploring the antioxidant, anti-diabetic, antimicrobial, and cytotoxic activities. Heliyon. 6: e05413.

Bangou, M.J., Kiendrebeogo, M., Meda, N.T., Coulibaly, A.Y., Compaoré, M., Zeba, B., Milogo-Rasolodimby, J. and Nacoulma, O.G. 2011. Evaluation of enzymes inhibition activities of medicinal plant from Burkina Faso. Pak. J. Biol. Sci. 14: 99–105.

Barreca, D., Lagana, G., Ficarra, S., Tellone, E., Leuzzi, U., Galtieri, A. and Bellocco, E. 2011. Evaluation of the antioxidant and cytoprotective properties of the exotic fruit Annona cherimola Mill. (Annonaceae). Food Research International 44: 2302–2310. DOI: 10.1016/j.foodres.2011.02.031.

Becker, M.M., Mandaji, C.M., Catanante, G., Marty, J.L. and Nunes, G.S. 2018. Mineral and bromatological assessment and determination of the antioxidant capacity and bioactive compounds in native Amazon fruits. Brazilian Journal of Food Technology 21. DOI:10.1590/1981-6723.02218.

Bermejo, A., Figadere, B., Zafra-Polo, M.C., Barrachina, I., Estornell, E. and Cortes, D. 2005. Acetogenins from Annonaceae: recent progress in isolation, synthesis and mechanisms of action. Natural Product Reports 22(2): 269–303. https://doi.org/10.1039/b500186m.

Berto, A., da Silva, A.F., Visentainer, J.V., Matsushita, M. and de Souza, N.E. 2015. Proximate compositions, mineral contents and fatty acid compositions of native Amazonian fruits. Food Research International 77: 441–449. DOI:10.1016/j.foodres.2015.08.018.

Bomfim, D.S., Ferraz, R.P.C., Carvalho, N.C., Soares, M.B.P., Pinheiro, M.L.B., Costa, E.V. and Bezerra, D.P. 2013. Eudesmol isomers induce caspase-mediated apoptosis in human hepatocellular carcinoma HepG2 cells. Basic & Clinical Pharmacology & Toxicology 113: 300–306.

Botrel, D.A., Rodrigues, I.C.B., de Souza, H.J.B. and de Fernandes, R.V.B. 2016. Application of inulin in thin-layer drying process of araticum (Annona crassiflora) pulp. LWT - Food Science and Technology 69: 32–39. DOI: 10.1016/j.lwt.2016.01.018.

Boyon, F.F., Ngouana, V., Zollo, P.H.A., Menut, C., Bessiere, J.J., Gut, J. and Rosenthal, P.J. 2003. Composition and anti-plasmodial activities of essential oils from some Cameroonian medicinal plants. Phytochemistry 64(7): 1269–1275.

Branches, A.D.S., Costa, R.A., Junior, E.S.A., Bezzera, D.P., Soares, M.B.P., Costa, E.V. and Oliveira, K.M.T. 2019. Theoretical and experimental study by DFT, molecular docking calculations and cytotoxicity assay of 7,7-dimethylaporphine alkaloids type isolated from *Guatteria friesiana* (Annonaceae). Journal of Molecular Structure 1177: 347–362. DOI: 10.1016/j.molstruc.2018.09.060.

Brannan, R.G., Salabak, D.E. and Holben, D.H. 2012. Sensory analysis of pawpaw (Asimina triloba) pulp puree: consumer appraisal and descriptive lexicon. J. Food Res. 1(1): 179. DOI: 10.5539/jfr.v1n1p179.

Brindza, J., Grygorieva, O., Klymenko, S., Vergun, O., Marecek, J. and Ivanisová, E. 2019. Variation of fruits morphometric parameters and bioactive compounds of Asimina triloba (L.) Dunal germplasm collection. Potr. S. J. F. Sci. 13(1): 1–7. DOI: 10.5219/1019.

Brito, M.T., Ferreira, R.C., Beltrão, D.M., Moura, A.P.G., Xavier, A.L., Pita, J.C.L.R., Batista, T.M., Longato, G.B., Ruiz, A.L.T.G., de Carvalho, J.E. et al. 2018. Antitumor activity and toxicity of volatile oil from the leaves of *Annona leptopetala*. Braz. J. Pharmacogn. 28: 602–609.

Bruginski, E.R.D. 2016. Identificação de alcaloides e acetogeninas diretamente em tecidos de folhas e sementes de *Annona rugulosa* (Annonaceae) por DESI-MSI. Curitiba, dissertação.

Cascaes, M.M., Silva, S.G., Cruz, J.N., Santana de Oliveira, M., Oliveira, J., de Moraes, A.A.B., da Costa, F.A.M., da Costa, K.S., Diniz do Nascimento, L. and de Aguiar Andrade, E.H. 2021. First report on the *Annona exsucca* DC. Essential oil and *in silico* identification of potential biological targets of its major compounds. Nat. Prod. Res. 35: 1–4.

Chabattula, S.C., Gupta, P.K., Tripathi, S.K., Gahtori, R., Padhi, P., Mahapatra, S., Biswal, B.K., Singh, S.K., Dua, K., Ruokolainen, J., Mishra, Y.K., Jha, N.K., Bishi, D.K. and Kesari, K.K. 2021. Anticancer therapeutic efficacy of biogenic Am-ZnO nanoparticles on 2D and 3D tumor models. Materials Today Chemistry 22: 100618.

Coelho, M.B., Souza, I.A., Freire, M.G., Marangoni, S., Antunes, E. and Macedo, M.L. 2006. Neutrophil migration in mice induced by a mannose-binding lectin isolated from *Annona coriacea* seeds. Toxicon 48: 529–535.

Coelho, M.B., Marangon, S. and Macedo, M.L. 2007. Insecticidal action of *Annona coriacea* lectin against the flour moth *Anagasta kuehniella* and the rice moth *Corcyra cephalonica* (Lepidoptera: pyralidae). Comp. Biochem. Physiol. C Toxicol. Pharmacol. 146: 406–414.

Costa, E.V., Pinheiro, M.L.B., Xavier, C.M., Silva, J.R.A., Amaral, A.C.F., Souza, A.D.L., Barison, A., Campos, F.R., Ferreira, A.G., Machado, G.M.C. and Leon, L.L.P. 2006. A pyrimidine-β-carboline and other alkaloids from *Annona foetida* with antileishmanial activity. Journal of Natural Products 69(2): 292–294.

Costa, E.V., Pinheiro, M.L.B., Marques, F.A., Braga, R.M. and Sales Maia, B.H.L.N. 2009. First report of alkaloids in the genus *Guatteriopsis* (Annonaceae). Biochemical Systematics and Ecology 37: 43–45.

Costa, E.V., Dutra, L.M., Jesus, H.C.R., Nogueira, P.C.L., Moraes, V.R.S., Salvador, M.J., Cavalcanti, S.C.H., Santos, R.C. and Prata, A.P.N. 2011. Chemical composition and antioxidant, antimicrobial, and larvicidal activities of the essential oils of *Annona salzmannii* and *A. pickelii* (Annonaceae). Nat. Prod. Commun. 6: 907–912.

Costa, E.V., Da Silva, T.B., D'Souza Costa, C.O., Soares, M.B.P. and Bezerra, D.P. 2016. Chemical composition of the essential oil from the fresh fruits of *Xylopia laevigata* and its cytotoxic evaluation. Nat. Prod. Commun. 11: 417–418.

Costa, R.A., Barros, G.A., Silva, J.N., Oliveira, K.M., Bezerra, D.P., Soares, M.B.P. and Costa, E.V. 2021. Experimental and theoretical study on spectral features, reactivity, solvation, topoisomerase I inhibition and *in vitro* cytotoxicity in human HepG2 cells of guadiscine and guadiscidine aporphine alkaloids. Journal of Molecular Structure 1229: 129844. DOI: 10.1016/j.molstruc.2020.129844.

Costa, R.G.A., Anunciação, T.A.D., Araujo, M.D.S., Souza, C.A., Dias, R.B., Sales, C.B.S., Rocha, C.A.G., Soares, M.B.P., Silva, F.M.A.D., Koolen, H.H.F. et al. 2020. *In vitro* and *in vivo* growth inhibition of human acute promyelocytic leukemia HL-60 cells by *Guatteria megalophylla* Diels (Annonaceae) leaf essential oil. Biomed. Pharmacother. 122: 109713.

da Silva, E.P., Siqueira, H.H., do Lago, R.C., Rosell, C.M. and de Vilas Boas, E.V.B. 2014. Developing fruit-based nutritious snack bars. Journal of the Science of Food and Agriculture 94(1): 52–56. 10.1002/jsfa.6282.

da Silva, M.A.P., Vieira, N.F., Martins, Y.A.A., Araujo, V.F.P., Goncalves, D.S., de Moura, L.C. and Carvalho, B.S. 2015. Sensory profile of fermented milk drinks flavored with fruits from the Brazilian Cerrado. African Journal of Food Science 9(6): 379–389. 10.5897/AJFS2015.1293.

Damiani, C., Silva, E.P., Becker, F.S., Endrigo, D.E., Asquieri, E.R., da Silva, F.A. and de Boas, E.V.B.V. 2017. Antioxidant potential of marolo jam (Annona crassiflora Mart) during storage. Open Access Library Journal 4(5). DOI: 10.4236/oalib.1103158 e3158.

de Cardoso, L.M., Oliveira, D.S., de Bedetti, S.F., Martino, H.S.D. and Pinheiro-Sant'Ana, H.M. 2013. Araticum (*Annona crassiflora* Mart.) from the Brazilian Cerrado: Chemical composition and bioactive compounds. Fruits 68(2): 121–134. DOI: 10.1051/fruits/2013058.

de Lima, A.V.S.C., Nicolau, E.S., Rezende, C.S.M., Torres, M.C.L., Novais, L.G. and Soares, N.R. 2016. Characterization and sensory preference of fermented dairy beverages prepared with different concentrations of whey and araticum pulp. Semina: Ciências Agrárias 37(6): 4011–4026. 10.5433/1679-0359.2016v37n6p4011.

De Sousa Orlando, V., Glauciemar, D.V.V., Bruna, C.S.S., Ceacutelia, H.Y., Santos de Matos Arajo, A.L., da Luz, A., Aparecida de Oliveira Pinto, M., Pereira Rodarte, M. and Alves, M.S. 2016. *In-vivo* and *in vitro* bioactivities of the essential oil of *Duguetia lanceolata* branches. Afr. J. Pharm. Pharmacol. 10: 298–310.

De Souza Araújo, C., De Oliveira, A.P., Lima, R.N., Alves, P.B., Diniz, T.C. and Da Silva Almeida, J.R.G. 2015. Chemical constituents and antioxidant activity of the essential oil from leaves of *Annona vepretorum* Mart. (Annonaceae). Pharmacogn. Mag. 11: 615–618.

Della Lucia, F., Ferreira, E.B., dos Santos, R.C., Moreira, M.A.M., Corrêa, S.C. and de Vilas Boas, E.V.B. 2012. Acceptability of yoghurt with marolo pulp (Annona crassiflora Mart.) in the traditional and diet/light formulations. Revista da Universidade Vale do Rio Verde 10(2): 85–92. DOI:10.5892/615.

El shafey, A.M. 2020. Green synthesis of metal and metal oxide nanoparticles from plant leaf extracts and their applications: a review. Green Process. Synth. 9: 304e339. DOI:10.1515/gps-2020-0031.

Eyog-Matig, O., Ndoye, O., Kengue, J. and Awono, A. 2006. Les Fruitiers Forestiers Comestibles du Cameroun. International Plant Genetic Resources Institute (IPGRI), Cotonou, Benin, p. 220. Regional Office for West and Central Africac/o ITTA 08 BP. 0932.

Feitosa, C.M., Costa Júnior, J.S., Costa, E.V., Cavalcante, N.A. and Silva, V.L. 2015. Antioxidant compounds in fruits and Alzheimer's disease. pp. 103–122. *In*: Chistiane Mendes Feitosa (ed.). Medicinal Plants and Alzheimer's Disease. Campinas: Editora Átomo.

Fontes, J.E.N., Ferras, R.P.C., Britto, A.C.S., Carvalho, A.A., Moraes, M.O., Pessoa, C., Costa, E.V. and Bezerra, D.P. 2013. Antitumor effect of the essential oil from leaves of *Guatteria pogonopus* (Annonaceae). Chem. Biodivers. 10: 722–729.

Formagio, A.S.N., Vieira, M.C., Volobuff, C.R.F., Silva, M.S., Matos, A.I., Cardoso, C.A.L. and Carvalho, J.E. 2015. *In vitro* biological screening of the anticholinesterase and antiproliferative activities of medicinal plants belonging to Annonaceae. Braz J. Med. Biol. Res. 48: 308–315.

Fournet, A., Angelo, B.A., Muñoz, V., Hocquemiller, R. and Cavé, A. 1993. Effect of some bisbenzylisoquinoline alkaloids on American *Leishmania* sp. Phytotherapy Research 7: 281–284.

Fournier, G. and Leboeuf, M. 1999. Cavé A. Annonaceae essential oils: A review. J. Essent. Oil Res. 11: 131–142. doi: 10.1080/10412905.1999.9701092.

Giner-Casares, J.J., Henriksen-Lacey, M., Coronado-Puchau, M. and Liz-Marzan, L.M. 2016. Inorganic nanoparticles for biomedicine: where material scientists meet medical research, Mater. Today 19: 19e28. DOI:10.1016/j.mattod.2015.07.004.

Gomes, J.V.D., Borges, A.S., Athaydes, B.R., Gomes-Copeland, K.K.P., Silveira, D., Magalhaes, P.O., Gonçalves, R.C.R., Kitagawa, R.R., Fonseca-Bazzo, Y.M. and Jamal, C.M. 2022. Anti-*Helicobacter pylori* potential, antioxidant capacity, and anti-inflammatory activity of *Xylopia sericea* A. St.-Hil. (Annonaceae) leaves. Phytomedicine Plus. 2: 100214. DOI: 10.1016/j.phyplu.2021.100214.

Gontijo, D.C., do Nascimento, M.F.A., Brandão, G.C. and de Oliveira, A.B. 2019. Phytochemistry and antiplasmodial activity of *Xylopia sericea* leaves. Nat. Prod. Res. 6419.

González-Trujano, M.E., Navarrete, A., Reyes, B., Cedillo-Portugal, E. and Hong, E. 2001. Anticonvulsant properties and bio-guided isolation of palmitone from leaves of *Annona diversifolia*. Planta Medica. 67(2): 136–141.

Han, J., Zhou, X., Fu, J., Gao, G., Zuo, C., Guo, Y., Han, M. and Wang, X. 2021. Annonaceous acetogenins nanosuspensions stabilized by poloxamer 188: preparation, properties and *in vivo* evaluation. Journal of Drug Delivery Science and Technology 66: 102676.

Hidalgo, E.M.P. 2017. Biological activities of alkaloids isolated from *Annona crassiflora* Mart. Dissertation of Master's Degree, Instituto de Biociências, Universidade de São Paulo, São Paulo. DOI:10.11606/D.41.2018.tde-22032018-135520.

Ibrahim, A., Umar, I.A., Aimola, I.A. and Mohammed, A. 2019. Inhibition of key enzymes linked to diabetes by *Annona senegalensis* Pers (Annonaceae) leaf *in vitro*. Journal of Herbal Medicine 16: 100248. DOI: 10.1016/j.hermed.2018.11.004.

Isas, A.S., Celis, M.S.M., Correa, J.R.P., Fuentes, E., Rodríguez, L., Palomo, I., Mozzi, F. and Nieuwenhove, C.V. 2020. Functional fermented cherimoya (Annona cherimola Mill.) juice using autochthonous lactic acid bactéria. Food Research International 138: 109729.

Ismael Montero, F., Chagas, E.A., Melo Filho, A.A.D.E., Saravia, S.A.M., Santos, R.C , Jamkhande, P.G. and Wattamwar, A.S. 2015. *Annona reticulata* Linn. (Bullock's heart): plant profile, phytochemistry and pharmacological properties. J. Tradit. Complement. Med. 5: 144–152.

Jamkhande, P.G. and Wattamwar, A.S. 2015. *Annona reticulata* Linn. (Bullock's heart): plant profile, phytochemistry and pharmacological properties. Journal of Traditional and Complementary Medicine 5(3): 144–152.

Joly, A.B. 2002. Botânica: introdução à taxonomia vegetal. Nacional. 13 Ed: 286.

Kessler, P.J.A. 1993. Annonaceae. pp. 93–129. *In*: Kubitzki, K., Rohwer, J.G. and Bittrich, V. (ed.). The Families and Genera of Vascular Plants. II. Flowering Plants. Dicotyledons. Magnoliid, Hamamelid and Caryophyllid families. Berlin: Springer Verlag.

Khallouki, F., Haubner, R., Ulrich, C.M. and Owen, R.W. 2011. Ethnobotanical survey, chemical composition, and antioxidant capacity of methanolic extract of the root bark of *Annona cuneata* Oliv. J. Med. Food 11: 1397–1402.

Kittakoop, P., Mahidol, C. and Ruchirawat, S. 2013. Alkaloids as important scaffolds in therapeutic drugs for the treatments of cancer, tuberculosis, and smoking cessation. Current Topics in Medicinal Chemistry 14(2): 239–252.

Kral, R. 1960. A revision of *Asimina* and *Deeringothannus* (Annonaceae). Brittonia 12(4): 233–278. DOI: 10.2307/2805119.

Krinski, D., Massaroli, A. and Machado, M. 2014. Potencial inseticida de plantas da família Annonaceae. Rev. Bras. Frutic. 36: 225–242.

Lavagnini, C.G., Di Carne, C.A.V., Correa, F., Henrique, F., Tokumo, L.E., Silva, M.H. and Santos, P.C.S. 2014. Fisiologia Vegetal - hormônio giberelina. Garça: Revista Científica de Agronomia. 25(1): 48–52.

Layne, D.R. 1996. The pawpaw [Asimina triloba (L.) Dunal]: a new fruit crop for Kentucky and the United States. HortScience 31(5): 777–784.

Lee, S.S., Wu, D.Y., Tsai, S.F. and Chen, C.K. 2015. Anti-acetylcholinesterase alkaloids from *Annona glabra* leaf. Nat. Prod. Commun. 10: 891–893.

Leite, D.O.D., Camilo, C.J., Nonato, C.F.A., Carvalho, N.K.G., Salazar, G.J.T., de Morais, S.M., and Costa, J.G.M.D. 2021. Chemical profile and evaluation of the antioxidant and anti-acetylcholinesterase activities of *Annona squamosa* L. (Annonaceae) Extracts. Foods (Basel, Switzerland) 10(10): 2343. https://doi.org/10.3390/foods10102343.

Li, T.M., Li, W.K. and YU, J.G. 1997. Flavonoids from *Artabotrys hexapetalus*. Phytochemistry 45: 831–833.

Liaw, C.C., Liou, J.R., Wu, T.Y., Chang, F.R. and Wu, Y.C. 2016. Acetogenins from annonaceae. pp. 113–228. *In*: Kinghorn, A.D., Falk, H., Gibbons, S. and Kobayashi, J. (eds.). Progress in the Chemistry of Organic Natural Products. DOI: 10.1007/978-3-319-05275-5.

Likhitwitayawuid, K., Angerhofer, C.K., Cordell, G.A. and Pezzuto, J.M. 1993. Cytotoxic and antimalarial bisbenzylisoquinoline alkaloids from *Stephania erecta*. Journal of Natural Products 56: 30–38.

Lim, T.K. 2012. *Rollinia mucosa*. In Edible Medicinal and Non-Medicinal Plants.

Lima, L.A., Pimenta, L.P. and Boaventura, M.A. 2010. Acetogenins from *Annona cornifolia* and their antioxidant capacity. Food Chem. 122: 1129–1138.

Lima, L.A., Johann, S., Cisalpino, P.S., Pimenta, L.P. and Boaventura, M.A. 2011. *In vitro* antifungal activity of fatty acid methyl esters of the seeds of *Annona cornifolia* A. St.- Hil. (Annonaceae) against pathogenic fungus *Paracoccioides brasiliensis*. Rev. Soc. Bras. Med. Trop. 44: 777–780.

Lopes, S., Santos, L. and Choupina, A. 2016. A extrusão em tecnologia alimentar: aplicações, caraterísticas dos produtos, composição e tendências futuras. Revista de Ciências Agrárias. 39(1): 4–14.

Lúcio, A.S.S.C., da Silva Almeida, J.R.G., da-Cunha, E.V.L., Tavares, J.F. and Barbosa Filho, J.M. 2015. Alkaloids of the Annonaceae: Occurrence and a compilation of their biological activities. In The Alkaloids: Chemistry and Biology; Elsevier: Amsterdam, The Netherlands 74: 233–409. ISBN 1099–4831.

Luzia, D.M.M. and Jorge, N. 2013. Bioactive substance contents and antioxidant capacity of the lipid fraction of Annona crassiflora Mart. seeds. Industrial Crops and Products 42: 231–235. DOI: 10.1016/j.indcrop.2012.05.027.

Maciel, A.G.S., Trindade, R.C.P., Basílio Júnior, I.D., Santana, A.E.G., Silva, J.P., Santos, L.A.T., Silva, E.S., Freitas, J.D. and Nascimento, T.G. 2019. Microencapsulation of *Annona squamosa* L. (Annonaceae) seed extract and lethal toxicity to *Tetranychus urticae* (Koch, 1836) (Acari: Tetranychidae). Industrial Crops & Products 127: 251–259.

Mahiou, V., Roblot, F., Fournet, A. and Hocquemiller, R. 2000. Bisbenzylisoquinoline alkaloids from *Guatteria boliviana* (Annonaceae). Phytochemistry 54: 709–716.

Meira, C.S., Guimarães, E.T., Macedo, T.S., Da Silva, T.B., Menezes, L.R.A., Costa, E.V. and Soares, M.B.P. 2015. Chemical composition of essential oils from *Annona vepretorum* Mart. and *Annona squamosa* L. (Annonaceae) leaves and their antimalarial and trypanocidal activities. J. Essent. Oil Res. 27: 160–168.

Melo, C.H.S. 2017. Análises fitoquímica e farmacológica das espécies *Eugenia dysenterica* DC. (Cagaiteira) e *Annona crassiflora* Mart. (Araticunzeiro) direcionadas para doenças neurodegenerativas. Dissertation of Master's Degree, Federal University of Piaui.

Morzelle, M.C., de Souza, E.C., Assumpção, C.F., Vilas Boas, B.M. 2011. Development and sensory evaluation of mixed nectar of passion fruit (Passiflora edulis Sims) and araticum fruit (Annona crassiflora). Revista Brasileira de Produtos Agroindustriais. 13(2): 131–135.

Morzelle, M.C., Lamounier, M.L., Souza, E.C., Salgado, J.M. and de Vilas-Boas, E.V.B. 2012. Nutritional and sensory characteristics of ice cream from savana fruits. Journal of Candido Tostes Dairy Institute 67(387): 70–78. 10.5935/2238-6416.20120052.

Moura, A.P.G., Beltrão, D.M., Pita, J.C.L.R., Xavier, A.L., Brito, M.T., de Sousa, T.K.G., Batista, L.M., de Carvalho, J.E., Ruiz, A.L.T.G., Della Torre, A. et al. 2016. Essential oil from fruit of *Xylopia langsdorffiana*: antitumour activity and toxicity. Pharm. Biol. 54: 3093–3102.

Munshi, C.P., Vaidya, P.M., Buranpuri, J.J. and Gulati, O.D. 1972. Kala-azar in Gujarat. Journal of the Indian Medical Association 59: 287–293.

Murilo, J. and Restrepo, D. 2000. Las anonáceas de la región de Araracuara: estudios en La Amazonía Colombiana XX. Soporte Editorial: Bogotá.

Nguiam, M.P., Wouamba, S.C.N., Longo, F., Kamkumo, R.G., Foweda, L.D.K., Djomeni, P.D.D., Ndjakou, B.L., Sewald, N., Boyom, F.F., Fogue, S.K. and Dimo, T. 2021. Antibacterial and antishigellosis activity of *Xylopia staudtii* (engl. & diels), Annonaceae. Journal of Ethnopharmacology 280: 114406. DOI:10.1016/j.jep.2021.114406.

Nkunya, M.H.H., Waibel, R. and Achenbach, H. 1993. Three flavonoids from the stem bark of the antimalarial Uvaria dependens. Phytochemistry 34(3): 885–888.

Omar, S., Chee, C.L., Ahmad, F., Ni, J.X., Jaber, H., Huang, J. and Nakatsu, T. 1992. Phenanthrene lactams from *Goniothalamus velutinus*. Phytochemistry 31: 4395–4397.

Osorio, E., Arango, G.J., Jiménez, N., Alzate, F., Ruiz, G., Gutiérrez, D., Paco, M.A., Giménez, A. and Robledo, S. 2007. Antiprotozoal and cytotoxic activities *in vitro* of Colombian Annonaceae. Journal of Ethnopharmacology 111: 630–635.

Paredes, A., Hasegawa, M., Prieto, F., Mendez, J., Rodríguez, M. and Rodríguez-ortega, M. 2001. Biological activity of *Guatteria cardoniana* fractions. Journal of Ethnopharmacology 78: 129–132.

Pimenta, A.C., Rego, S.S., Zuffellato-Ribas, K.C., Nogueira, A.C. and Koelher, H.S. 2013. Morphological characterization of fruits, seeds and seedlings of araticum plant (*Annona crassiflora* Mart - Annonaceae). Journal of Seed Science 35: 524–531.

Rejón-Orantes, J.C., González-Esquinca, A.R., Mora, M.P., Roldan, G. and Cortes, D. 2011. Annomontine, an alkaloid isolated from *Annona purpurea*, has anxiolytic like effects in the elevated plus-maze. Planta Med. 77: 322–327.

Ribeiro, L.P., Gonçalves, G.L.P., Bicalho, K.U., Fernandes, J.B. and Vendramim, J.D. 2020. Rolliniastatin-1, a bis-tetrahydrofuran acetogenin: the major compound of *Annona mucosa* Jacq. (Annonaceae) has potent grainprotective properties. Journal of Stored Products Research 89: 101686.

Rocha, R.S., Kassuya, C.A., Formagio, A.S., Mauro, M.O., Andrade-Silva, M., Monreal, A.C., Cunha-Laura, A.L., Vieira, M.D. and Oliveira, R.J. 2015. Analysis of the anti-inflammatory and chemopreventive potential and description of the antimutagenic mode of action of the *Annona crassiflora* methanolic extract. Pharm. Biol. 17: 1–13.

Rodrigues, A.M., De Paula, J.E., Degallier, N., Molez, J.E. and Espindola, L.S. 2006. Larvicidal activity of some Cerrado plant extracts against *Aedes aegypti*. J. Am. Mosq. Control Assoc. 22: 314–317.

Rojas de arias, A., Inchausti, A., Ascurrat, M., Fleitas, N., Rodriguez, E. and Fournet, A. 1994. *In vitro* activity and mutagenicity of bisbenzylisoquinolines and quinones against *Trypanosoma cruzi* trypomastigotes. Phytotherapy Research 8: 141–144.

Santos, R.C., Melo Filho, A.V. and Chagas, E.A. 2015. Fatty acid profile and bioactivity from *Annona hypoglauca* seeds oil. Afr. J. Biotechnol. 14: 2377–2382.

Schiassi, M.C.E.V., de Souza, V.R., Lago, A.M.T., Campos, L.G. and Queiroz, F. 2018. Fruits from the Brazilian Cerrado region: Physico-chemical characterization, bioactive compounds, antioxidant activities, and sensory evaluation. Food Chemistry 2453: 05–311. DOI: 10.1016/J.FOODCHEM.2017.10.104.

Selvam, K., Allam, A.A., Ajarem, J.S., Sudhakar, C., Selvankumar, T., Senthilkumar, B. and Kim, W. 2022. *Annona reticulata* leaves-assisted synthesis of zinc oxide nanoparticles and assessment of cytotoxicity and photocatalytic impact. Materials Letters 309: 131379.

Senadeera, S.S., Prasanna, P.H.P., Jayawardana, N.W.I.A., Gunasekara, D.C.S.P. and Chandrasekara, A. 2018. Antioxidant, physicochemical, microbiological, and sensory properties of probiotic yoghurt incorporated with various Annona species pulp. Heliyon. e00955. DOI: 10.1016/j.heliyon.2018.e00955.

Shaaban, H.A., El-Ghorab, A.H. and Shibamoto, T. 2012. Bioactivity of essential oils and their volatile aroma components: Review. J. Essent. Oil Res. 24: 203–212.

Silva, A. 2019. Aproveitamento tecnológico de resíduo sólido da graviola (*Annona muricata* L.) na elaboração de barra de cereal e avaliação do seu potencial antioxidante. Cuité: Universidade Federal de Campina Grande, Paraíba, Brasil.

Silva, J.S., Damiani, C., Silva, E.P., Ruffi, C.R.G., Asquieri, E.R., Silva, T.L.L.E. and de Vilas Boas, E.V.B. 2018. Effect of marolo (Annona crassiflora Mart.) pulp flour addition in food bars. Journal of Food Quality. 1–12. DOI: 10.1155/2018/8639525.

Silva, M.V., Costa, T.R., Ferreira, E.C., Fernandes, L.F.L., Santos, S.C., Liao, L.M., Ferri, P.H., Jaula, J.R., Ferreira, H.D. and Silva, M.R.R. 2001. Growth inhibition effect of Brazilian cerrado plant extracts on Candida species. Pharm. Biol. 39: 138–141.

Tsai, S.F. and Lee, S.S. 2010. Characterization of acetylcholinesterase inhibitory constituents from *Annona glabra* assisted by HPLC microfractionation. J. Nat. Prod. 73: 1632–1635.

Valentin, A., Benoit-vical, F., Moulis, C., Stanislas, E., Mallié, M., Fouraste, I. and Bastide, J.M. 1997. *In vitro* antimalarial activity of penduline, a bisbenzylisoquinoline from *Isopyrum thalictroides*. Antimicrobial Agents and Chemotherapy 41: 2305–2307.

Villela, P., Batista, Â.G. and Dessimoni-Pinto, N.A.V. 2013. Nutritional composition of Annona crassiflora pulp and acceptability of bakery products prepared with its flour. Food Science and Technology 33(3): 417–423. DOI: 10.1590/S0101-20612013005000082.

Wanke, T., Philippus, A.C., Zatelli, G.A., Vieira, L.F.O., Lhullier, C. and Falkenberg, M. 2015. C15 acetogenins from the Laurencia complex: 50 years of research—an overview. Braz. J. Pharmacogn. 25: 569–587. DOI:10.1016/j.bjp.2015.07.027.

Willer, H. and Lernoud, J. 2019. The world of organic agriculture: statistics and emerging trends, twentieth ed. Research Institute of Organic Agriculture FiBL and IFOAM Organics International, Frick and Bonn.

Zafra-Polo, M.C., Figadère, B., Gallardo, T., Tormo, J.R. and Cortes, D. 1998. Natural acetogenins from annonaceae, synthesis and mechanisms of action. Phytochemistry 48: 1087–1117. DOI:10.1016/S0031-9422(97)00917-5.

Chapter 14

Potential Exploitation of Residual Avocado (*Persea americana*) Seeds in the Development of Functional Foods with Glycemia- and Cholesterol-lowering Properties

Mercedes V. Urquiza Martínez,[1] *José Pérez Villarreal,*[2] *Zaira I. Bedolla Valdez,*[2] *José Venegas González,*[3] *Luz Torner,*[4] *Raul Manhães de Castro,*[5,6] *Ana E. Toscano*[5,7,8] and *Omar Guzmán Quevedo*[1,4,8,*]

Introduction

Metabolic diseases currently represent the main health problem in the world (Saklayen, 2018). The modern diet, characterized by a high content of saturated

[1] Instituto Tecnológico Superior de Tacámbaro, Tacámbaro, Michoacán, Mexico;
[2] Instituto Tecnológico Superior de Uruapan, Uruapan, Michoacán, Mexico;
[3] Centro Interdisciplinario de Investigación para el Desarrollo Integral Regional (CIIDIR), Unidad Michoacán, Instituto Politécnico Nacional, Jiquilpan, Michoacán, Mexico;
[4] Centro de Investigación Biomédica de Michoacán, Instituto Mexicano del Seguro Social, Morelia, Michoacán, Mexico;
[5] Unidade de Estudos em Nutrição e Plasticidade Fenotípica do Departamento de Nutrição, Universidade Federal de Pernambuco, Recife, Pernambuco, Brazil;
[6] Departamento de Nutrição, Universidade Federal de Pernambuco, Recife, Pernambuco, Brazil;
[7] Departmento de Enfermagem, Universidade Federal de Pernambuco, Vitória de Santo Antão, Pernambuco, Brazil;
[8] Pós-Graduação em Neuropsiquiatria e Ciências do Comportamento, Universidade Federal de Pernambuco, Recife, Pernambuco, Brazil.
* Corresponding author: omar.guzman-quevedo@itstacambaro.edu.mx

fats and refined sugars along with a poor consumption of healthy foods, has led to a high prevalence of obesity, cardiovascular diseases (CVD) and Type 2 Diabetes (T2D) (Mozaffarian, 2016). CVD and T2D are intimately related to high levels of cholesterol (hypercholesterolemia) and blood glucose (hyperglycemia), respectively. Both medical conditions, largely related to each other, are of great relevance in public health as they are the world's leading cause of death. In 2019, 17.9 million people died from CVD, representing 32% of all global deaths (WHO, 2021). T2D affects ~ 9% of the world's population and the occurrence increases every year (Einarson et al., 2018). CVD affects 32.2% of people with diabetes, and is the main cause of mortality in these patients (Einarson et al., 2018).

The development of nutritional therapies has been the focus of the scientific community in recent years, trying to resolve the lack of long-term effectiveness of pharmacological treatments against metabolic disorders including hyperglycemia and hypercholesterolemia. Inspired by the Hippocratic precept "let food be thy medicine" the concept of functional foods appeared in the 1980s. This was used for the first time by the Japanese Ministry of Health and Welfare, establishing the consumption of foods with possible health benefits. Since then, research in the field of functional foods has grown greatly, leading to the development and consolidation of the industry around them. Functional foods are based on the content of biologically active compounds, which are capable of favorably modifying some functions of the organism promoting health. Bioactive compounds can be present naturally or be added to food from another source. Thus, to ensure the supply of bioactive compounds to the increasingly demanding industry of functional foods, new sources must be exploited.

The avocado (*Persea americana*), originally from Mexico and Central America, is a largely consumed and cultivated fruit, with Mexico as the main producer. The avocado fruit is composed of an edible pulp, the peel and the seed. The latter represents one by-product of the avocado fruit and constitutes around 13% of the total weight of the fruit (Rodriguez-Carpena et al., 2011). Literature reports indicate that the avocado seed, despite being a waste fraction, has a high nutritional value, in addition to having a high content of bioactive compounds. Considering the above and the ecological impact that its non-use implies, it is important to develop strategies for its exploitation in the food industry. This chapter discusses the functional properties of the avocado seed, emphasizing its high content of bioactive compounds and its potential exploitation as a source of bio-active compounds for the production of functional foods.

Avocado

Avocado is an evergreen tree, which belongs to the family *Lauraceae* and the *Persea* genus, whose most cultivated species in the world is *Persea americana* (Fig. 14.1A). Also known as alligator pear, avocado is a treasured climacteric fruit with origin in Mexico and Central America, consumed as far back as 8000–7000 BCE (Chen et al., 2009). The name avocado is derived from the Nahuatl word "*ahuacatl*" which means "testicle", this is due to its anatomical shape. It has a buttery flesh, a large single seed, and its peel, when ripe, is a black-purple color (Fig. 14.1B) (Chen et al., 2009;

Figure 14.1. Avocado. (A) Avocado tree and (B) anatomy of the avocado fruit.

Weschenfelder et al., 2015). Due to the large popularity of avocado consumption, its production increases year after year. Annual avocado production in 2013 was estimated at 4.7 million tonnes, of which Mexico contributed 31.1% of global production (Hurtado-Fernández et al., 2018).

The high nutritional quality and healthy properties attributed to avocado have greatly contributed to the high popularity of avocado consumption. Avocado is a rich source of monounsaturated fatty acids, protein, fiber, minerals (calcium, iron, magnesium, phosphorus, potassium, sodium, copper, zinc, etc.), vitamins (vitamin C, thiamine, riboflavin, niacin, pantothenic acid, vitamin B-6, vitamin A, vitamin K_1 and vitamin E), phenolic compounds and phytosterols (Dreher and Daverpont, 2013). The uses of avocado are wider than its use as food; it is used along with parts of its tree as traditional medicine in several cultures, exploiting their antioxidant, anti-inflammatory, hepatoprotective, antihypertensive, antimicrobial and hypolipidemic properties (Nwaoguikpe et al., 2011; Noorul et al., 2017).

Avocado seeds

In recent years, the avocado industry has begun to transform, going from its sale as a simple commodity to its commercialization as a processed product known as 'guacamole'. While making guacamole, a high volume of peel and seed is thrown away (Fig. 14.2). This market trend allows access to large masses of these by-products, which can be used to produce other foods. The food industry has taken a most sustainable approach nowadays, focusing on the development and study of discarded by-products of food like avocado seed or peel, which comprise ~ 34% (18% seed and 16% peel) of total fruit weight (Calderón-Oliver et al., 2016). This means that of the 4.7 million tonnes produced annually, around 1.6 million tonnes of waste are discarded, of which ~ 846 thousand tonnes correspond to seed (Rodriguez-Carpena et al., 2011; Hurtado-Fernández et al., 2018; Salazar-López et al., 2020).

In addition to the ecological and economic impact of the waste of the avocado seed, it is regrettable that this by-product is not used since it represents an important source of nutrients and bioactive compounds. The general composition of the avocado

Figure 14.2. Avocado by-products generated during the making of 'guacamole'. (A) Transport belt of avocado peel and seed discarded during the process; (B) Container for the waste generated.

seed consists of 12.70% moisture, 2.50% ash, 18.44% protein, 18.63% crude fat, 3.25% crude fiber and 48.10% carbohydrate. The avocado seed is a rich source of phenolic and flavonoid compounds promoting health benefits as well. Among them are catechins, hydroxycinnamic acids, flavonols and procyanidins, showing a strong antioxidant capacity, backing up its research and use on health issues (Nwaoguikpe et al., 2011; Pahua-Ramos et al., 2012; Noorul et al., 2017; Tabeshpour et al., 2017).

Health effects of avocado seed-derived bioactive compounds

About 80% of people in developing countries depend on traditional medicine as the main treatment for several diseases because of its low cost and minor secondary effects. Due to this, there is a great interest from the scientific community to study their uses (processing, dose, route of administration) and effects (beneficial for health and toxicity) of them. As mentioned before, avocado is highly valued as a food, but also used as traditional medicine along with other parts of its tree such as leaves, peel or seeds (Kasole et al., 2019).

The avocado seeds are promising therapeutic alternative to several health problems, mostly due to its high content in phenolic compounds and antioxidant capacity (Pahua-Ramos et al., 2012). Use of by-products of the avocado such as the peel and seed allows to increase its exploitation and may lead to novel product development. The avocado seed is a probed rich source of health promoter polyphenols such as protocatechuic acid, kaempferide, vanillic acid, catechins, hydroxycinnamic acids, flavonols and procyanidins. These compounds show a strong antioxidant capacity, backing up its research and use on health issues (Nwaoguikpe et al., 2011; Pahua-Ramos et al., 2012; Noorul et al., 2017; Tabeshpour et al., 2017).

The therapeutics studied, attributed to the avocado seed, reported in literature include the prevention and treatment of cancer, inflammation, microbial, diabetes and cardiovascular diseases. Table 14.1 summarizes the principal therapeutic effects of avocado seed investigated for several health conditions. There are a number of nutritious and bioactive compounds in the avocado seed underlying the aforementioned health benefits. Among them can be highlighted unsaturated fatty acids (mainly linolenic and oleic acids), sterol compounds (responsible for lowering

Table 14.1. Therapeutic effect of avocado seed on different diseases.

Therapeutic effect	Mechanism	Model	Reference
Anticancer	Cell proliferation arrest. ↓ Cyclin D1 and E2. ↑ Apoptosis mediated by Caspase 3. ↑ PARP cleavage. ↓ Nuclear translocation of NF-κB. ↑ Oxidative stress-dependent apoptosis via mitochondrial membrane depolarization. ↑ Transcription factor p53, protease caspase-3 and APAF.	Cancer cell lines of breast (MCF7, MDA-MB-231), prostate (LNCaP), lung (H1299, A549), colon (HT29, HCT116), epidermal (HaCaT), gastric (BGC823), liver (HePG2) and Jurkat lymphoblastic leukemia cells	Bhuyan et al., 2019
Antimicrobial	Antimicrobial activity against both Gram-positive and Gram-negative bacteria (except *E. coli*). Aqueous extracts had antimicrobial activity against *L. monocytogenes* and *S. epidermidis*. The antimicrobial effect by increased membrane permeability. Peel and seed extracts did not present antimicrobial activity against: *Listeria innocua* *E. coli* *Lactobacillus sakei* *Weissella viridescens* *Leuconostoc mesenteroides*	*Listeria monocytogenes* *Staphylococcus epidermidis* *Staphylococcus aureus* *Enterococcus faecalis* *Escherichia coli* *Salmonella enteritidis* *Citrobacter freundii* *Pseudomonas aeruginosa* *Salmonella typhimurium* *Enterobacter aerogenes*	Egbuonu et al., 2018 Bahru et al., 2019 Bhuyan et al., 2019
Antiinflammatory	↓ Pro-inflammatory cytokines including IL-6, TNF-α, and IL-1β along with NO (nitric oxyde), translocation of NF-κB and production of Prostaglandin E2	RAW264.7 murine macrophage cells	Alkhalaf et al., 2019 Dabas et al., 2019
Antilipidemic	↓ TC, TG, LDL-C and prediction of atherogenic index, along with an increase in HDL-C. Attributed to ↑ LDL receptor expression, ↓ Hepatic lipid synthesis, ↓ AST and ALT activities and lipoprotein secretion and ↑ TC elimination via bile acids.	Hyperlipidemic mice model	Imafidon and Amaechina, 2010 Mudassir et al., 2020 Pahua-Ramos et al., 2012 Shehata and Soltan 2013

Table 14.1 contd. ...

...Table 14.1 contd.

Therapeutic effect	Mechanism	Model	Reference
Antihyperglycemic	↓ Blood glucose levels (fasting and postprandial). ↓ Fasting insulin resistance index, percent glycemic change and HbA1c. Attributed to the modulation of the α-amylase and α-glucosidase expression. ↓ Blocking gluconeogenesis. ↑ Glucose utilization, presence of insulin stimulatory substances. ↓ Intestinal glucose absorption. ↑ Peripheral glucose metabolism. ↑ Insulin secretion by protecting pancreatic β-cells	Alloxan-induced diabetic rats	Azantsa et al., 2018 Ejiofor et al., 2018 Ezejiofor et al., 2013 Kasole et al., 2019
Antihypertensive	↓ Blood pressure of hypertensive rats, along with related biomarkers as TC, LDL-C and TG. Attributed to ↑ content of potassium (promotes vasodilatation and sodium regulation). ↓Cholesterol levels (by beta sitosterol and tocopherols)	Albino rats	Imafidon and Amaechina, 2010

APAF = Apoptosis-inducing factor, ALT = Alanine transaminase, AST = Aspartate transaminase, HbA1c = Glycated hemoglobin, HDL-C = high-density lipoprotein cholesterol, IL-1β = Interleukin-1β, IL-6 = Interleukin-6, LDL = Low-density lipoprotein, LDL-C = Low-density lipoprotein cholesterol, NF-κB = Nuclear factor kappa B, PARP = Poly (ADP-ribose) polymerase, TC = Total cholesterol, TG = Triglycerides, TNF-α = Tumor necrosis factor-α, ↓ = Decrease, and ↑ = Increase.

cholesterol levels), tocopherols, carotenoids, vitamin C, minerals (magnesium, calcium, zinc, potassium, sodium, and phosphorus), phytochemicals (flavonoid, tannin, saponin and total phenolics). Moreover, an antioxidant capacity of 44.65%, evaluated by the 2,2-diphenyl-1-picrilhidrazilo technique, has been reported (Ifesan et al., 2015; Talabi et al., 2016; Alkhalaf et al., 2019; Flores et al., 2019).

Hypoglycemic and hypocholesterolemic properties of the avocado seed-derived bioactive compounds

Among the most relevant therapeutic use of the avocado seed, is the timely and strong effect on dyslipidemia. This alteration over circulating lipids, includes high levels of Total Cholesterol (TC), triglycerides (TG), low and very low-density lipoprotein cholesterol (LDC-C and VLDL-C) and low levels of High-Density Lipoprotein Cholesterol (HDL-C) (Kopin and Lowenstein, 2017). These combinations of altered lipid levels lead to coronary artery disease and peripheral artery disease considered the modern silent killers because of their close relation to CVD including heart attacks and stroke (number 1 cause of death in developed and developing countries) (WHO, 2017). The mechanism responsible for it involves the formation of the atheromatous plaque, defined as the abnormal accumulation of lipids, macrophages, calcium and fibrous connective tissue in the inner layer of an artery wall that leads to the narrowing and restricted blood flow (Sun, 2014). As mentioned earlier, therapeutics are getting a more complex perspective widening the options in it. Typically, pharmaceutics was the only option, but with the increase in scientific knowledge, a much richer approach is willing to include natural compounds in its list of treatments.

Several studies on animal models of hypercholesterolemia and dyslipidemia using different extractions from the avocado seed [e.g., aqueous (hot and cold), hydroethanolic and methanolic extracts] have found significant reduction of TC, TG and LDL-C, while HDL-C showed a significant increase (Fig. 14.3). It results in a reduction of the atherogenic index due to the upregulation of LDL receptor expression, inhibition of hepatic lipid synthesis (reducing the activities of the enzymes aspartate transaminase and alanine transaminase) and lipoprotein secretion. In addition, this leads to an increase in cholesterol elimination through bile acids given the presence of phenolic compounds, monounsaturated fatty acids, carotenoids, phytosterols and dietary fiber (Pahua-Ramos et al., 2012; Shehata and Soltan, 2013; Tabeshpour et al., 2017).

T2D is a disease that is manifested by high levels of glucose in the blood. This hyperglycemia results from a process of insulin resistance. T2D is a highly prevalent disease in the world with a marked socioeconomic impact. In this context, the avocado seed has shown hypoglycemic and antihyperglycemic effects by significantly decreasing blood glucose levels. Among the list of examined effects on glucose metabolism, it may be referred to as a decrease in fasting blood glucose, decreased postprandial glucose levels, improvement in fasting insulin resistance index, percent glycemic change and HbA1c (Fig. 14.3), to mention the most important ones. The effect was linked to enzymes involved in T2D including α-amylase and α-glucosidase. Furthermore, the avocado seed contains important trace elements such as sodium, potassium, calcium, magnesium, zinc and chromium which

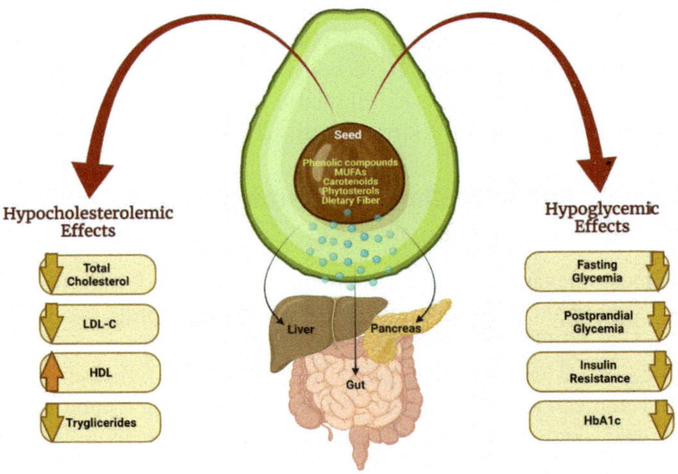

Figure 14.3. Hypocholesterolemic and hypoglycemic effects of the avocado seed. Avocado seed-derived bioactive compounds can act on different tissues including the liver, pancreas and gut to promote metabolic benefits. LDL-C: low-density lipoprotein cholesterol; HDL: high-density lipoprotein; MUFAs: monounsaturated fatty acids; HbA1c: glycated hemoglobin. Created with BioRender.com.

have key inhibition over enzymes involved in hepatic gluconeogenesis including glucose-6-phosphatase, fructose-1,6-biphosphatase and phosphoenolpyruvate carboxykinase. This inhibition results in a blockage of gluconeogenesis and enhances glucose utilization in the body, promoting glucose homeostasis. Other mechanisms involved in glucose modulation are the presence of insulin stimulatory substances, decreasing its intestinal absorption, increasing its peripheral metabolism or inducing insulin secretion by surviving pancreatic β-cells (Ezejiofor et al., 2013; Azantsa et al., 2018; Ejiofor et al., 2018; Bhuyan et al., 2019; Nardi et al., 2020; Mudassir et al., 2020). In the same line, an anti-diabetic effect of an aqueous extract of *Persea americana* seeds was observed in alloxan-induced diabetic rats. The mechanism of such an effect involved the activation of PI3K/AKT signaling pathway, which is closely related to insulin (Ojo et al., 2022).

Functional foods

The Functional Food Center (FFC) in 2018 defined functional food as a "natural or processed foods that contain biologically active compounds, which, in defined, effective and non-toxic amounts, provide a clinically proven and documented health benefit utilizing specific biomarkers for the prevention, management, or treatment of chronic disease or its symptoms" (FFC, 2018). The so-called "biologically active compounds" include the nutritious and non-nutritious constituents of foods (usually in small amounts) that act as antioxidants, as chemo- and cardio-preventive, and prevent or reduce the risk of developing certain diseases. The properties attributed to them are mostly incorporated in having an effect, causing a reaction or triggering a response in tissues of the human body (Gur et al., 2018).

According to Henry (2010), functional foods can be classified as (a) naturally occurring bioactive compounds in conventional food, (b) foods that have been enriched with bioactive compounds, and (c) synthesized food ingredients (could be: removal of constituents, adjustment of components, modifying bioavailability or a combination of them) (Henry, 2010). The two final options (parenthesis b and c) require a process, proposed by the FFC, for the development of functional food. It begins with the examination of the relation between a particular food and health benefits. Second, researchers must find the biologically compound responsible for the observed health benefit. Third, the determination by *in vitro* and *in vivo* studies of the mechanism and the daily dose needs to be carried out in order to avoid toxicity. Fourth, tests on humans must be conducted to monitor side effects with a determined dose of bioactive compounds. Finally, an appropriate food vehicle for the bioactive compounds must be developed that guarantees its conservation and then, is finalized the functional food (Gur et al., 2018).

Functional food ingredients can be found naturally in vegetable products like seeds, fruits, vegetables, cereals, leaves, beverages like tea, coffee, wine, etc. Among the wide variety of ingredients considered to have beneficial actions are vitamins, minerals, fiber (soluble and insoluble), β-glucans, prebiotics, probiotics and synbiotics, phenolic compounds, monounsaturated fatty acids, essential fatty acids and certain proteins, among others (Helkar et al., 2016).

In this context, the food industry has become concerned lately about the disposal of by-products that are rich in many of the aforementioned products and can be utilized in food, pharmaceutics or other industries and lower the amount of waste and environmental damage.

Sources of functional compounds for the food industry

Among the main sources of natural bioactive compounds are vegetables, fruits, cereals, some non-vegetarian leaves (tea *Camélia sinensis*), fermented products, sprouts, etc. However, agro-industrial wastes are of great interest because studies are examined to find the richest in such bioactive compounds. The list of such compounds includes polyphenols, glucosinolates, carotenoids, terpenoids, alkaloids, saponins, vitamins, minerals, fatty acids and fibers (Helkar et al., 2016; Câmara et al., 2021).

From the grain processing industry, rice bran is rich in vitamins, minerals, key unsaturated fats, dietary fiber and different sterols, antioxidant compounds like polyphenols, vitamin E, tocotrienols and carotenoids. It has been used in bread, cakes, noodles, pasta and ice creams without influencing the functional and textural properties. It also contains oryzanol and psitosterol, with examined cholesterol-lowering properties, anti-tumorous action (inducing natural killer activity), inactivation of macrophages and the inhibition of angiogenesis. In summary, rice bran is a very good example of a hypoallergenic by-product, antioxidant and anti-cancer activity, very reliable to be used in the food industry as a source of bioactive compounds (Helkar et al., 2016). From the brewery, three types of wastes have been obtained: the spent grain, the hot trub and the residual yeast, which can be used as flavoring agents, enzymes, single-cell protein and substrate for microalgae cultivation. Sugars, proteins, acids and antioxidants can also be extracted. From winery waste, materials include vine prunings, grape stalks, grape pomace and grape seeds, yeast lees, tartrate, carbon dioxide and wastewater, characterized by the presence of natural safer antioxidants vs synthetic (Helkar et al., 2016). Finally, the vegetable and fruit industry provide a very large list of by-products usable as bioactive compounds. One of them is xoconostle, rich in antioxidants, glucose, citric and linoleic acids, tocopherols and isorhamnetin-O-(di-deoxyhexosyl-hexoside); it has been studied for properties such as antihypoglycemic, antihyperlipidemic, hypocholesterolemic, anti-inflammatory, antidiuretic, antiulcerogenic, immunostimulating activity and the prevention of some cancers (Arias-Rico et al., 2020). Another example is apple pomace, with a content of nearly 74% of dietary fiber (an important role in improvement and management of human health, particularly gastrointestinal system), consisting of pectin, cellulose, hemicellulose, lignin and gums and phenolic compounds bound with skin, i.e., dihydrochalcones, flavonols and phenolic acids (Lyu et al., 2020).

Avocado by-products offer a very wide and promising variety of bioactive compounds. Both peels and seeds have high contents of bioactive phytochemicals such as phenolic acids, condensed tannins and flavonoids including procyanidins, flavonols, hydroxybenzoic and hydroxycinnamic acids. These bioactive compounds have shown various biological activities such as antioxidant and anti-inflammatory properties. The anti-inflammatory activity of phenolic compounds is largely related to their ability to scavenge oxidative radicals, which is important for cell and oxidative stress regulation (Helkar et al., 2016; Tremocoldi et al., 2018). Compared with the other sources of bioactive compounds mentioned, the avocado seed represents much higher quantities in mass, allowing a greater amount of obtaining the aforementioned compounds.

Plausibility of avocado seed exploitation in functional food design

The use of by-products to develop functional food is now a reality. The trends in investigation show the increasing interest of the scientific community in making up the exploitation of these types of products in the food industry, given the examined health beneficial effects and the reduction to the environment impact. The avocado seed, which has many functional ingredients in its composition, is a perfect example of it. From the total weight of the fruit, the seed represents ~ 18% and concentrates 70% of the total antioxidants of avocados (Calderón-Oliver et al., 2016). Moreover, it has more soluble fiber than oatmeal, and a large amount of monounsaturated fatty acids and minerals (like phosphorus and potassium). Its rich composition allows it to demonstrate pro-health effects such as hypolipidemic, hypoglycemic, anti-inflammatory, antimicrobial and anti-aging. Thus, the next step in this equation is to examine its uses in the food industry to keep these properties and develop functional foods that are attractive to the consumer. Some products made with avocado seed flour show emerging strategies for the exploitation of this by-product. Some examples include (1) candy from avocado seed (Ifesan et al., 2015), which was well accepted in a 20 member group; the formula preferred was 85% seed flour, 15% ginger and sugar; (2) barren cakes from Indonesia—usually made with rice flour—

have a low cost and good taste, according to the study the most accepted formula was with 75% of avocado seed flour (Terati et al., 2021); (3) biscuits for diabetic patients were made with avocado seed flour (Ervianingsih et al., 2020) traditional dishes like moi-moi from Nigeria (bean-based steam-cooked pudding) included avocado seed flour to its preparation with better sensory characteristics at 10% of seed flour (Emelike et al., 2020).

One important aspect to be mentioned here is the consideration of the presence of anti-nutritional components (characterized to compete or reduce the absorption of minerals, metals, proteins, etc.) in avocado seed, among them are: saponin, phytate, oxalate, tannin, alkaloid and cyanogenic glycosides. However, investigations have provided different techniques based on boiling, defatting and fermentation with the probiotics process to improve its nutritious capacity and reduce toxicity risks (Egbuonu et al., 2018; Emelike et al., 2020; Kumari and Amarakoon, 2021).

Conclusions

In summary, the avocado seed, a waste product of this fruit, is a rich source of bioactive compounds with significant beneficial effects on health. The literature reviewed in this chapter shows a clear effect of avocado seed-derived compounds on elevated blood glucose and cholesterol levels. Considering its high dietary value, the high amount of mass (13% of the fruit weight) and the low cost (as it is a by-product) that the avocado seed represents, its use in the preparation of functional foods with glycemia- and cholesterol-lowering properties can be highly profitable. Additional research focusing on its application in the food industry is necessary to establish the intake required to obtain health benefits and avoid potential toxicity.

References

Alkhalaf, M.I., Alansari, W.S., Ibrahim, E.A. and ELhalwagy, M.E. 2019. Anti-oxidant, anti-inflammatory and anti-cancer activities of avocado (*Persea americana*) fruit and seed extract. J. King Saud. Univ. Sc. 31(4): 1358–1362.

Arias-Rico, J., Cruz-Cansino, N., Cámara-Hurtado, M., López-Froilán, R., Pérez-Rodríguez, M.L., Sánchez-Mata, M.C. et al. 2020. Study of Xoconostle (Opuntia spp.) powder as source of dietary fiber and antioxidants. Foods (Basel, Switzerland) 9(4): 403.

Azantsa, B.G.K., Kuikoua, W.T., Takuissu, G.N., Takwi, E.A.M. and Tagne, V.M. 2018. Effects of polyherbal formulation of Allium sativum and *Persea americana* seeds' extracts on postprandial hyperglycemia and sucrose digestion in acute treatment of normoglycemic rats. Biol. Med. (Aligarh) 10(2): 432.

Bahru, T.B., Tadele, Z.H. and Ajebe, E.G. 2019. A review on avocado seed: functionality, composition, antioxidant and antimicrobial properties. Chem Sci. Int. J. 27(2): 1–10.

Bhuyan, D.J., Alsherbiny, M.A., Perera, S., Low, M., Basu, A., Devi, O.A. et al. 2019. The odyssey of bioactive compounds in avocado (*Persea americana*) and their health benefits. Antioxidants 8(10): 426.

Câmara, J.S., Albuquerque, B.R., Aguiar, J., Corrêa, R.C., Gonçalves, J.L., Granato, D. et al. 2021. Food bioactive compounds and emerging techniques for their extraction: Polyphenols as a case study. Foods 10(1): 37.

Calderón-Oliver, M., Escalona-Buendía, H.B., Medina-Campos, O.N., Pedraza-Chaverri, J., PedrozaIslas, R. and Ponce-Alquicira, E. 2016. Optimization of the antioxidant and antimicrobial response of the combined effect of nisin and avocado byproducts. LWT-Food Sci. Technol. 65: 46–52.

Chen, H., Morrell, P.L., Ashworth, V.E., De La Cruz, M. and Clegg, M.T. 2009. Tracing the geographic origins of major avocado cultivars. J. Hered. 100(1): 56–65.

Dabas, D., Ziegler, G.R. and Lambert, J.D. 2019. Anti-inflammatory properties of a colored avocado seed extract. Adv. Food Technol. Nutr. Sci. Open J. 5: 8–12.

Dreher, M.L. and Davenport, A.J. 2013. Hass avocado composition and potential health effects. Crit. Rev. Food Sci. Nutr. 53(7): 738–750.

Egbuonu, A.C., Opara, I.C., Onyeabo, C. and Uchenna, N.O. 2018. Proximate, functional, antinutrient and antimicrobial properties of avocado pear (*Persea americana*) seeds. J. Nutr. Health Food Eng. 8(1): 00260.

Einarson, T.R., Acs, A., Ludwig, C. and Panton, U.H. 2018. Prevalence of cardiovascular disease in type 2 diabetes: a systematic literature review of scientific evidence from across the world in 2007–2017. Cardiovasc Diabetol. 17(1): 83.

Ejiofor, C.C., Ezeagu, I.E. and Ayoola, M. 2018. Hypoglycaemic and Biochemical effects of the aqueous and methanolic extract of *Persea americana*seeds on alloxan-induced albino rats. European J. Med. Plants 1–12.

Emelike, N.J.T., Ujong, A.E. and Achinewhu, S.C. 2020. Proximate and sensory properties of moi-moi developed from cowpea and avocado pear seed flour blends. J. Food Technol. Res. 7(2): 136–143.

Ervianingsih, Hurria, Astari, C., Mursyid, M. and Kadir, A.R. 2020. Avocado seed waste utilization into biscuits as diabetes mellitus patients' alternative foods. In IOP Conf Ser: Earth Environ. Sci. 575(1): 012004. IOP Publishing.

Ezejiofor, A.N., Okorie, A. and Orisakwe, O.E. 2013. Hypoglycaemic and tissue-protective effects of the aqueous extract of *Persea americana*seeds on alloxan-induced albino rats. Malays J. Med. Sci. 20(5): 31.

Flores, M., Ortiz-Viedma, J., Curaqueo, A., Rodriguez, A., Dovale-Rosabal, G., Magaña, F. et al. 2019. Preliminary studies of chemical and physical properties of two varieties of avocado seeds grown in Chile. J. Food Qual. 2019: 1–11.

Functional food center. 2018. Official site: https://www.functionalfoodscenter.net/

Gur, J., Mawuntu, M. and Martirosyan, D. 2018. FFC's advancement of functional food definition. Funct Food Health Dis. 8(7): 385–397.

Helkar, P.B., Sahoo, A.K. and Patil, N.J. 2016. Review: Food industry by-products used as a functional food ingredients. Int. J. Waste Resour. 6(3): 1–6.

Henry, C.J. 2010. Functional foods. Eur. J. Clin. Nutr. 64(7): 657–659.

Hurtado-Fernández, E., Fernández-Gutiérrez, A. and Carrasco-Pancorbo, A. 2018. Avocado fruit—Persea americana. pp. 37–48. Editor(s): Sueli Rodrigues, Ebenezer de Oliveira Silva, Edy Sousa de Brito, Exotic Fruits, Academic Press, ISBN 9780128031384.

Ifesan, B.O.T., Olorunsola, B.O. and Ifesan, B.T. 2015. Nutritional composition and acceptability of candy from avocado seed (*Persea americana*). Int. J. Agric Innov. Res. 3(6): 1631–1634.

Imafidon, K.E. and Amaechina, F.C. 2010. Effects of aqueous seed extract of *Persea americana* Mill. (avocado) on blood pressure and lipid profile in hypertensive rats. Adv. Biol. Res. 4(2): 116–121.

Kasole, R., Martin, H.D. and Kimiywe, J. 2019. Traditional medicine and its role in the management of diabetes mellitus: "patients' and herbalists' perspectives". Evid-based Complement Altern. Med. 1–12.

Kopin, L. and Lowenstein, C.J. 2017. Dyslipidemia. Ann. Intern. Med. 167(11): ITC81–ITC96.

Kumari, P.D.S.R. and Amarakoon, R. 2021. Fermentation of *Persea americana* seed flour using Lactobacillus plantarum and investigate its' effect on nutritional quality. European J. Agric Food Sci. 3(5): 55–59.

Lyu, F., Luiz, S.F., Azeredo, D.R.P., Cruz, A.G., Ajlouni, S. and Ranadheera, C.S. 2020. Apple Pomace as a functional and healthy ingredient in food products: a review. Processes 8(3): 319.

Mozaffarian D. 2016. Dietary and policy priorities for cardiovascular disease, diabetes, and obesity: a comprehensive review. Circulation 133(2): 187–225.

Mudassir, H.A., Khaliq, S., Azmi, M.B., Bano, M., Naheed, M. and Fatima, M. 2020. *Persea americana*seeds improve glycosylation and dyslipidemia in fructose-fed streptozotocin-injected type 2 diabetic male rats. Pak. J. Pharm. Sci. 33(6).

Nardi, L., Lister, I.N., Girsang, E. and Fachrial, E. 2020. Hypoglycemic effect of avocado seed extract (*persea americana* mill) from analysis of oral glucose tolerance test on Rattus norvegicus L. Am. Sci. Res. J. Eng. Tech. Sci. 65(1): 49–56.

Noorul, H., Mujahid, M., Khalid, M., Vartika, S., Nesar, A., Zafar, K. et al. 2017. Physico-phytochemical analysis and Estimation of total phenolic, flavonoids and proanthocyanidin content of *Persea americana* (avocado) seed extracts. World J. Pharm. Sci. 70–77.

Nwaoguikpe, R.N., Braide, W. and Ujowundu, C.O. 2011. Biochemical composition and antimicrobial activities of the seed extracts of Avocado (*Persea americana*). J. Microbiol. Antimicrob. 3(7): 184–190.

Ojo, O.A., Amanze, J.C., Oni, A.I., Grant, S., Iyobhebhe, M., Elebiyo, T.C. et al. 2022. Antidiabetic activity of avocado seeds (Persea americana Mill.) in diabetic rats via activation of PI3K/AKT signaling pathway. Sci. Rep. 12(1): 2919.

Pahua-Ramos, M.E., Ortiz-Moreno, A., Chamorro-Cevallos, G., Hernández-Navarro, M.D., Garduño-Siciliano, L., Necoechea-Mondragón, H. et al. 2012. Hypolipidemic effect of avocado (*Persea americana* Mill) seed in a hypercholesterolemic mouse model. Plant Foods Hum. Nutr. 67(1): 10–16.

Rodríguez-Carpena, J.G., Morcuende, D., Andrade, M.J., Kylli, P. and Estévez, M. 2011. Avocado (*Persea americana* Mill.) phenolics, in vitro antioxidant and antimicrobial activities, and inhibition of lipid and protein oxidation in porcine patties. J. Agric Food Chem. 59: 5625–5635.

Saklayen, M.G. 2018. The global epidemic of the metabolic syndrome. Curr. Hypertens Rep. 20(2): 12.

Salazar-López, N.J., Domínguez-Avila, J.A., Yahia, E.M., Belmonte-Herrera, B.H., Wall-Medrano, A., Montalvo-González, E. et al. 2020. Avocado fruit and by-products as potential sources of bioactive compounds. Food Res. Int. 138(Pt A): 109774.

Shehata, M.M.S.M. and Soltan, S.S. 2013. Effects of bioactive component of kiwi fruit and avocado (fruit and seed) on hypercholesterolemic rats. World J. Dairy Food Sci. 8(1): 82–93.

Sun, Z. 2014. Atherosclerosis and atheroma plaque rupture: normal anatomy of vasa vasorum and their role associated with atherosclerosis. Sci. World J. 285058.

Tabeshpour, J., Razavi, B.M. and Hosseinzadeh, H. 2017. Effects of avocado (*Persea americana*) on metabolic syndrome: A comprehensive systematic review. Phytother. Res. 31(6): 819–837.

Talabi, J.Y., Osukoya, O.A., Ajayi, O.O. and Adegoke, G.O. 2016. Nutritional and antinutritional compositions of processed Avocado (*Persea americana* Mill) seeds. Asian J. Plant Sci. Res. 6(2): 6–12.

Terati, M., Zikrullah, L., Puspitasari, M., Hasri, R.N. and Wahyuni, T. 2021. The substitution of avocado seed flour to rice flour in the manufacture of traditional palembang food gandus cake. pp. 37–40. In First International Conference on Health, Social Sciences and Technology (ICOHSST 2020). Atlantis Press.

Tremocoldi, M.A., Rosalen, P.L., Franchin, M., Massarioli, A.P., Denny, C., Daiuto, É.R. et al. 2018. Exploration of avocado by-products as natural sources of bioactive compounds. PloS One 13(2): e0192577.

Weschenfelder, C., dos Santos, J.L., de Souza, P.A.L., de Campos, V.P. and Marcadenti, A. 2015. Avocado and cardiovascular health. Open J. Endocr. Metab. Dis. 5(07): 77.

World Health Organization. 2017. Official site: https://www.who.int/es/news-room/fact-sheets/detail/cardiovascular-diseases-(cvds).

World Health Organization. 2021. Official site: https://www.who.int/news-room/fact-sheets/detail/cardiovascular-diseases-(cvds)

Chapter 15

Perspectives for the Reuse of Bacuri (*Platonia insignis*) for Nutritional and Biotechnological Applications

Ana Paula Rodrigues de Sousa,[1] *Pedro Vitor Oliveira Silva Furtado,*[1] *Maria Geovana Santos Ribeiro,*[1] *Antônio do Nascimento Cavalcante,*[2] *Mahendra Rai*[3] and *Chistiane Mendes Feitosa*[4,*]

Introduction

Since the dawn of civilization, plant species have been a source of resources within man's reach, allowing the accumulation and experience-based learning of information about the environment through constant observation of natural phenomena in order to improve feeding conditions, and consequently nutrition and in the permanent search for the cure of many diseases, thus demonstrating, a narrow relationship between the use of medicinal plants and their own evolution in the livelihood process.

In this context, Brazil having the largest and richest biodiversity on the planet, stands out as a nation with the greatest potential for research with various plant species (Nascimento, 2013). According to some researchers, such as Noldin and coworkers (2006) and Nualkaew and coworkers (2012), one of the families of medicinal plants, where fruits can be widely used in food and that has been attracting a wide interest in the scientific community due to its promising chemical and biological properties, is

[1] Department of Chemistry, Federal University of Piauí, 64049-550, Teresina, PI, Brazil.
[2] Institute of Education, Science, and Technology of Maranhão, Campus President Dutra, 65760-000, President Dutra, MA, Brazil.
[3] Department of Biotechnology. SGB Amravati University, Amravati-444 602, Maharashtra State, India.
[4] Federal University of Piaui, Department of Chemistry and Pharmacy, 64049-550, Teresina, PI, Brazil.
* Corresponding author: chistiane@ufpi.edu.br

the *Clusiaceae* family. This family includes approximately 1000 species belonging to 47 genera, distributed in mainly tropical and subtropical regions of the planet and one genus that reaches the temperate climate regions in nine of these genera, about 90 species, the fruits are edible (Barroso et al., 2002; Costa Júnior et al., 2011) and therefore, adding value to these vegetables. In addition, many of these species are commonly used in folk medicine to treat various diseases such as pain, infection, inflammation and to combat ulcers (Noldin et al., 2006), among others.

The bacurizeiro is a plant that belongs to the *Clusiaceae* family, subfamily *Clusioideae*, genus *Platonia* and species *Platonia insignis* Mart. (Braga, 1976). The term *Platonia* is an attribute of Plato, a Greek philosopher, and *insignis* means

Figure 15.1. Fruits of *Platonia insignis* Mart.

notable, important, great, which attracts a lot of attention, in reference to the physical structure and usefulness of the plant, as well as the size, flavor and aroma of the fruits (Barroso et al., 2002; Moura et al., 2007).

P. insignis Mart is a large pole fruit tree, with a height ranging from 15 to 25 meters and a diameter of up to 100 centimeters (Cavalcante, 1988). This tree is native to the Amazon region of Brazil and Guyana, but is also found in Colombia and Paraguay, always occurring in open areas of low vegetation, clearings and mainly in secondary vegetation, rarely found in dense primary forest (Cavalcante, 1988; Mourão and Beltrati, 1995; Chitarra and Chitarra, 2005). In Brazil, the center of dissemination is the state of Pará, where a wide variation in fruit shape and size, pulp yield and quality and other characteristics of economic interest are found. It can also be found in the states of Maranhão, Mato Grosso, Piauí and Goiás (Villachica et al., 1996; Aguiar et al., 2014).

Bacuri is a large, rounded fruit with a thick skin and a yellowish color, as can be seen in Fig. 15.1. The pulp is inside the bark, which is whitish in color, viscous, and has a very pleasant smell and taste (Fontenele et al., 2005). The fruit presents great potential, both from the point of view of its industrial processing, for the preparation of ice creams, creams, refreshments, jams and jellies and for consumption *in natura* (Silva et al., 2009). When it is in the ripening process, it exudes a soft and fragrant perfume, very rich in terpenes, a secondary metabolite of extreme commercial value (Alves and Jennings, 1979).

The bark of the fruit can also be used to make sweets, creams and ice cream, increasing the yield of the fruit and consequently adding value. This process must take place after the resin in this part of the fruit has been separated (Aguiar et al., 2014). The ever-increasing demand for fruits has stimulated producers to establish cultivation areas. The bacuri seeds have no use in food, however, they can be used in the production of soap or bacuri lard. The oil extraction process occurs with great difficulty since the seeds are soaked in water for a long time and then boiled, and the oil is removed from the surface of the boiling water. Bacuri butter gives a golden tone to the skin, a few minutes after its application, it is absorbed and the skin is left with a velvety touch (Ferreira, 2008; Costa Júnior et al., 2011).

Although most of the studies on the bacurizeiro tree focus on the use of its flowers and fruits, the wood also has wide applicability, being compact, resistant and of good quality. As its main physical characteristic, it presents a pinkish-beige and light beige color, and can be used mainly in carpentry, furniture, flooring, laths and civil and naval construction, among others (Aguiar, 2006).

In this chapter, the chemical composition and nutritional aspects of *P. insignis*, the pharmacological activities, the use of the various parts of *P. insignis* in cooking and the biotechnological applications are highlighted.

Chemical composition: the nutritional aspects of *Platonia insignis* Mart.

Platonia insignis Mart. popularly known as bacuri is very valued by the population, especially for the pulp of its fruit, which has a unique aroma and flavor making it the most consumed part (Santos et al., 2020). The fruits are not only used due to their smell and taste, but also because they are important sources of essential nutrients with high nutritional power, such as vitamins, proteins, lipids, carbohydrates, fiber, minerals and other important substances including natural antioxidants and phenolic compounds that should be consumed daily in order to maintain a healthy diet and prevent non-transmissible chronic diseases (Santos et al., 2019).

Thus, several studies have been developed seeking to evaluate the nutritional characteristics of these fruits (Ferreira and Melo, 2007), as well as the chemical composition, since the presence of some classes of compounds gives the specific medicinal properties of extreme importance for chemical, pharmaceutical and food industries (Freitas et al., 2017).

Rogez et al. (2004) evaluated the nutritional content of the pulp in relation to Dry Matter (DM), determining the presence of several constituents. The pulp is described as pasty and viscous, this aspect comes from the DM content, which in this study was about 21%. Its protein content was considered poor, and its carbohydrate content relatively low compared to other tropical fruits, with glucose, fructose and sucrose present in similar amounts, while the TDF had a high content, around 28%, which is interesting from a nutritional point of view due to its benefits related to intestinal and colon physiology.

Another highlight of Rogez et al. (2004) concerns the richness of the pulp in relation to the minerals obtained, some of which are essential to the body, such as potassium and zinc. Their presence is also related to the high content of TDF. Thus, this fruit can also be characterized by its typical tropical flavor, having a good amount of soluble sugar that can be used in various ways in cooking. Besides its use in cooking, the bacurizeiro tree also has its application in ethnopharmacology due to its rich composition in phytochemical compounds such as terpenes, xanthones and phenolic compounds that are some of the main constituents (Yamaguchi et al., 2014).

The current literature provides evidence of phytochemical studies conducted on the fruits (bark, pulp and seeds), flowers and branches (Castelo, 2018). Some pharmacologically important chemical constituents present in the pulp have already been identified using techniques such as Gas Chromatography, High Concentration Capacity Headspace techniques (HCC-HS) and Headspace Solid Phase Microextraction (HS-SPME). Linalool (Fig. 15.2) was the main compound obtained by Uekane et al. (2017) in their studies.

Rufino et al. (2010) reported the presence of carotenoids, vitamin C, total anthocyanins and flavonoids obtained from the ethanolic extract of the pulp. In another study, Viera et al. (2011) reported the presence of phenolic compounds in fruits from the study of aqueous and hydroalcoholic extracts, pointing out that water is more efficient in extracting these substances which leads to the presence of more polar constituents in tropical fruits, in agreement with other evidence in literature (Wu et al., 2004).

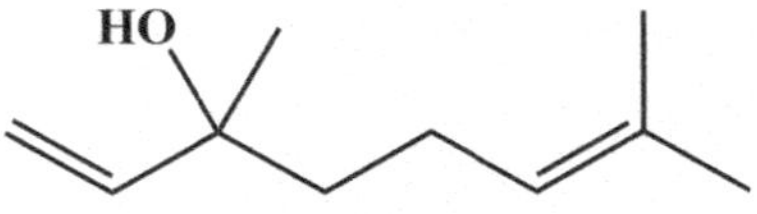

Linalol

Figure 15.2. Chemical structure of linalool.

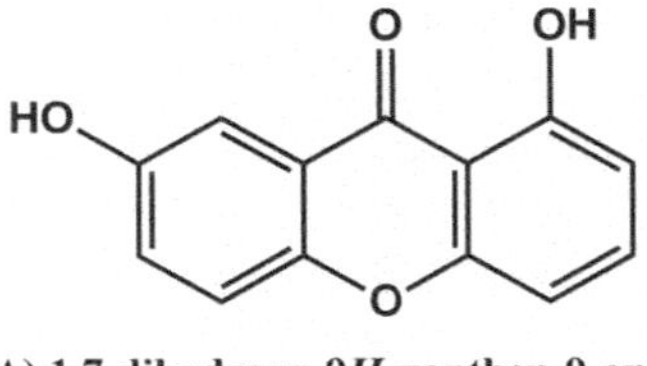

(A) 1,7-dihydroxy-9*H*-xanthen-9-one

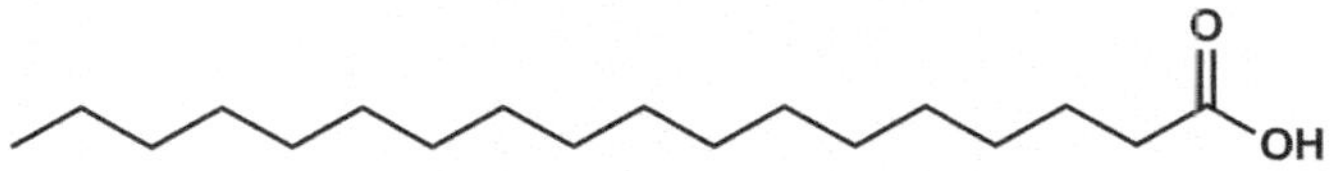

(B) stearic acid

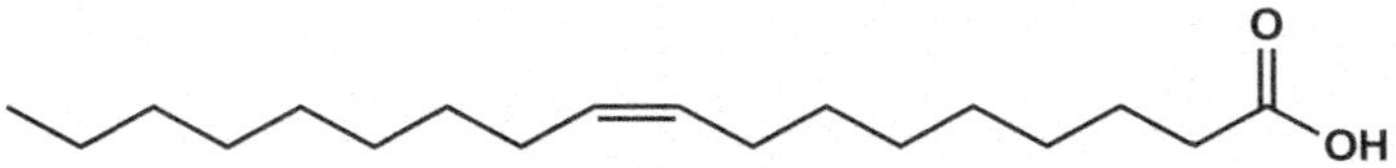

(C) oleic acid

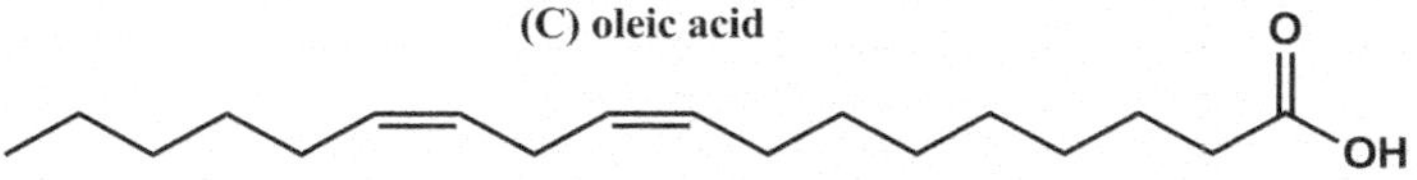

(D) linoleic acid

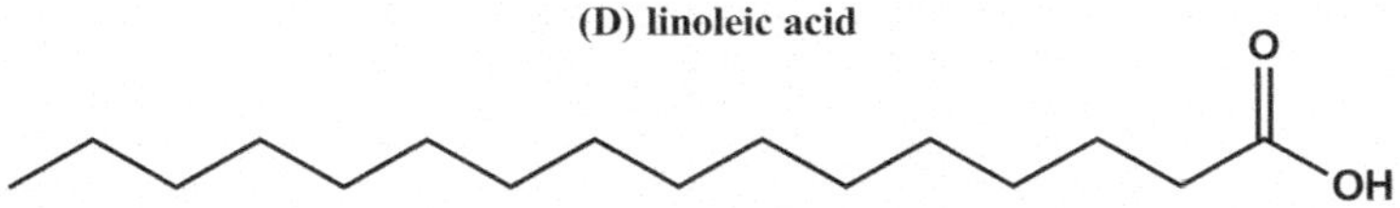

(E) palmitic acid

Figure 15.3. Compounds present in *Platonia insignis* Mart.

In studies carried out on the seeds, one can see the presence of fatty acids, triacylglycerols and metabolites such as xanthones that are responsible for some of the pharmacological activities of this species (Fig. 15.3a) (Santos et al., 2013). The seeds are rich in apolar compounds, being a source of oils containing fatty acids such as stearic, oleic, linoleic and palmitic (Fig. 15.3b, c, d, e) (Yamaguchi et al., 2014).

From fractions with lower polarity character obtained from hexanic extracts of this fruit part, compounds such as 2-oleyl-1,3-dipalmitoyl-glycero (ODG, Fig. 15.4a) which is a triacylglycerol (Cavalcante et al., 2020) and a 1,3-diastearoyl-2-oleoylglycerol (TG1) a triacylglyceride derived from trioleate of formula $C_{57}H_{108}O$ (Fig. 15.4b), characterized from spectroscopic techniques such as Infrared (IR), Carbon and Hydrogen Nuclear Magnetic Resonance (RMN C^{12} and RMN H^{1}) (Feitosa et al., 2016). The garcinielliptone FC (GFC) (Fig. 15.4c) was also isolated from less polar fractions and are polycyclic polyprenylated benzophenones, which constitute a tautomeric pair, these are compounds that have been investigated for

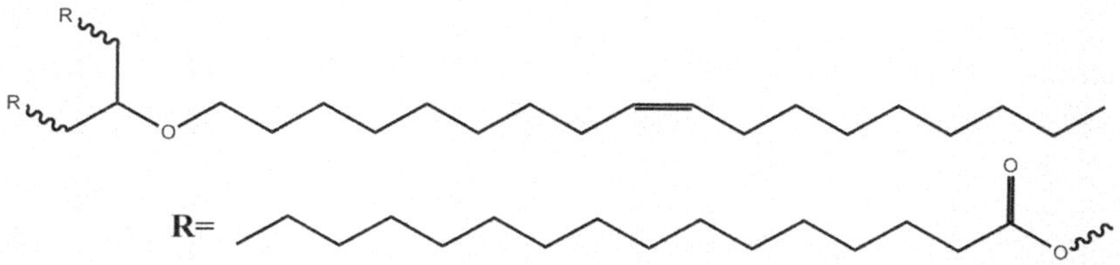

(a) 2-oleyl-1,3-dipalmitoyl-glycerol (ODG)

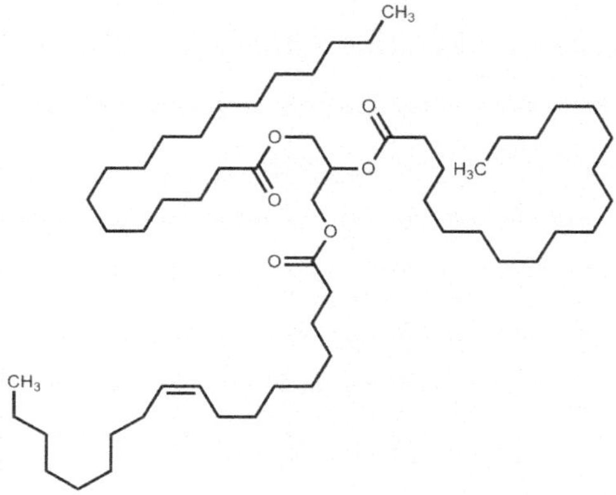

(b) 1,3-distearoyl-2-oleoylglycerol (TG1)

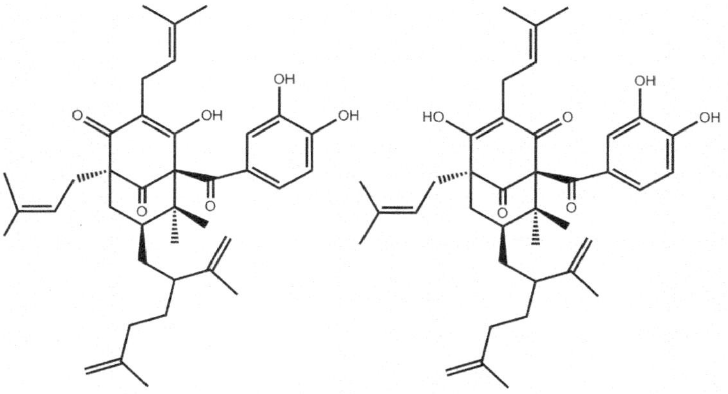

(c) **(d)**

(c) e (d) Garcinueliptona FC (GFC)

Figure 15.4. Substances isolated from the seeds of the *Platonia insignis* Mart.

their pharmacological application (Costa Junior et al., 2013; Arcanjo et al., 2014; Silva et al., 2014).

Investigations from more polar fractions derived from the ethanolic extract of the seeds were also performed, resulting in the isolation of substances obtained from the ethyl acetate (AcOet) and dichloromethane ($CHCl_2$) fraction, with 1,3,5,6-tetrahydroxy-2-(2-methylbut-2-en-2yl)-7-(3-methylbut-2-enyl)xanthen-9-one, the main constituent present in the fraction $CHCl_2$ with a yield of 47,82% and the alpha-mangostin and 1,3,5,6-tetrahydroxy-2-(2-methylbut-3-en-2-yl)-7-(3-methylbut-2-enyl)xanthen-9-one the compounds present in larger amounts in the fraction AcOet with 40,74% and 40,11%, respectively, the isolated substances are presented in Table 15.1 (Costa Junior et al., 2012).

Another part of the fruit that has been studied a lot is the bark, which corresponds to most of the fruit's composition (50 to 80%) and is generally discarded for culinary and industrial uses, however, it can be used both for obtaining olive oil and as a natural antioxidant that can be a technological innovation in the cosmetic industry, since its composition has the presence of bioflavonoids such as the morelloflavone that presents this activity, besides being rich in resins that can be extracted and used for other industrial uses (Bezerra et al., 2005).

Monteiro et al. (1997) compared the methods of extraction of liquid carbon dioxide (LCO_2), supercritical carbon dioxide, steam distillation, cold ethanol, Soxhlet extractions with ethanol, LCO_2 and ethanol and obtained the free fatty acids such as palmitic, linoleic and α-linolenic, oleic and stearic acids as the predominant constituents. Caprylic and myristic acids, the alcohols linalool and 3,7-dimethyloct-1-en-3,7-diol, methyl citrate, eugenol (a phenolic ether), in addition to hydrocarbons and oxides. Besides the fruits, studies with other parts of the plant have also presented interesting results, lupeol is a triterpenoid that has relevance in pharmacology (Fig. 15.5) and can be isolated from the hexanic fraction obtained from the ethanolic extract of stem barks of this species (Souza et al., 2017). The branch extracts, on the other hand, have the presence of secondary metabolites such as bioflavonoids (majority constituents), as well as benzophenones and xanthone. Silva et al. (2021) conducted a study on branch extract of the plant and used fractions and subfractions obtained from the ethanolic extract, resulting in the isolation of 11 compounds, 10 of which were characterized and presented in Table 15.1.

The compounds were analyzed in the mass spectrometer and characterized by comparing the spectra and fragmentations with literature data, the compounds **2, 6** and **8** were classified as bioflavonoids of the morelloflavone type, **3, 7** and **9** bioflavonoids of the genus *Garcinia* (of type GB1a), **4** and **10** GB1, these classifications are made based on the binding pattern of their monomers (Silva et al., 2021). The constituents morelloflavone **(6)** and garcinylipton FC (Fig. 15.4c), were also identified in the flowers of *P. insignis* Mart. along with two other phytochemical constituents, quinic acid **(12)** and volkensiflavona **(11)**. Fractions of different polaritis (hexane, AceOEt and $CHCl_2$) were obtained from the hydroalcoholic extract and allowed the identification of more compounds that are also presented in Table 15.1. The constituent **13** was a mixture of campesterol, stigmasterol and sitosterol obtained from the hexanic fraction after a methylation reaction allowed the identification of

Table 15.1. Main constituents isolated from *Platonia insignis* Mart.

Part	Extract	Fraction	N°	Constituents	Reference
branches	Extract ethanolic	Fraction AcOEt	1	α-Mangostin	Silva et al. (2021)
			2	Morelloflavona	
			3	GB-2a	
			4	GB-2	
			5	Garciniliptona FC	
			6	Morelloflavona-7″-*O*-sulfato	
			7	GB-2ª-sulfato	
			8	Fukugisídio	
			9	GB-2ª-*O*-glicopiranosídio	
			10	GB2-7"-*O*-glicopiranosídeo	
flowers	Hydroalcoholic	–	11	Volkensiflavona	Pinheiro (2016)
			12	Ácido Quínico	
		Hexanic	13	Campesterol, estigmasterol e sitosterol	
			14	Hexadecanoato de metila C16:0	
			15	Octadeca-9-12-dienoato de metila C18:2	
			16	(9Z)-octadec-9-enoato de metila C18:1	
			17	Octadecanoato de metila C18:0	
			18	Sitosterol	
		Dichloromethane	19	Catequina	
			20	Sinfonona F	
			21	Garcinoma B	
			22	3,5-dihidroxi-1,2-dimetoxixantona	

Table 15.1 contd. ...

...Table 15.1 contd.

Part	Extract	Fraction	N°	Constituents	Reference
Seeds	Ethanolic extract	Dichloromethane fraction	23	Trimethyl citrate	Junior et al. (2012)
			24	Hexadecanoic acid methyl ester	
			25	6-Octadecenoic acid methyl ester	
			26	11,14-Eicosadienoic acid, methyl ester	
			27	1-hidroxy-3,5-6-trimethoxy-xanthen-9-one	
			28	1,3,5,6-tetrahydroxy-2-(2-methylbut-3-en-2-yl)-7-(3-methylbut-2-enyl)xanthen-9-one (gamma-mangostin)	
		Ethyl acetate fraction	29	Hexadecanoic acid methyl ester	
			30	Heptadecanoic acid methyl ester	
			31	10-Octadecenoic acid methyl ester	
			32	9-Octadecenoic acid methyl ester	
			33	1,3,6-trihydroxy-7-methoxy-2,8-bis(3-methylbut-2-enyl)xanthen-9-one (alpha-mangostin)	
			34	1,3,5,6-tetrahydroxy-2-(2-methylbut-3-en-2-yl)-7-(3-methylbut-2-enyl)xanthen-9-one (gamma-mangostin)	

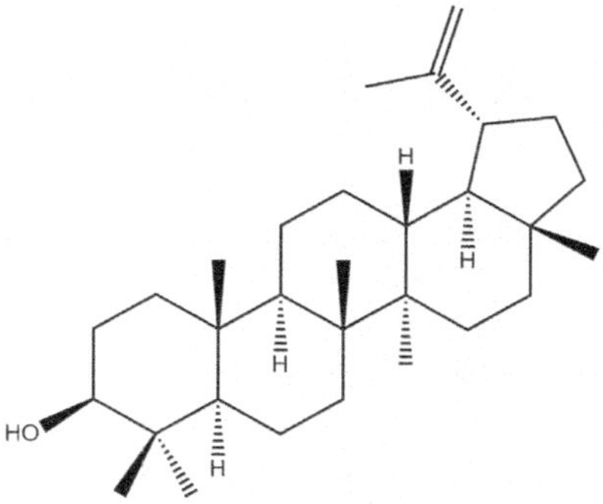

Lupeol

Figure 15.5. Chemical structure of Lupeol.

the compounds **14, 15, 16, 17** and **18**. Already the fraction CH_2Cl_2 provided the constituents **2, 5, 19, 20 and 11,** which after a silylation reaction also provided the compounds **21 and 22** (Pinheiro, 2016).

Pharmacological activities of *Platonia insignis* Mart.

P. insignis is an important medicinal plant and is widely used in Brazilian popular medicine, mainly for the treatment of eczema, herpes, gastrointestinal diseases, dermatitis, schistosomiasis, leishmaniasis and malaria. Studies involving biological-pharmacological activities of this species can be found in literature, and most of these studies focus mainly on extracts of its seeds. Extracts from parts of this plant have demonstrated important pharmacological activity as anti-inflammatory, antimalarial, antihypertensive, antidiabetic, immunomodulatory, antiviral, antitumor, antidepressant, antiallergic, antimutagenic and antioxidant effects (Bilanda et al., 2010; Costa Júnior et al., 2013). Additionally, these extracts are widely used in folk medicine for the treatment of several diseases such as diarrhea, skin problems, earaches, spider and snake bites, rheumatism, arthritis and even as a cicatrizing agent (Costa Júnior et al., 2013; Mendes et al., 2015).

One of the by-products obtained from the oil of the seeds is "bacuri lard" this by-product is widely used in popular medicine as a cicatrizing agent for burns and to treat dermatological diseases (Agra et al., 2007). In *in vivo* tests, lard has shown the potential to accelerate the healing of skin wounds in rats (Santos et al., 2013; Feitosa et al., 2016). The seed decoction is often used to treat diarrhea, while the seed oil is used both against spider or snake bites and to treat skin problems, otitis, rheumatism and arthritis (Agra et al., 2007; Mendes et al., 2015).

Pharmacological studies have shown that the ethyl acetate fraction of the seeds of *P. insignis* possesses potential anticonvulsant activity in pilocarpine-induced rats (Costa Júnior et al., 2011). The hexanic, dichloromethane, ethyl acetate extracts and especially isolated tautomeric substances, such as garcinielliptona, showed potential leishmanicidal activity against promastigotes of *Leishmania amazonensis* (Costa Júnior et al., 2013).

The *in vitro* antioxidant property of bacurizeiro has been usually reported, through the sequestration of free radicals. Thus, it presents a possible protective action against the onset and/or development of degenerative processes associated with various types of diseases (Rufino et al., 2010; Vieira et al., 2011; Costa Júnior et al., 2013).

Mendes et al. (2015) tested the cicatrizing activity, in *Wistar* rats, of a paste pharmaceutical formulation containing the compound 1,3-diestearoyl-2-oleyl-glycerol, isolated from the hexanic extract of the seeds of *P. insignis*. Macroscopic and histological analyses of the rat wounds demonstrated possible healing activity of the compound. Other biological activities reported in literature for the seeds of *P. insignis* are antimicrobial against strains of *Saccharomyces cerevisiae*, cytotoxicity in *Artemia salina*, genotoxic effect on Chinese *hamster* lung fibroblasts (V79), a pilocarpine-induced anticonvulsant effect, healing, anti-inflammatory activity and stimulatory effect on the Central Nervous System (CNS) in mice (Santos Júnior et al., 2010; Costa Júnior et al., 2011; Costa Junior et al., 2012; Costa Júnior et al., 2013).

Culinary use of bacurizeiro parts

The Brazilian food industry, throughout its production chain, generates an excessive amount of waste, especially fruit processing, from the seeds, pulp and bark. This issue has sparked the interest of researchers in studies that seek to minimize the waste of these inputs from the reuse of materials that have a nutritional richness, physiological functions and a technological application, minimizing the generation of waste during the processing of products such as the pulp (Sousa et al., 2011a; Sousa et al., 2011b). As mentioned earlier, the nutritional value and applicability of the fruit of *P. insignis* are considerably vast in both popular and industrial use. Thus, studies have been developed in order to further expand the diversity of uses of the fruit as the development of new types of foods (Aguiar, 2006).

In this context, one of the main factors that also contribute to the formulation of new foods is related to the growing number of consumers in the market who are concerned about their healthy lifestyle. Over the years, scientific findings disseminated by the media such as the incidence of cardiovascular accidents, cancer, cerebrovascular accident and arteriosclerosis, among others, caused in part by the individual's poor eating habits, have directly affected the food choices of the general public. In this way, with the growing interest in healthy foods, a need was generated in the market to meet this demand, and furthermore, studies still prove that bacuri has enormous potential to delay most of these diseases mentioned, due to its numerous antioxidants and the amounts of vitamins, minerals and fiber, making it an indispensable food in a healthy diet (Avello and Suwalsky, 2006; Canuto et al., 2010).

Thus, an example of the studies developed in the food area is in the research developed by Mendoza et al. (2017) that elaborate analysis of the formulation of granola with the addition of bacuri pulp and nuts for consumption in yogurt, through sensory tests it was possible to determine an improvement in the properties of color, texture and nutritional value, proving that its consumption has become a practical and healthy option. Additionally, the work of Rodrigues (2012) presents the possibility of preparing a soy-based bacuri pulp ice cream for lactose intolerant people, obtaining results according to the legislation in force with microbiological aspects, the product achieved good acceptance in the tests further proving the versatility of the fruit in food composition.

The bark and seeds of the fruit are also used, although the pulp is the main product, they are used in the manufacture of sweets and creams, however, the use of the bark is difficult due to the large concentration of resin (Aguiar, 2006). The research of Vasconcelos et al. (2018) is important for the manufacture and characterization of a cookie produced from the bacuri bark, even with the high concentration of resin present in the fruit bark. The product still acquired a good acceptance by consumers with reasonable percentages in the market.

Another problem in the Brazilian food industry concerns the issue of post-harvest preservation in regions far from large commercial centers, as is the case of Pará. The waste of *in natura* food has also driven the creation of new technologies and methodologies that could expand the time of preservation of fruit while maintaining sensory, microbiological and nutritional quality (Fontenele, 2007). Fruit structuring can be mentioned as one of the main preservation techniques, has shown promising results, and occurs by means of gelling which allows the final product to present significant retention of its nutritional properties. The structured fruit is an example of a low-cost raw material with great yield since the bacuri bark is also used, as in the case of bacuri. The use of bacuri's bark as a structured fruit represents an alternative utilization besides a chance of income for small producers (Vijayanand et al., 2000).

Biotechnological activities and future perspectives

As described earlier, *P. insignis* Mart. is one of the most profitable plants in several areas of study, whether pharmaceutical, food or industrial (Fig. 15.6), driving the creation of new methodologies and formulations that further explore the properties that the bacurizeiro tree presents. The species resonates with modern trends seeking to meet the needs of the population (Chirife and Favetto, 1992; Aguiar, 2006).

In the food area, modern technologies have been used for the development of new products, to replace costly preservation methods, such as heat treatments, with other methods that do not require as much energy, but still guarantee quality and a stable product. Due to the increasing trends of healthy food consumption, and the marketing of products with minimal industrialized processing, with a significant reduction of additives, natural and preserved foods, there has been interest in using the combined methods in the food industry (Alzamora et al., 1993; Argaiz et al., 1993).

Some studies with bacuri pulp demonstrated significant preservation through the use of combined methods, the barriers employed guaranteed microbiological stability, sensory characteristics, nutritional quality and convenience, therefore being suitable for the consumption of the product as a self-stable food. In addition, this methodology has proved to be an alternative for reducing post-harvest losses, extending the shelf life of the pulps and using few obstacles that do not compromise the final product (Chirife and Favetto, 1992; Aguiar, 2006).

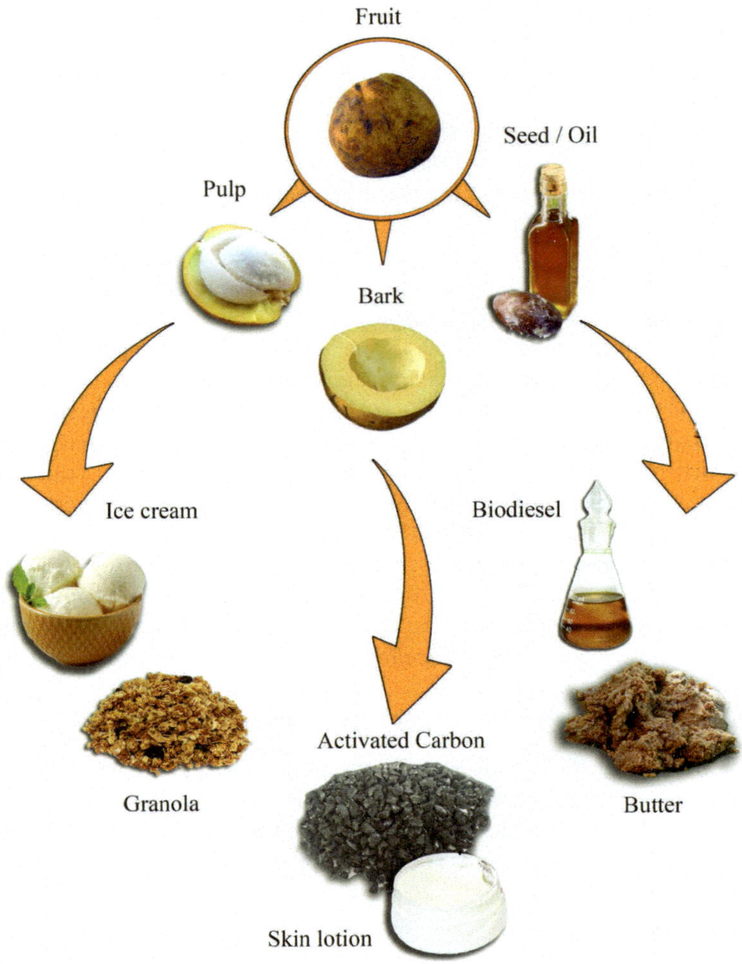

Figure 15.6. Prospects for use of the bacuri fruit.

In pharmacology, from the use of isolated chemical species, biotechnologies such as the use of inclusion complexes may be an option to improve the physicochemical characteristics of substances promoting a better interaction with the organisms (Fraceto et al., 2007). Thus, Cavalcante et al. (2019) developed a complex using ODG (Fig. 15.4) (a substance that can be isolated from bacurizeiro) with β-cyclodextrin, being the first work to make this type of encapsulation with a substance isolated from this species, they obtained the effective formation of the complex causing a modification in its release profile and in its solubility, improving its bioavailability and promoting a promising approach for its clinical application.

The properties of this species also enabled the study of a formulation with amphotericin B (AmB) which is an antimicrobial agent with a molecular structure that has hydrophilic and hydrophobic regions, widely used in the treatment of fungal and parasitic diseases such as leishmaniasis (Falci and Pasqualotto, 2015). Coelho et al. (2018) conducted an *in vitro* study where a topical AmB emulgel was developed with bacuri butter, which demonstrated antileishmanial action and a promising potential for topical use. Lustosa et al. (2021) also developed a formulation with the same action based on bacuri butter and applied it in an *in vivo* test, obtaining satisfactory results when compared to a butter-free formulation, thus proving the effectiveness of formulations based on the use of this material.

Another line of studies related to the use of bacuri properties uses compounds such as pectins that can be obtained from the bacuri bark (Bezerra et al., 2005). These compounds are widely used by the food industry to make jellies, but they also have a non-edible application by the cosmetics industry as a natural texturizing agent in creams and by the pharmaceutical industry due to their biological effects (Fani, 2012). The cosmetics industry can also take advantage of the grease composition of the seeds to use the oil as an emollient and humectant. This oil has high absorption due to the presence of tripalmitin which allows rapid penetration into the skin (Morais and Gutjah, 2012).

In the biofuels industry, one of the future applications of the species that stands out the most is linked to the production of biodiesel, the use of this fuel has shown energy performance similar to common diesel, derived from petroleum, and maybe a more ecological alternative because its synthesis is derived from the reaction of transesterification of vegetable/animal oils and fats, being biodegradable and its burning releases much less carbon monoxide and hydrocarbons. Some research shows that bacuri oil is promising for the production of biofuels, due to its superior thermal stability and particular composition, which further adds value to the extracts of the fruit (Soares, 2010; Yamaguchi et al., 2014).

Additionally, even if it is not used as a matrix material for biodiesel production, *P. insignis* oil can still be used as an antioxidant for the biofuel, increasing its resistance to oxidation, due to the presence of active metabolites such as phenolic compounds, flavonoids, fatty acids and carotenoids among others (Chendynski et al., 2020).

While on the topic of sustainability, in Brazil wood is mostly used for charcoal production, however, less polluting alternatives have been used that are able to take advantage of byproducts from the agricultural industry, as is the case of activated charcoal produced from the bacuri bark. Activated carbon proves to be an excellent substitute because besides being less aggressive to the environment, it also has the ability to adsorb liquids and gases, this property is widely applicable in the industry because it can be used both in water treatment, gas purification, molecular sieves, catalysts and pollutant emission control (Sousa et al., 2015).

Conclusion

As can be observed in this chapter, the nutritional aspects of the bacuri pulp show richness in relation to the presence of essential minerals in the body and satisfactory amounts of soluble sugars, along with the presence of fiber, carbohydrates and proteins. Its chemical composition is a source of carotenoids, vitamin C and phenolic compounds, while other parts of the fruit, such as the seeds and bark, are rich in more apolar compounds such as fatty acids, and can be used to obtain oil, lard, butter and olive oil, which are used for various purposes. From the flowers, branches and stems, constituents such as biflavanoids and xanthones can be extracted, in addition to substances such as lupeol, which has an important application from the pharmacological point of view.

In literature, there are several studies on the pharmacological activities of extracts and compounds extracted from this species, such as anti-inflammatory, antimalarial, antihypertensive, antidiabetic, immunomodulatory-respiratory, antiviral, antitumor, antidepressant, antiallergic, antimutagenic effects, antioxidant, leishmanicidal activity, antimicrobial, anticonvulsant, cicatrizant and stimulant effect on the central nervous system. Moreover, *P. insignis* is a medicinal plant widely used in Brazilian popular medicine, mainly for the treatment of eczema, herpes, gastrointestinal diseases, dermatitis and schistosomiasis.

This chapter demonstrated that the *Platonia insignis* Mart. species has a diversified application due to its physical and sensory characteristics that instigate not only the production of traditional foods such as sweets and derivatives but also the development of new foods such as cookies produced from the bark. Besides its culinary use, the species is also being evaluated for its use in the energy sector to manufacture biodiesel or as an oxidizing agent of it, in the production of activated carbon, saving burning of wood in the production of traditional charcoal, and in the cosmetic industry, acting in the formulation of emulsifiers.

Finally, the use of this species for various purposes have been highlighted. More and more studies are being carried out about its application, hoping that parts not used by the industries can be reused, reducing the generation of waste, contributing to the environment and the total use of the species.

References

Agra, M.F., Freitas, P.F. and Barbosa-Filho, J.M. 2007. Synopsis of the plants known as medicinal and poisonous in Northeast of Brazil. Rev. Bras Farmacogn. 17: 114–140.

Aguiar, L.P. 2006. Qualidade e potencial de utilização de Bacuri (*Platonia insignis* Mart.) Oriundos da Região Meio-Norte. Dissertação (Mestrado). 122 f.

Aguiar, U.N., Lima, S.G., Rocha, M.S., Freitas, R.M., Oliveira, T.M., Silva, R.M., Moura, L.C.B. and Almeida, L.T.G. 2014. Preparação e caracterização do complexo de inclusão do óleo essencial de *Crton zehntneri* com β-ciclodextrina. Quim Nova. 37: 50–55.

Alves, S.M. and Jennings, W.G. 1979. Volatiles composition of certain Amazonian fruits. Food Chem. 4: 149–159.

Alzamora, S.M., Tapia, M.S., Argaiz, A. and Welti, J. 1993. Application of combined methods technology in minimally processed fruits. Int. Food Res. J. 26: 125–130.

Arcanjo, D.D.R., Costa-Junior, J.S., Moura, L.H.P., Ferraz, A.B.F., Rossatto, R.R., David, J.M., Quintans-Junior, L.J., Oliveira, R.C.M., Citó, A.M.G.L. and Oliveira, A.P. 2014. Garcinielliptone FC, a polyisoprenylated benzophenone from *Platonia insignis* Mart., promotes vasorelaxant effect on rat mesenteric artery. Nat. Prod. Res. 28(12): 923–927. http://dx.doi.org/10.1080/14786419.2014.889136.

Argaiz, A., Vergara, F., Welti, J. and Lõpez-Malo, A. 1993. Durazano conservado por factores combinados. Bol. Int. Divulg. 1: 22–30.

Avello, M. and Suwalsky, M. 2006. Radicales libres, antioxidantes naturales y mecanismos de proteccion. Atena. 2(494): 161–172.

Barroso, G.M., Peixoto, A.L., Ichaso, C.L.F., Costa, C.G., Guimarães, E.F. and Lima, H.C. 2002. Sistemática de Angiospermas do Brasil Minas Gerais. 309 p.

Bezerra, G.S.A., Maia, G.A., Figueiredo, R.W. and Souza Filho, M.S.M. 2005. Potencial agroeconomico do bacuri: revisão. B.Ceppa. 23: 47–58.

Bilanda, D.C., Dimo, T., Dzeufiet Djomeni, P.D., Bella, N.M.T., Aboubakar, O.B.F., Nguelefack, T.B., Tan, P.V. and Kamtchouing, P. 2010. Antihypertensive and antioxidant effects of Allanblackia floribunda Oliv. (Clusiaceae) aqueous. J. Ethnopharmacol. 128(3): 634–640.

Braga, R. 1976. Plantas medicinais do Nordeste, especialmente do Ceará: Mossoró. ESAM. 540.

Canuto, G.A.B., Xavier, A.A.O., Neves, L.C. and Benassi, M.T. 2010. Caracterização físico-química de polpas de frutos da Amazônia e sua correlação com a atividade anti-radical livre. Ver. Bras. Frutic. 32(4): 1196–1205.

Castelo, K.F.A. 2018. Estudo Químico dos extratos ativos de bacuri *(Platonia insignis*). Dissertação (Mestrado). 126 f.

Cavalcante, A.N., Feitosa, C.M., Santos, F.P.S., Sousa, A.P.R., Sousa, R.S. Souza, A.A., Pinto, B.F., Araujo, C.M. and Rashed, K. 2019. Elaboration and characterization of the inclusion complex between β-cyclodextrin and the anticholinesterase 2-oleyl-1,3-dipalmitoyl-glycerol extracted from the seeds of *Platonia insignis* MART. J. Mol. Struct. 1177: 286–301. http://dx.doi.org/10.1016/j.molstruc.2018.09.067.

Cavalcante, A.N., Lima, L.K.F., Araujo, C.M., Santos, F.P.S., Nascimento, M.O., Sousa, J.M.C., Rai, M. and Feitosa, C.M. 2020. Toxicity, cytotoxicity, mutagenicity and in vitro antioxidant models of 2-oleyl-1,3-dipalmitoyl-glycerol isolated from the hexane extract of *Platonia insignis* MART seeds. Toxicol. Rep. 7: 209–216. http://dx.doi.org/10.1016/j.toxrep.2020.01.014.

Cavalcante, P.B. 1988. Frutas comestíveis da Amazônia. Belém: Museu Paraense Emílio Goeldi. 4. Ed: 48–50.

Chendynski, L.T., Cordeiro, T., Messias, G.B., Mantovani, A.C.G., Spacino, K.R., Zeraik, M.L. and Borsato, D. 2020. Evaluation and application of extracts of rosemary leaves, araçá pulp and peel of bacuri in the inhibition of the oxidation reaction of biodiesel. Fuel. 261: 116379.

Chirife, J. and Favetto, G.J. 1992. Some physico-chemical basis of food preservation by combined methods. Int. Food Res. J. 25(5): 389–396.

Chitarra, A.B. and Chitarra, M.I.F. 2005. Pós-colheita de frutos e hortaliças. Lavras. 256 p.

Coêlho, E.S., Lopes, G.L.N., Pinheiro, I.M., Holanda, J.N.P., Alves, M.M.M., Nogueira, N.C., Carvalho, F.A.A. and Carvalho, A.L.M. 2018. Emulgel based on amphotericin B and bacuri butter (*Platonia insignis* Mart.) for the treatment of cutaneous leishmaniasis: characterization and *in vitro* assays. Drug Dev. Ind. Pharm. 44(10): 1713–1723. http://dx.doi.org/10.1080/03639045.2018.1492610.

Costa júnior, J.S., Ferraz, A.B.F., Filho, B.A.B., Feitosa, C.M., Cito, A.M.G.L. and Freitas, R.M. 2011. Evaluation of *in vitro* antioxidant effects of garcinielliptona fc (GFC) isolated from *Platonia insignis* Mart. J. Med. Plant. Res. 5(2): 293–299.

Costa Júnior, J.S., Ferraz, A.B.F., Sousa, T.O., Silva, R.A.C., Lima, S.G., Feitosa, C.M., Citó, A.M.G.L., Cavalcante, A.A.C.M., Freitas, R.M. and Sperotto, A.R.M. 2012. Investigation of biological activities of dichloromethane and ethyl acetate fractions of *Platonia insignis* Mart. seed. Basic Clin. Pharmacol. Toxicol. 112: 34–41, http://dx.doi.org/10.1111/j.1742-7843.2012.00924.x.

Costa Júnior, J.S., Almeida, A.A.C., Ferraz, A.B.F., Rossatto, R., Silva, T.G., Silva, P.B.N., Militão, G.C.G., Citó, A.M.G.L., Santana, L.C.L.R. and Carvalho, F.A.A. 2013. Cytotoxic and leishmanicidal properties of garcinielliptone FC, a prenylated benzophenone from *Platonia insignis*. Nat. Prod. Res. 27(4–5): 470–474. http://dx.doi.org/10.1080/14786419.2012.695363.

Falci, D.R. and Pasqualotto, A.C. 2015. Anfotericina B: uma revisão sobre suas diferentes formulaões, efeitos adversos e toxicidade. Clin. Biomed. Res. 35(2): 65–82. http://dx.doi.org/10.4322/2357-9730.56021.

Fani, M. 2012. Pectina ação e utilização nos alimentos. Revista Aditivos Ingredientes. 86: p. 40.

Feitosa, C.M, Santos, P.R.P., Freitas, R.M., Rodrigues, A.M.X., Oliveira, G.A.L., Costa Junior, J.S. and Cavalcante, A.N. 2016. Ensaios pré-clínicos em ratos tratados com 1,3-diestearil-2-oleil-glicerol, constituinte isolado de *Platonia insiginis*. ConScientiae Saúde. 14(4): 555–567. http://dx.doi.org/10.5585/conssaude.v14n4.5658.

Ferreira, M.S.G. 2008. Manejo da espécie *Platonia insignis* Mart – Bacurizeiro, em florestas secundárias da Amazônia Oriental: proposta para uma produção sustentável. Tese de Doutorado. 246 p.

Ferreira, M.S. and Melo, M. 2007. *Platonia insignis* Mart. species richness in secondary forests of north-eastern Pará, Brazil. Bois & Forets Des Tropiques - Revues Du Cirad. 4(294): 21–28.

Fontenele, M.A., Figueiredo, R.W., Maia, G.A., Alves, R.E., Sousa, P.H.M.D. and Souza, V.A.B.D. 2005. Conservação pós-colheita de bacuri (*Platonia insignis* Mart.) sob refrigeração e embalado em PVC. Ver. Ceres. 57: 292–296.

Fontenele, M.A. 2007. Conservação pós-colheita do bacuri sob condições ambiente e refrigeração associada à atmosfera modificada. Dissertação (mestrado). 138f.

Fraceto, L.F., Gonçalves, M.M., Moraes, C.M., Araújo, D.R., Zanella, L. and Paula, E. 2007. Caracterização do complexo de inclusão ropivacaína: beta-ciclodextrina. Quim. Nova. 30(5): 1203–1207.

Freitas, F.A., Araujo, R.C., Soares, E.R., Nunomura, R.C.S., Silva, F.M.A., Silva, S.R.S., Souza, A.Q.L., Souza, A.D.L., Franco-Montalbán, F. and Acho, L.D.R. 2017. Biological evaluation and quantitative analysis of antioxidant compounds in pulps of the Amazonian fruits bacuri (*Platonia insignis* Mart.), ingá (*Inga edulis* Mart.), and uchi (*Sacoglottis uchi* Huber) by UHPLC-ESI-MS/MS. J. Food Biochem. 42: 1–24. http://dx.doi.org/10.1111/jfbc.12455.

Lustosa, A.K.M.F., Coêlho, A.G., Santos, A.A., Barros, Y.S.O., Rodrigues, K.A.F., Amorim, L.V., Alves, M.M.M., Carvalho, A.L.M., Mendes, A.N. and Carvalho, F.A.A. 2021. Formulações tópicas à base de manteiga das sementes de *Platonia insignis Mart.* para o tratamento de lesões relacionadas à leishmaniose cutânea experimental. Res., Soc. Dev. 10(4): e52310413665.

Mendes, m.C.S., Lacerda, j.S., Junior, M.R., Silva, M.L.D.G., Coelho, M.L., Tome, A.R. and Freitas, R.M. 2015. Evaluation of the cicatrizant activity of a semisolid pharmaceutical formulation obtained from *Platonia insignis* mart. Afr. J. Pharmacy Pharmacol. 9(6): 154–164.

Mendoza, V.S., Santos, L.L. and Sanjinez-Argadoña, E.J. 2017. Elaboração de granola com adição de polpa e castanha de Bacuri para consumo com iogurte. Evidência – Ciênc Biotec. 16(2): 83–100.

Morais, L.R.B. and Gutjah, E. 2012. Química de oleaginosas: valorização da biodiversidade amazônica. Ed. do Autor, 78 p.

Monteiro, A.R., Meireles, M.A.A., Marques, M.O.M. and Petenate, A.J. 1997. Extraction of the soluble material from the shells of the bacuri fruit (*Platonia insignis* Mart) with pressurized CO_2 and other solvents. J. Supercrit. Fluids. 1(11): 91–102.

Moura, M.C.C., Homma, A.K.O., Menezes, A.J.E.A. et al. 2007. Bacuri: Agrobiodiversidade. Interamerican Institute for Cooperation on Agriculture, 1 ed. 210p.

Mourão, k.S.M. and Beltrati, C.M. 1995. Morfologia dos frutos, sementes e plântulas de *Platonia insignis* mart. (clusiaceae): aspectos anatômicos dos frutos e sementes em desenvolvimento. Acta Amazônica 25(1/2): 11–32.

Nascimento, J.L. 2013. Complexo de inclusão do extrato hexânico de *Platonia insignis* Mart e β-ciclodextrina: caracterização e avaliação das atividades gastroprotetora e antioxidante *in vitro*. Dissertação (Mestrado). 137 p.

Noldin, v.F., Isaias, D.B. and Cechinel Filho, V. 2006. Gênero *calophyllum*: importância química e farmacológica. Quím. Nova. 29(3): 549–554.

Nualkaew, N., Morita, H., Shimokawa, Y., Kinjo, K., Kushiro, t., Deeknamkul, W., Ebizuka, Y. and Abe, I. 2012. Benzophenone synthase from *Garcinia Mangostana* L. pericarps. Phytochemistry (77): 60–69.

Pinheiro, E.E.A. 2016 Contribuição ao estudo fitoquímico e potencial biológico das flores de *Platonia insignis* Mart. Dissertação (Mestrado). 122 f.

Rodrigues, T.A. 2012. Elaboração e avaliação de sorvete de bacuri a base do extrato aquoso de soja. TCC (Graduação). 63 f.

Rogez, H., Buxant, R., Mignolet, E., Souza, J.N.S., Silva, E.M. and Larondelle, Yvan. 2004. Chemical composition of the pulp of three typical Amazonian fruits: araça-boi (*Eugenia stipitata*), bacuri (*Platonia insignis*) and cupuaçu (*Theobroma grandiflorum*). Eur. Food Res. Technol. 218(4): 380–384. Http://dx.doi.org/10.1007/s00217-003-0853-6.

Rufino, M.S.M., Alves, R.E., Brito, E.S., Pérezjiménez, J., Saura Calixto, F. and Mancini Filho, J. 2010. Bioactive compounds and antioxidant capacities of 18 non-traditional tropical fruits from Brazil. Food Chemistry 121: 996–1002.

Santos Júnior, R.Q., Soares, C., Maia Filho A.L.M., Araújo, K.S., Santos, I.M.S.P., Costa Júnior, J.S. and Saffi, J. 2010. Estudo histológico da cicatrização de feridas cutâneas utilizando a banha de bacuri (*Platonia insignis* Mart.). Conscientiae saúde. 9(4): 575–581.

Santos, B.A., Teixeira, F., Amaral, L.A., Randolpho, G.A., Schwarz, K., Santos, E.F., Resende, J.T.V. and Novello, D. 2019. Caracterização química e nutricional de polpa de frutas armazenadas sob congelamento. Rev. Univ. Vale Rio Verde. 1(17): 1–13.

Santos, D.B., Aguiar, R.O., Cruz, W.P., Bernardino, P.D.L.S., Martins, L.H.S., Carvalho, F.I.M., Bichara, C.M.G. and Silva, P.A. 2020. Desenvolvimento e caracterização de doces de leite bubalino pastosos saborizados com doces de bacuri e cupuaçu. Braz. J. Dev. 6(8): 56917–56935. Http://dx.doi.org/10.34117/bjdv6n8-195.

Santos, P.R.P., Carvalho, R.B.F., Costa Júnior, J.S., Freitas, R.M. and Feitosa, C.M. 2013. Survey of physicochemical and pharmacological properties of extracts and compounds isolated from *Platonia insignis* Mart. a perspective for developing phytomedicines. Braz. J. Pharm. Sci. 94: 161–168.

Silva, A.P.S.C.L., Lopes, J.S.L., Vieira, P.S., Pinheiro, E.E.A., Silva, M.L.G., Silva Filho, José Carlos C.L., Costa Júnior, J.S., David, J.M. and Freitas, R.M. 2014. Behavioral and neurochemical studies in mice pretreated with garcinielliptone FC in pilocarpine-induced seizures. Pharmacol. Biochem. Behav. 124: 305–310. http://dx.doi.org/10.1016/j.pbb.2014.05.021.

Silva, A.G.A., Moreira, R.A., Sousa, R.P.S., Evaldo Filho, S.M., Veras, M.D.A., Chaves, M.H. and Freitas, S.D.L. 2021. Composição Quimica e atividades fitoprotetoras e antiradicalar *in vitro* dos galhos de *Platonia insignis* (clusiacae) Quím. Nova. 44(8): 954–962. http://dx.doi.org/10.21577/0100-4042.20170761.

Silva, L.C., Ferreira, B. Cromatografia em Coluna (C). 2009. Universidade Federal de Goiás. Goiás. Site: http://www.ebah.com.br/content/ABAAAAjJwAK/apostila-cromatografia-coluna (Acesso em 29/06/2015).

Soares, A.G. 2010. Caracterização físico-química do resíduo agroindustrial dos frutos do bacurizeiro (*Platonia Insignis* Mart.) com objetivo de produção de insumos para indústria de alimentos e química. Tese (Doutorado). 101 f.

Sousa, M.S.B., Vieira, L.M., Silva, M.J.M., Lima, A. 2011a. Caracterização nutricional e compostos antioxidantes em resíduos de polpas de frutas tropicais. Ciên. Agrotec. 35(3): 554–559.

Sousa, M.S.B., Vieira, L.M. and Lima, A. 2011b. Fenólicos totais e capacidade antioxidante in vitro de resíduos de polpas de frutas tropicais. Braz. J. Food Technol. 14(03): 202–210.

Sousa, S.F.Z., Soares, A.G., Lima, E.C.S. and Srur, A.U.O.S. 2015. Produção e avaliação de carvões ativados de cascas de bacuri (*platonia insignis mart.*) e coco (*cocos nucifera*) - Uma alternativa de exploração sustentável. J. Fruits Vegetables 1: 81–87.

Souza, A.C., Alves, M.M.M., Brito, L.M., Oliveira, L.G.C., Sobrinho-Junior, E.P.C., Costa, I.C.G., Freitas, S.D.L., Rodrigues, K.A.F., Chaves, M.H. and Arcanjo, D.D.R. 2017. *Platonia insignis* Mart., a Brazilian Amazonian Plant: the stem barks extract and its main constituent lupeol exert antileishmanial effects involving macrophages activation. Evid. Base Complement Alternat. Med. 2017: 1–12. http://dx.doi.org/10.1155/2017/3126458.

Uekane, T.M., Nicolotti, L., Griglione, A., Bizzo, H.R., Rubiolo, P., Bicchi, C., Rocha-Leão, M.H.M. and Rezende, C.M. 2017. Studies on the volatile fraction composition of three native Amazonian-Brazilian fruits: murici (*Byrsonima crassifolia* L., malpighiaceae), bacuri (*Platonia insignis* M., clusiaceae), and sapodilla (*Manilkara sapota* L., sapotaceae). Food Chemistry 219: 13–22. http://dx.doi.org/10.1016/j.foodchem.2016.09.098.

Vasconcelos, K.M., Santos, I.L., Medeiros, E.M., Souza, F.C.A., Vasconcelos, K.M.M. and Oliveira, W.W.C. 2018. Desenvolvimento e caracterização de biscoito elaborado a partir da farinha de bacuri (*Platonia insignis* Mart.). Scientia Amazonia 1: 6–20.

Vieira, L.M., Sousa, M.S.B., Mancini-Filho, J. and Lima, A. 2011. Total phenolics and antioxidant capacity "*in vitro*" of tropical fruit pulps. Rev. Bras. Frutic. 33(3): 888–897.

Vijayanand, P., Yadav, A.R., Balasubramanyam, N. and Narasimham, P. 2000. Storage stability of guava fruit bar prepared using a new process. LWT 33: 132–137.

Villachica, H., Carvalho, J.E.U., Müller, C.H., Diaz, S.C. and Almanza, M. 1996. Frutales y hortaliças promossoras de la amazônia. Lima, tratado de cooperación amazônica. Secretaria pró-tempore. 152–156.

Wu, X., Beecher, G.R., Holden, J.M., Haytowitz, D.B., Gebhardt, S.E. and Prior, R.L. 2004. Lipophilic and hidrophilic antioxidant capacities of common foods in the United States. J. Agric. Food Chem. 52(12): 4026–4037.

Yamaguchi, K.K.L., Pereira, C.V.L., Lima, Emerson, S. and Veiga Junior, V.F. 2014. Química e farmacologia do bacuri (*Platonia insignis*). Sci. Amazon. 3(2): 39–46.

Chapter 16

Red Pitaya (*Hylocereus polyrhizus*) as a Functional Food

Nutritional Value, Phytochemical Content and Health-Promoting Effects

Chayane Gomes Marques,[1,*] *Bruno Bezerra da Silva,*[1]
Sandra Machado Lira,[2] *Ana Paula Dionísio,*[3]
Maurício Fraga van Tilburg,[4] *Maria Izabel Florindo Guedes*[1] and
Claisa Andrea Freitas Rabelo[1]

Introduction

The genus Hylocereus is grown in western Mexico, Southeast Asia and Central and South America and comprises 14 species, but only four of them are economically significant: *H. megalanthus* (yellow pitaya), *H. undatus* (white pitaya), *H. costaricensis* and *H. polyrhizus* (red pitaya) (Tsai et al., 2019; Abirami et al., 2021). The *Hylocereus polyrhizus* is a perennial plant from the Cactaceae family (Ramli et al., 2014). This plant has triangular, succulent and spiny cladodes 2 to 4 mm wide, and it naturally grows on trees or stones, where its fibrous and adventitious roots are responsible for its support and nutrient uptake (Almeida et al., 2016). *H. polyrhizus* was introduced in Brazil around the 1990s through imports from

[1] State University of Ceara, Department of Nutrition, Fortaleza, CE, 60714-903, Brazil.
[2] Federal Institute of Education, Science and Technology of Ceara, Iguatu, CE, 63500-000, Brazil.
[3] Embrapa Agroindustria Tropical, Dr[a]. Sara Mesquita Street, Fortaleza, CE, 2270, 60511-110, Brazil.
[4] Federal University of the Semi-Arid Region, Mossoró-RN.
* Corresponding author: chaygmarq@gmail.com

Colombia, and its cultivation has increased, especially in the Brazilian Southeast and Northeast regions (Lima, 2013; Jeronimo et al., 2017).

The tropical fruit from *H. polyrhizus*, also known as dragon fruit, red pitaya and *buah naga,* has a sweet mesocarp interspersed with edible black seeds (Tran et al., 2015). Its main characteristic is the extraordinary red color of its pulp, mainly due to pigments denominated betacyanins (Abirami et al., 2021). *H. polyrhizus* mesocarp can be consumed fresh (raw or juice) or as jellies or wine. Its pigments may be used as a coloring agent for foods like ice cream, yogurt and other industrial products (Choo and Yong, 2011; Yong et al., 2017).

Red pitaya has nutraceutical properties and health benefits attributed to prebiotics, polyunsaturated fatty acids, polyphenols and antioxidant vitamins (Joshi and Prabhakar, 2020). Thus, when included in the human diet, this fruit may prevent chronic diseases such as atherosclerosis, liver steatosis, cancer, cardiovascular disease, diabetes and metabolic syndrome (Luo et al., 2014; Song et al., 2016; Ravichandran et al., 2021; Al-Radadi, 2022). The peel, the primary byproduct of this fruit, can also be explored by the food and pharmaceutical industries as a pectin source, a polysaccharide with a hypolipidemic effect (Zaid et al., 2019).

Earlier studies on the Hylocereus genus and other genera from the Cactaceae family, identified different phytoconstituents in flowers, leaves, stems, cladodes and pericarp (Wu et al., 2011; Yi et al., 2012; Fathordoobady et al., 2016; Cupido et al., 2022). Based on the potential of red pitaya for sustainable production of therapeutic compounds, this chapter focuses on the nutritional and phytochemical contents of red pitaya as a functional food for human health.

Approach

The chapter was conducted on the scientific literature available at the PubMed, Science Direct, and Scielo databases, where the following descriptors were used: ("*Hylocereus polyrhizus*" OR "red pitaya" OR "red pitahaya"), ("chemical characterization" OR "bioactive compounds" OR "phytoconstituents"), ("nutritional composition" OR "nutritional quality" OR "nutritional value"), ("health benefits" OR "biological activity" OR "bioactivity" OR "biological effects"). Publications from the last 20 years in English and Portuguese were evaluated and carefully revised to include *in vivo* and *in vitro* studies on the chemical and nutritional analysis of red pitaya and its biological activities. Systematic reviews and gray literature have also been evaluated.

Nutritional value, bioactive compounds and biological effects

Nutritional composition

The nutrient content of red pitaya has been the subject of intense scientific research in the last years (Jamilah et al., 2011; Moo-Huchin et al., 2014; Nurul and Asmah, 2014; Cordeiro et al., 2015; Arivalagan et al., 2021). This interest relates to its potential as a functional food, its economic value and the bioprospection of pharmacological compounds in non-edible parts. It is known that edaphoclimatic conditions contribute to the nutritional diversity of crops, justifying the necessity to determine their nutritional composition in distinct world regions (Tien et al., 2022).

Red pitaya consists of approximately 22% of peel and 65% of mesocarp. The peel has a high level of total fiber (69.3% approximately) with a proportion of 3.8:1.0 of insoluble to soluble fiber, besides pectin (10.79 ± 0.01%) and betacyanin (150.46 ± 2.19 mg/100 g) (Jamilah et al., 2011).

The pulp is also a source of insoluble fiber (37.70 ± 1.27 g/100 g) with a total content equal 44.58 ± 0.77 g/100 g (Moo-Huchin et al., 2014). It has a high carbohydrate content (85.70 ± 1.12 g/100 g) (total sugar, reducing and non-reducing sugars corresponding to 8.79, 5.56, and 3.07%, respectively) (Cordeiro et al., 2015), and lower protein (7.33 ± 0.85 g/100 g) and lipids content (2.19 ± 0.10 g/100 g) (Dias, 2016). Its water content (85.58 ± 0.15 g/100 g) is similar to the peel (87.14 ± 1.58 g/100 g) (Cordeiro et al., 2015).

The minerals found in pulp are phosphorus (2.3 g/kg), potassium (12.6 g/kg), calcium (8.0 g/kg), magnesium (3.6 g/kg), copper (21.71 g/kg), iron (337.58 g/kg), manganese (113.93 g/kg) and zinc (116.26 g/kg) (Cordeiro et al., 2015). In addition, it contains a large amount of vitamin A (85.22 ± 3.21 μg/100 g), moderate vitamin C content (24.66 ± 0.33 μg/100 g) (Nurul and Asmah, 2014), vitamin E (0.26 μg/100 g), betacarotene (1.4 μg/100 g) and lycopene (3.4 μg/100 g) (Wichienchot et al., 2010). It has been also reported as a source of thiamin (B1), riboflavin (B2) and niacin (B3) (Le Bellec et al., 2006).

A recent study identified and quantified water and fat-soluble vitamins in different pitaya clones from India. Vitamin C (2.94–5.64 mg/100 g), pantothenic acid (40.1–49.3 μg/100 g) and niacin (7.10–12.55 μg/100 g) showed the highest amount within the hydrosoluble vitamins, while the highest fat-soluble vitamins were vitamin E (75–182 μg/100 g), vitamin K1 (9.43–31.4 μg/100 g) and vitamin D2 (0.534–0.746 μg/100 g) (Arivalagan et al., 2021). The authors observed variation in the nutritional values between the same species clones and between different cultivars, reflecting the effect of environmental factors on their nutritional content.

The red pitaya seeds correspond to 27 or even 46 g/kg in Hylocereus genus (Esquivel et al., 2007) and contain carbohydrates (352 ± 15.0 g/kg seeds), fibers (302 ± 19.0 g/kg), lipids (296 ± 6.0 g/kg), proteins (206 ± 6.0 g/kg), moisture (126 ± 6.0 g/kg) and ash (21 ± 1.0 g/kg) (Villalobos-Gutiérrez et al., 2012). The lipids present in the red pitaya oil seeds are mostly polyunsaturated (49.30 ± 0.28 g/100 g oil) and monounsaturated (27.92 ± 0.09 g/100 g oil), among which linoleic (48.00 ± 0.25 g/100 g), oleic (26.80 ± 0.07 g/100 g) and linolenic acids (1.30 ± 0.03 g/100 g) stand out (Lim et al., 2010).

The bark and pulp of red pitaya are also a source of oligosaccharides (raffinose, stachyose, maltotriose, maltotetraose, maltopentaose, maltohexaose, maltoheptaose and fructooligosaccharides) (Khalili et al., 2014; Lira et al., 2020), which are carbohydrates of low molecular weight that can work as prebiotics. These can modulate the composition of the intestinal microbiota, subsequently modifying bile acids (Catry et al., 2018), lipids and carbohydrates metabolism with health-promoting effects (Liu et al., 2017).

Bioactive compounds

The consumption of bioactive compounds or secondary metabolites from fruits and vegetables is related to preventing human diseases. In plants, they contribute to the color, flavor, odor, growth, oxidative stability and pollinator attraction for seed dispersion and pollination. They are also a mechanism of the plant defense system against multiple factors (Pandey and Rizvi, 2009; Cartea et al., 2010).

The fruits and subproducts from red pitaya contain several bioactive compounds such as flavonoids, phenolic acids, betalains, terpenes, steroids and functional lipids. These compounds have also been reported in flowers, leaves, stems and cladodes provided by other genera of the Cactaceae family (Wu et al., 2011; Yi et al., 2012; Cupido et al., 2022).

The processes of extraction, isolation and identification of bioactive compounds from fruits and vegetable extracts are defined according to the botanic part chosen for analysis (Altemimi et al., 2017; Gadioli et al., 2018). In these processes, the compounds identified, as well as the accuracy of the analysis, could be influenced by the characteristics of the sample (crop type, variety, germination, maturity, environmental conditions, location, processing and storage), the sample preparation, the nature and polarity of the solvent, and also the technique of extraction (Hidalgo and Almajano, 2017; Gadioli et al., 2018). Table 16.1 shows the chemical composition of red pitaya from different locations using distinct extraction methods.

The extraction of the red pitaya seeds had the highest extract yield using chloroform, followed by hexane and ethanol 95%, according to Adnan et al., 2011. However, the antioxidant activity is not related to yield but depends on the recovery of antioxidant compounds from extracts (Adnan et al., 2011; Hidalgo and Almajano, 2017). In the study, a different antioxidant ability according to the polarity of samples were observed.

The pigment profile of the peel and pulp extract was found to be similar in both the solvent extraction and SFE method by Fathordoobady et al. (2016). In this study, the different ratios of EtOH/water showed to be more appropriate and efficient for extracting bioactive compounds. The highest total betacyanins contents (120.28 mg/100 ml) of pulp extract were detected in the 7:3 EtOH/water (v/v) ratio, as well as the best yield. A ratio of 1:1 EtOH/water (v/v) recovered 28.44 mg/100 ml for the peel extract. Similarly, the best total phenol content was found for the peel and pulp in these same proportions, without significant difference for the SFE method.

Betacyanins belong to the betalains class and have a red-violet color. Those identified in red pitaya are betanin, betanidin and phyllocactin, which have an effective antioxidant capacity in low concentrations, particularly betanidin, which has a high affinity for Nitric Oxide (NO) due to the presence of a catechol group in its structure (Taira et al., 2015). In *in vitro* studies, betanin performed a hepatoprotective and anticarcinogenic effect by inducing the expression of the nuclear factor sensitive to oxidative stress (*Nrf2*), which is an essential element in the antioxidant responses in living beings, by directing the expression of phase II detoxification enzymes (Krajka-Kuzniak et al., 2013).

Tenore et al. (2012) identified eight betacyanins, six flavonoids and seven phenolic acids in the red pitaya peel and pulp using the fractionation extraction

Table 16.1. Bioactive compounds reported in red pitaya (*H. polyrhizus*) from different countries.

Fruit part	Extraction/identification method	Chemical constituents	Reference/Country
Seeds	Ethanol 95% Chloroform Hexane HPLC analysis	Catequin, epicatequin, rutin, quercetin, mirecetin and kaempferol.	Adnan et al. (2011). Malaysia
Pulp	Juice HPLC-DAD-MS analysis	Gallic acid, malonylglycoside, tyrosine, malonyl-glycoside cyclo-dopa, 2-decarboxy-cyclo-dopa, ellagic acid, betalamic acid, indicaxanthin, acetylcoumarin, betanin (betanidin-5-O-β-glycoside), isobetanin, phyllocactin I (6'-O-malonyl-betanin), isophyllocactin I, phyllocactin II (4'-O-malonyl-betanin), neobetanidin decarboxy, decarboxylated neobetanin, 5.5 ', 6, 6'-tetrahydroxy-3,3'-biindolyl, dehydrogenated decarboxylated neobetanidin.	Esquivel et al. (2007). Israel
Peel	Ethanol 60% HPLC-PDA analysis	Betanin, and isobetanin.	Faridah et al. (2015). Indonesia
Pulp Peel	$SFE\text{-}CO_2$ plus EtOH/water as the co-solvent. LC/MS/MS analysis HPLC analysis (positive mode)	Betanin, isobetanin, phyllocactin, butyrylbetanin, isophyllocactin and isobutyrylbetanin. 2′-Apiosyl-phyllocactin (peel), and 2′- Apiosyl-isophyllocactin (peel).	Fathordoobady et al. (2016). Malaysia
Pulp Seeds	Freeze-dried material UPLC-QTOF-MS^E analysis (positive and negative mode)	Caffeic acid, ferulic acid, caffeic acid hexoside, citramalic acid, citric acid, glucose-malic acid, tartaric ester, phyllocactin, isophyllocactin, betanin, isobetanin, apiosyl-malonyl-betanin, 6'-O-malonyl-2-descarboxy-betanin isomer, isoharmnetin glucoxyl-rhamnosyl-rhamnoside, indicaxanthin, quercetin hexoside, glucose, sucrose, maltotriose, maltotetraose.	Holanda et al. (2021). Brazil
Pulp Peel	Methanol 80%. LC/MS/MS analysis (positive mode)	Gallic acid, hydroxycinnamic acid derivatives, trans-caftaric, tyrosine, phloretin 2'-O-glucoside, mirecetin-3-O-galactopyranoside, betalamic acid, hylocerenin, betanin, isobetanin, isophyllocactin-I, phyllocactin-II, neobetanin, 5,5 ', 6,6'-tetrahydroxy 3,3'-biindolyl, dehydrogenated neohilocerenin.	Kim et al. (2011) Korea
Seeds	Oil Gas chromatographic analysis	α, γ-Tocopherol. Gallic acid, protocatechuic, *p*-hydroxybenzoic, vanillic acid, caffeic, syringic acid, *p*-coumaric. Cholesterol, campesterol, stigmasterol, β-sitosterol.	Lim et al. (2010). Malaysia

Pulp Peel	Freeze-dried material UPLC-QTOF-MS^E analysis (positive mode)	Phyllocactin, betanin, isobetanin, isophyllocactin, 2′-O-Apiosyl-phyllocactin, 6′-*O*-malonyl-2-descarboxy-betanin, isorhamnetin triglycosides isomer, quercetin-3-*O*-hexoside.	Lira et al. (2020) Brazil
Pulp Peel	Methanol 70% HPLC-DAD and HPLC MS–MS analysis (positive and negative mode)	Isorhamnetin triglycoside; quercetin-3-O-rutinoside; flavonol glycoside; kaempferol-3-O-rutinoside; isorhamnetin-3-O-rutinoside; isorhamnetin-3-O-glucoside. Betanidin-5-O-β-sophoroside; betanidin-5-O-β-glucoside (betanin); isobetanidin-5-O-β-glucoside (isobetanin); betanidin-5-O-(6'-O-malonyl)-β-glucoside (phyllocactin); isobetanidin-5-O-(6'-O-malonyl)-β-glucoside (isophyllocactin); betanidin-5-O-(6'-O-3-hydroxy-3-methyl-glutaryl)-β-glycoside; Iso-betanidin-5-O-(6'-O-3-hydroxy-3-methyl-glutaryl)-β-glucoside; Betanidin-5-O-(6'-O-3-hydroxy-butyryl)-β-glucoside. Gallic acid; protocatechuic acid; *p*-hydroxybenzoic acid; vanillic acid; caffeic acid; syringic acid; *p*-coumaric acid.	Tenore et al. (2012). Taiwan
Pulp	Ethanol 80% HPLC–PDA and ESI–MS/MS analysis	Betanin, isobetanin, phyllocactin, isophyllocactin, hylocerenin, isohylocerenin.	Wybraniec et al. (2001). Israel

SFE-CO_2: Supercritical Fluid Extraction; EtOH: Ethanol; HPLC-PDA: High Performance Liquid Chromatography coupled to a Photodiode Array Detector; ESI-MS: Electrospray Ionization Mass Spectrometry; HPLC-DAD: High-Performance Liquid Chromatography with Diode-Array Detection; LC/MS/MS: Liquid Chromatography with tandem Mass Spectrometry; UPLC-QTOF-MS^E: Ultra-high-performance Liquid Chromatography-quadrupole time-of-flight Mass Spectrometry.
Source: Elaborated by the authors.

method. The first fraction of the peel showed a large amount of total phenols, with a number almost 10 times higher of betacyanins than in the pulp. In the second and third fractions, the amounts of flavonoids and phenolic acids were comparable, both in the peel and pulp.

The characterization of 100 g of red pitaya seed oil revealed the presence of functional lipids, such as α–tocopherol (31.90 ± 0.70 mg) and γ-tocopherol (11.60 ± 0.20 mg), phytosterols such as cholesterol (6.0 ± 0.40 mg), campesterol (252 ± 3.10 mg), stigmasterol (106 ± 1.20 mg) and β-sitosterol (676 = 9.10 mg). Seven phenolic acids were also found: gallic acid (0.25 ± 0.15 mg), protocatechuic acid (0.93 ± 0.11 mg), ρ-hydroxybenzoic (0.66 ± 0.05 mg), vanylic acid (0.64 ± 0.11 mg), caffeic (0.08 ± 0.01 mg), syringic acid (0.08 ± 0.01 mg) and ρ-cumárico (0.78 ± 0.06 mg) (Lim et al., 2010). Phytosterols consumption has been associated with reduced serum levels of total cholesterol and Low-Density Lipoprotein cholesterol (LDL) in humans (Alvarez-Sala et al., 2018).

Bioactivity

The consumption of fruits and vegetables has shown preventive and attenuating effects against many human diseases (Brazil, 2015; Mozaffarian, 2016). The increase in the incidence and prevalence of chronic diseases, partly due to unhealthy eating habits and aging of the population, promotes the need for research on strategies for preventing and treating morbidities associated with these conditions.

For most fruits and vegetables, only the pulp is consumed. However, studies have shown significant amounts of biologically active compounds and essential nutrients in seeds, bark, stems and other components that are usually wasted (Rudra et al., 2015; Roriz et al., 2022). The extraction of bioactive compounds from the non-edible portion and their application for the generation of bioproducts or in the food industry, for instance, represents a sustainable form of reuse (Sagar et al., 2018).

The biological activities of bioactive compounds found in the red pitaya are described next, and Fig. 16.1 summarizes these activities.

Antioxidant activity

Natural antioxidants are secondary metabolites, with an aromatic ring and one or more hydroxyl groups in their structure, widely known for their human health-promoting effects (Denardin et al., 2015; Hidalgo and Almajano, 2017).

The antioxidant activity of foods and plants can be determined by chromatographic methods, such as High-Performance Liquid Chromatography (HPLC) coupled with DAD, MS or fluorescent detector and colorimetric methods. In the DPPH, ABTS and FRAP methods, a colorimetric change is measured by spectrophotometry at a given wavelength. In ORAC, the antioxidant compounds present in the sample inhibit the reduction of fluorescence caused by the fluorescein reaction with the peroxyl radicals (Hidalgo and Almajano, 2017). The antioxidant properties of red pitaya have been evaluated by some of the methods previously mentioned and correlated with the content of total phenols, flavonoids and betacyanins (Tenore et al., 2012; Ramli et al., 2014; Abirami et al., 2021).

Red pitaya
(Hylocereus polyrhizus)

Biological effects	Compounds
Antioxidant action	Phenols, flavonoids, betanin, betanidin, phyllocactin.
Anti-inflammatory effect	Kaempferol, betacyanins, quercetin, apigenin.
Antiproliferative action	Mirecetin, gallic acid, baicalein, 5-cedranone, β-selinene, eucalyptus, terpinolene and α-terpineol.
Immunomodulatory activity	Lupeol (a terpenoid).
Antiviral activity	Betacyanins, tannins, and terpenes.
Wound healing action	Phenolic acids and flavonoids, such as quercetin.
Anti-anxiety effect	Betalains and quercetin.
Hypoglycemic and hypolipidemic properties	Betacyanins, quercetin, betanin, antocyanins, unsaturated fatty acids, soluble fiber.

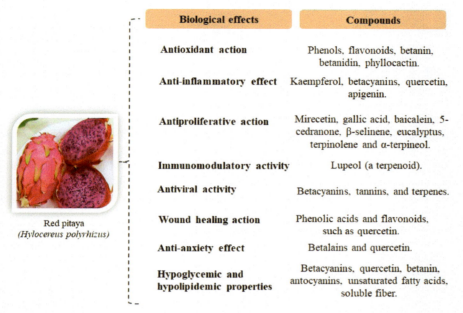

Figure 16.1. Biological activities of red pitaya and associated bioactive compounds. These compounds were identified in red pitaya edible and non-edible parts based on the papers analyzed. Source: Elaborated by the authors.

The antioxidant activity determined by the DPPH method and the reducing potential analyzed by the FRAP method of the red pitaya peel and aqueous pulp extracts ranged according to the extraction method used. In Conventional Extraction (CE), the peel was more effective than the pulp 6.06 ± 0.03% and 0.85 ± 0.55%, respectively, for DPPH. In the Ultrasound-Assisted Extraction (UAE), the pulp showed a higher value (5.73 ± 0.33%) than that observed in the peel (3.04 ± 0.36%). The reducing potential by the FRAP method of the pulp extract was higher than that of the peel, regardless of the extraction method used: 620 ± 54.08 µmol Fe^{2+}/g and 609.17 ± 54.51 µmol Fe^{2+}/g dry extract (pulp) and 255 ± 49.94 µmol Fe^{2+}/g and 200.83 ± 38.3 µmol Fe^{2+}/g dry extract (peel) (Ramli et al., 2014).

Betanin, phyllocactin and betanidin isolated from red pitaya (Japan) showed effective antioxidant potential at low concentrations, with IC_{50} values corresponding to 167.87 nM, 160.91 nM and 64.42 nM, respectively, when evaluated using the ORAC method. The Nitric Oxide (NO) reduction capacity of these compounds was also evaluated, and the IC_{50} values were 24.48 µM, 17.51 µM and 6.81 µM, respectively, for betanin, phyllocactin and betanidin. In both tests, betanidin showed a high antioxidant capacity compared to the other compounds (Taira et al., 2015).

The radical scavenging activity of polyphenols or bioactive compounds depends on both the number of hydroxyl groups available in the chemical structure and the proton donor capacity of these groups and their relative concentration in the cell matrix (Rice-Evans et al., 1996). Furthermore, phenolic compounds can inhibit reactive species in a synergistic, additive or antagonistic way (Denardin et al., 2015).

Anti-inflammatory activity

Many studies have reported anti-inflammatory and antinociceptive action to flavonoids, mainly glycosylates, such as kaempferol 3-O-rutinoside and kaempferol 3-O-glucoside (Wang et al., 2014). The bioactivity of these compounds is associated with inhibiting cyclooxygenase, lipooxygenase, phospholipases, prostaglandins, histamine and NF-κB transcription (Wang et al., 2014; Corrêa and Rogero, 2019).

In addition to these pathways, studies have pointed out the action of some flavonoids on the NLRP3 (NOD-like receptor family, pyrin domain-containing 3) inflammasome activity (Miao et al., 2011; Lim et al., 2018). The NLRP3 is a protein complex that belongs to the NOD-Like Receptor family (NLR) and is responsible for activating pro-inflammatory cytokines in a caspase-dependent pathway. NLRP3 is formed by the NRL receptor, an adapter protein associated with apoptosis-associated speck-like protein containing a caspase-recruitment domain – CARD (ASC) and a pro-caspase 1 (Miao et al., 2011).

Flavonoids kaempferol and apigenin inhibited the activation of NLRP3 inflammasome via inhibition of ASC oligomerization, and apigenin inhibited phosphorylation of splenic protein tyrosine-kinase (SYK)/tyrosine-kinase protein 2 beta (PYK2) in an animal model of induced peritonitis. Furthermore, apigenin also promoted a reduction in the expression of interleukin-1 β (IL-1β) (Lim et al., 2018).

Studies with betalains, which are compounds derived from betalamic acid, present in beet (*Beta vulgaris*), in the fig tree [*Opuntia ficus-indica* (L. Miller)] and in the red pitaya (*H. polyrhizus*) have demonstrated the anti-inflammatory effect of these compounds *in vivo*. This effect occurs via suppression of Nuclear Factor kappa B (NF-κB) and consequently, the expression of IL-1β, Tumor Necrosis Factor-alpha (TNF-α), inducible Nitric oxide synthase (iNOS) and cyclooxygenase 2 (COX-2) (Allegra et al., 2014; Adhikari et al., 2017). Betalains have also been able to reduce leukocyte recruitment by inhibiting myeloperoxidase (MPO) activity and promoting increased levels of interleukin-10 (IL-10), an anti-inflammatory cytokine (Martinez et al., 2015).

Antiproliferative action

In vitro studies with extracts from different parts of red pitaya have been conducted in the last years to test their potential against cancer cells. The peel and pulp acetonic extracts (80%) were tested *in vitro* against human melanoma (B16F10). The peel extract significantly inhibited (IC_{50} 25.0 μg) cellular proliferation after 72 hr of exposure, and its effects were attributed to flavonoids myricetin, baicalein and gallic acid (Wu et al., 2006).

The methanolic extracts from red pitaya peel and pulp were tested against three human carcinoma cell lines (gastric adenocarcinoma, AGS; cervical adenocarcinoma, HeLa; and mammary adenocarcinoma, MCF-7). In this study, only the peel extract showed inhibition in the cell growth of AGS and MCF-7, although it was not significant. The authors suggested that these results could be related to cell lines and phytochemicals concentrated in the extracts. As a whole, they observed no correlation between the antiproliferative effect and the content of total phenols in the extracts and suggested that this effect could be the result of the synergism between total polyphenols and flavonoids (Kim et al., 2011).

Ismail et al. (2017) reported the antiproliferative effect of red pitaya cladodes methanolic extract (95%) against MCF-7 and HepG-2 cells (liver carcinoma), with an IC_{50} of 2.8 and 4.8 μg, respectively, after 48 hr of exposure. The compounds identified in the extract were 5-cedranone, β-selinene, eucalyptol, terpinolene and α-terpineol.

The supercritical carbon dioxide extracts from red pitaya peel and pulp have pentacyclic triterpenoids and steroids, which showed anticancer activity in cells of Prostate Cancer (PC3), breast cancer (Bcap-37), and gastric cancer (MCG-803). IC_{50} values were 0.61, 0.45 and 0.43 mg/mL, respectively, for the three cell types (Luo et al., 2014). The authors did not report the action mechanism but related this finding to the identified compounds.

In another study, gold nanoparticles (AuNPs) produced using the red pitaya seed oil extract, called AuNPs@D.pulp_seed oil, were tested on the breast cancer cell line (MCF-7), human colon carcinoma (HCT-116) and human liver cancer (HepG2) (Al-Radadi, 2022). The AuNPs@D.pulp_seed oil exhibited higher growth inhibition rates towards HCT-116 cell line (92.89%), followed by HepG2 (80.29%) and MCF-7 (74.83%) at 500 μg/mL. The author points out that the physical characteristics, surface chemistry and concentration of the gold nanoparticles influenced the total absorption, elimination and the anticancer activity. It has also been suggested that the AuNPs effects may be associated with DNA damage by Reactive Oxygen Species (ROS) and therefore cancer cell death, although the mechanism of action has not been fully clarified.

It is observed that some classes of compounds present in red pitaya, such as flavonoids, terpenoids and steroids are involved in the positive modulation of mutagenic, carcinogenic and cellular stress factors, reinforcing the benefits of consuming red and purple fruits and vegetables for cancer prevention and treatment (Altemimi et al., 2017).

Immunomodulatory activity

Lupeol (1-isopropenyl 3a, 5a, 5b, 8, 8, 11a-hexamethyl-eicosahydrocyclopenya [α] chrysen-9-o1), a terpenoid isolated from peel methanolic extract of red pitaya, showed an immunomodulatory action by an enhancement of macrophage phagocytosis (Wahdaningsih et al., 2020). A recent study using different concentrations of this compound identified its ability to increase lymphocyte proliferation and nitric oxide production (Wahdaningsih et al., 2021). These results revealed the high potential of lupeol to be an immunomodulator (Wahdaningsih et al., 2021) and reinforce the applicability of compounds isolated from red pitaya peel for pharmaceutical use (Faridah et al., 2015; Zaid et al., 2019).

Antiviral potential

Preliminary phytochemistry of red pitaya revealed hydroquinone, phenols, saponins, tannins, alkaloids, triterpenoids and steroids (Manihuruk et al., 2017; Kanchana et al., 2018). Tannins and diterpenes have been identified as antiviral compounds on *in vitro* assays against the acquired human immunodeficiency virus type 1 and 2 (HIV) and the Simian Immunodeficiency Virus (SIVmac251) (Gyuris et al., 2009).

The antiviral potential of red pitaya compounds has been less investigated. More recently, betacyanins extracted from red pitaya pulp and red spinach were tested against dengue virus type 2 (DENV-2). Betacyanin fraction from red pitaya demonstrated a direct virucidal effect against DENV-2 with an IC_{50} of 126.70 µg/ml and with 95.0% of virus inhibition at the maximum non-toxic betacyanin concentration (379.5 µg/ml) (Chang et al., 2020). The authors highlighted that the mechanism of action needs to be confirmed by further experiments, but they suggest that the betacyanin fractions might inactivate the extracellular DENV-2 particles, binding to the non-structural proteins and suppressing the viral infectivity.

Wound healing action

The topical use of red pitaya cream in wounded rats promoted significant wound diameter reduction on d three post-treatment with maximum effect on the 7th d. This effect was ascribed to flavonoids and phenols as anti-inflammatory agents of the pitaya, as well as tannins. Flavonoids stimulate the production of type III collagen, and quercetin promotes the angiogenesis and proliferation of epithelial cells and fibroblasts. According to the authors, these effects together may have contributed to the process of wound healing (Tahir et al., 2020).

An earlier study evaluating the effect of topical red pitaya extract at 7.5% polyphenols and flavonoids on Diabetes Mellitus (DM) wound and non-DM wound healing in Wistar rats verified a similar pattern, reducing the diameter size of the wound in both groups. Nevertheless, the extract demonstrated better results in the granulation and epithelialization tissue process in the Wistar non-DM group (Tahir et al., 2017).

Three different extracts from red pitaya stem, peel and flower were evaluated on cell proliferation and viability in the NIH-3T3 fibroblast cell line at distinct concentrations after 24 hr. The results revealed that the stem and flower extracts in 95% aqueous ethanol at 1000 µg/mL concentration exhibited significant cell migration (Tsai et al., 2019). All extracts also showed optimal antioxidant potential and excellent DNA damage protection effects due to their phenolics and flavonoid composition (Tsai et al., 2019).

Anti-anxiety effect

Emerging findings from intervention studies suggest that both diet modification and lifestyle can potentially prevent and treat mental health (Adan et al., 2019). Fruits and vegetables are a significant source of nutritional and non-nutritional components that have importance for mental health (Costa et al., 2020). The brain physiology and function are dependent on the availability of amino acids, lipids, vitamins, minerals and polyphenols (Bazinet and Laye, 2014; Lepinay et al., 2015; Castro et al., 2018).

Pulp and peel of red pitaya exhibited anxiolytic-like effects in zebrafish using the Light and Dark test (Lira et al., 2020). The animals received red pitaya pulp (0.5 and 1.0 mg/mL) and remained from 82.8 to 85.2% in the clear zone. Similar behavior was observed by red pitaya peel (0.1, 0.5 and 1.0 mg/mL), where animals remained from 78.10 to 95.57% in the clear zone. Both results were comparable to diazepam effect (5.0 mg/mL; 95.1% in the clear zone). The authors ascribed those anxiolytic properties to betalains, quercetin and other flavonoids found in the samples.

Hypoglycemic, hypolipidemic properties and other metabolic effects

Advanced Glycation End products (AGEs) are stable and irreversible end products produced non-enzymatically by condensation between carbonyl groups of reducing sugars and free amine groups of nucleic acids, proteins or lipids (Twarda-Clapa et al., 2022). Despite this, AGEs are involved in the development and progression of insulin resistance, diabetes, cancer, cardiovascular diseases and neurological disorders (Martin-Ventura et al., 2017; Twarda-Clapa et al., 2022). Thus, glycemia and oxidative stress control can reduce the endogenous production of these products (Twarda-Clapa et al., 2022).

In a recent study, red pitaya extracts exhibited efficient antiglycation potential against food *in vitro* assays, whose effects were associated with the chemical structure of flavonoids and other phenolic compounds. Therefore, this points to the potential use of red pitaya to prevent or treat glycation associated with diabetes (Ravichandran et al., 2021).

Many studies conducted in rodents showed promising results related to glycemic control and lipids profile by *H. polyrhizus* (Sani et al., 2009; Khalili et al., 2009; Omidizadeh et al., 2011; Omidizadeh et al., 2014; Song et al., 2016; Werdiningsih and Suhartati, 2018; Coelho et al., 2020; Holanda et al., 2021). Furthermore, this fruit could protect against liver damage and reduce heart stiffness (Ramli et al., 2014; Song et al., 2016; Holanda et al., 2021). These effects were ascribed to polyphenols, antioxidant content, unsaturated fats, betacyanins and soluble dietary fiber, which acted synergistically (Khalili et al., 2009; Omidizadeh et al., 2014).

Administration of betacyanins from red pitaya (200 mg/kg/day, for 14 wk) prevented hyperglycemia, hypertriglyceridemia and showed effects on weight loss, attenuated visceral adiposity and hepatic steatosis in obese mice fed a high-fat diet. These effects were associated with modulations of the gut microbiome exerted by betacyanins, probably with its ability to increase the relative abundance of *Akkermansia* and reduce the rate of *Firmicutes* and *Bacteroidetes* in the animals (Song et al., 2016).

A study evaluated the effect of the consumption of 400 g and 600 g per day of red pitaya for 7 wk in the blood glucose and lipid profile of healthy and type 2 diabetic subjects ($n = 28$). A reduction in blood glucose, LDL-cholesterol and triglycerides and a significant increase in HDL-cholesterol was reported for the group with diabetes receiving 400 g of pitaya. There were no significant differences in the group's analyzed parameters receiving 600 g, but an increase in the HDL-cholesterol level and decreased LDL-cholesterol, triglycerides and total cholesterol could also be observed (Abd Hadi et al., 2012). Although a small sample, this study confirms some observations that have already been described in animals.

A systematic review and meta-analysis about the effect of pitaya on glycemic control in pre-diabetes and type 2 diabetes *mellitus* showed better results in diabetes prevention, with a trend towards more significant blood glucose reduction with higher doses in the latter (Poolsup et al., 2017).

Some mechanisms of action reported to the red pitaya metabolic effects are illustrated in Fig. 16.2 and include: (1) Hypocholesterolemic effect due to soluble fibers, polyphenols, flavonoids and saponins which increase excretion of bile acids by direct molecular interaction, modification of intestinal microbiota and microbial metabolism (Partiff et al., 1994; Khalili et al., 2006; Kanchana et al., 2018; Chambers et al., 2019). Quercetin can induce bile acid excretion through increases in cholesterol 7 α-hydroxylase (CYP7A1) expression, a cytochrome P450 enzyme, involved in the conversion of cholesterol into bile acids (Chambers et al., 2019). Quercetin induces ATP-binding cassette transporter A1 (ABCA1) expression and apolipoprotein A-I (apoA-I)-mediated cholesterol efflux by activating of the p38 signaling cascade via stimulating phosphorylation of transforming growth factor β-activated kinase 1 (TAK1) and mitogen-activated protein kinase 3/6 (MKK3/6) in macrophages (Chang et al., 2012; Holanda et al., 2021). This molecular mechanism is essential for the biogenesis of HDL (Zannis et al., 2015). It has also been proved that quercetin regulates hepatic scavenger receptor class B type I (SR-BI) expression subsequently contributing to cholesterol homeostasis (Ren et al., 2018); (2) Inhibitory Cholesteryl Ester Transfer Protein (CETP) activity, promoting LDL decrease by flavonoids, especially anthocyanin (Werdiningsih et al., 2018); (3) Betacyanins increase the concentrations of IL-10, an anti-inflammatory cytokine and the serum levels of adiponectin, which has a substantial role in the glucose and lipid metabolism (Song et al., 2016); (4) Anti-hyperglycemic action related to betanin—a glucoside form of betanidin—through inhibition of alpha glucosidase and alpha-amylase, key digestive enzymes in the control of type 2 diabetes mellitus (Rivera et al., 2020); (5) Hypoglycemic activity of flavonoids (e.g., quercetin) mediated by a decreased oxidative stress, and inhibition of gut mucosal GLUT-2 (Song et al., 2002; Ajie et al., 2015); (6) Gut microbiota modulation by oligosaccharides prebiotic, such as maltotriose, coming from red pitaya are capable to stimulate the growth of *Lactobacillus* and *Bifidobacterium* (Wichienchot et al., 2010); (7) Betanin exhibits hepatoprotective activity by promoting translocation of *Nrf2* to the nuclear compartment. This factor regulates the expression of target genes related to antioxidant response, detoxification, lipid and glucose metabolism, cellular survival and indirect anti-inflammatory response by the negative feedback of inflammatory cytokines, thus blocking NF-κB activation (Krajka-Kuzniak et al., 2013).

Limitations and perspectives

The literature reviewed supports the premise that the chemical constituents and nutrients from red pitaya play crucial roles in metabolic disorders and health promotion. However, there is a limited amount of evidence suggesting that red pitaya may induce similar beneficial effects in humans as in animals. Furthermore, this is not a comprehensive review as some complete papers are not affordable and data from clinical trials are still scarce.

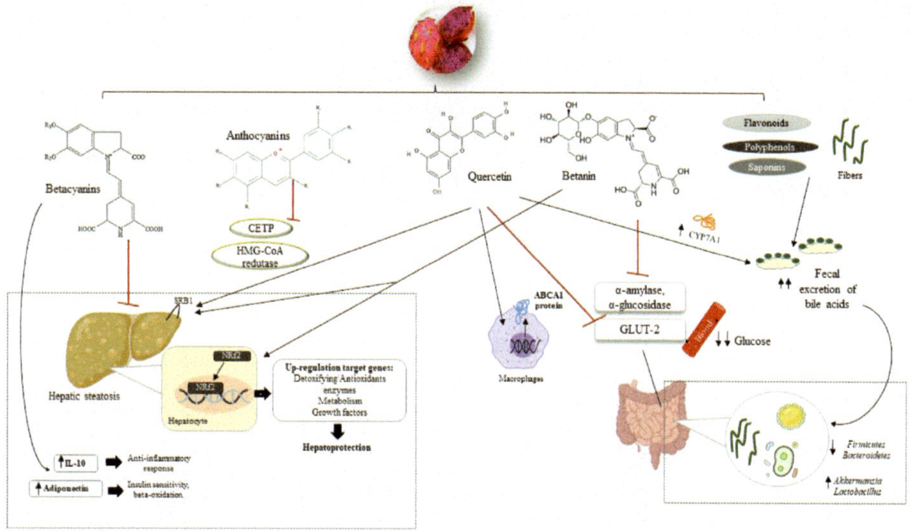

Figure 16.2. Proposed mechanism for compounds from red pitaya in the prevention and treatment of hyperglycemia, hypertriglyceridemia and liver steatosis. Betacyanins decrease liver inflammation by enhancing serum level of IL-10, adiponectin and decreasing lipid accumulation in hepatocytes. This effect can also be associated with the action of anthocyanins, a compound found in the red pitaya peel. Anthocyanins inhibit CETP and HMG-CoA redutase, decreasing the production of small and dense LDL-c and cholesterol synthesis, respectively. On the other hand, quercetin promotes cellular cholesterol efflux and the biogenesis of HDL through ABCA1 expression, and it may promote bile acids excretion via induction of CYP7A1, which predominantly converts cholesterol into bile acids; and it also regulates SR-BI expression, which is involved in cholesterol exchange. In addition, quercetin decreases glucose uptake by GLUT-2. Betanin inhibits the digestive enzymes alpha-glucosidase and alpha-amylase, which regulate postprandial glycemia, an important role in the management of type 2 diabetes. Although betanin is a kind of betacyanin, we described its isolated actions reported by earlier studies. Fibers, polyphenols, flavonoids and saponins promote the excretion of bile acids via direct molecular association with bile acids or interaction with gut microbes. Betacyanins and oligosaccharides induce shifts in the gut microbiome, for instance, reducing the ratio of pathogenic bacteria members of Firmicutes/ Bacteroidetes phylum, while increasing the relative abundance of Akkermansia and Lactobacillus. Red bars indicate blocking; black arrows indicate activation/stimulus. ABCA1, ATP-binding cassette transporter A1; CETP, Cholesteryl ester transfer protein; CYP7A1, cholesterol 7 α-hydroxylase; GLUT-2, Glucose transporter 2; HDL, High-density lipoprotein; HMG-CoA, 3-hidroxi-3-methyl-glutaril-CoA redutase; IL-10, Interleukin 10; LDL-c, Low-density lipoprotein cholesterol; Nrf2, Nuclear factor erythroid 2-related factor 2; SR-BI, Scavenger receptor class B type I. Chemical structures from ChemSpider and PubChem. Some icons used in this figure were from BioRender.com. Source: Elaborated by the authors.

In vivo tests regarding the action of red pitaya on metabolic syndrome, hepatic steatosis and insulin resistance should be continued in the form of clinical trials to assess the effect of consumption of this fruit and/or the supplementation of specific compounds in concomitance with a healthy diet and drug treatment when required. Research on the antiviral, immunomodulatory, wound healing and anxiolytic actions of pitaya seem to be promising subjects and the mechanistic insights for these activities are still not well understood. Furthermore, the use of computational analysis to guide the development of drugs based on pitaya compounds with promising actions is of interest. In these researches, it is also important to follow how the compounds present in the pitaya modulate the gut microbiota. For instance, betacyanins and magnesium have been reported to have beneficial effects on the composition of the gut microbiota (Song et al., 2016; Kuang et al., 2021), even so there is still much to be studied about the mechanism of action, such as which microbial metabolites are affected.

Conclusions

The red pitaya is an exotic fruit that has been introduced into Brazilian food culture, making it relevant to know its nutritional and functional features. Thus, this chapter provided background and updated information on the biological activities and the nutritional and phytochemical composition of the red pitaya.

Although there is a lack of studies analyzing the effect of *H. polyrhizus* consumption in humans, this fruit has phytochemicals recognized as physiologically active compounds based on experimental studies. Considering the health-promoting effects of fruits and vegetables, the consumption of pitaya for benefiting health and preventing chronic diseases is encouraged. Besides, this chapter reinforces pitaya non-edible portions as good sources of phytochemicals, which can be used in the food industry and the health industry for pharmaceuticals and nutraceuticals, contributing to the full use of this fruit and environmental sustainability.

Acknowledgment

The authors would like to thank National Coordination for the Improvement of Higher Education Personnel (CAPES), National Council for Scientific and Technological Development (CNPq), Biotechnology and Molecular Biology Laboratory (LBBM), Embrapa Tropical Agroindustry, for their support and for granting research scholarships. This paper was completed without input from funders and expresses the views of the authors and not necessarily those of the CAPES, CNPq, LBBM and Embrapa. This chapter includes content that appears in Chayane G. Marques dissertation, as fulfillment to obtain a Master Degree at State University of Ceara, Fortaleza, Brazil. The authors declare no conflicts of interest.

References

Abd Hadi, N., Mohamad, M., Rohin, M.A.K. and Mohd Yusof, R. 2012. Effects of red pitaya fruit (*Hylocereus polyrhizus*) consumption on blood glucose level and lipid profile in type 2 diabetic subjects. Borneo Sci. J. 31(2): 113–129.

Abirami, K., Swain, S., Baskaran, V., Venkatesan, K., Sakthivel, K. and Bommayasamy, N. 2021. Distinguishing three Dragon fruit (*Hylocereus* spp.) species grown in Andaman and Nicobar Islands of India using morphological, biochemical and molecular traits. Sci. Rep. 11: 2894. https://doi.org/10.1038/S41598-021-81682-X.

Adan, R.A.H., van der Beek, E.M., Buitelaar, J.K., Cryan, J.F., Hebebrand, J., Higgs, S., Schellekens, H. and Dickson, L.S. 2019. Nutritional psychiatry: Towards improving mental health by what you eat. Eur. Neuropsychopharmacol. 29(12): 1321–1332. doi:10.1016/j.euroneuro.2019.10.011.

Adhikari, A., Saha, A., Rania, I., Sur, T.K. and Das, A.K. 2017. Evaluation of anti-inflammatory effect of beetroot extract in animal models. Int. J. Basic Clin. Pharmacol. 6(12): 2853–2858. doi:10.18203/2319-2003.ijbcp20175207.

Adnan, L., Osman, A. and Hamid, A.A. 2011. Antioxidant activity of different extracts of red pitaya (*Hylocereus polyrhizus*) seed. Int. J. Food Prop. 14(1): 1171–1181. doi:10.1080/10942911003592787.

Ajie, R.B. 2015. White dragon fruit (*Hylocereus undatus*) potential as diabetes mellitus treatment. J. Majority. 4(1): 69–72.

Allegra, M., Ianaro, A., Tersigni, M., Panza, E., Tesoriere, L. and Livrea, M.A. 2014. Indicaxanthin from cactus pear fruit exerts anti-inflammatory effects in carrageenin-induced rat pleurisy. J. Nutr. 144(1): 185–192. doi:10.3945/jn.113.183657.

Almeida, E.I.B., Corrêa, M.C.M., Cajazeira, J.P., Queiroz, R.F., Barroso, M.M.A. and Marques, V.B. 2016. Cultivo de Hylocereus sp. com enfoque na propagação vegetativa, sombreamento e adubação mineral. R. Agro@mbiente On-line. 10(1): 65–76.

Al-Radadi, N.S. 2022. Biogenic proficient synthesis of (Au-NPs) via aqueous extract of red dragon pulp and seed oil: characterization, antioxidant, cytotoxic properties, anti-diabetic anti-inflammatory, anti-alzheimer and their anti-proliferative potential against cancer cell lines. Saudi J. Biol. Sci. 29(4): 2836–2855.

Altemimi, A., Lakhssassi, N., Baharlouei, A., Watson, D.G. and Lightfoot, D.A. 2017. Phytochemicals: extraction, isolation, and identification of bioactive compounds from plant extracts. Plants 6(42): 1–23. doi:10.3390/plants6040042.

Alvarez-Sala, A., Blanco-Morales, V., Cilla, A., Silvestre, R.A., Hernández-Álvarez, E., Granado-Lorencio, F., Barberá, R. and Garcia-Llatas, G. 2018. A positive impact on the serum lipid profile and cytokines after the consumption of a plant sterol-enriched beverage with a milk fat globule membrane: a clinical study. Food Funct. 9(10): 5209–5219. doi:10.1039/c8fo00353j.

Arivalagan, M., Karunakaran, G., Roy, T.K., Dinsha, M., Sindhu, B.C., Shilpashree, V.M., Satisha, G.C. and Shivashankara, K.S. 2021. Biochemical and nutritional characterization of dragon fruit (Hylocereus species). Food Chem. 353(15): 129426. doi:10.1016/j.foodchem.2021.129426.

Bazinet, R.P. and Laye, S. 2014. Polyunsaturated fatty acids and their metabolites in brain function and disease. Nat. Rev. Neurosci. 15(12): 771–785. doi:10.1038/nrn3820.

Brazil. Ministry of Health of Brazil. Secretariat of Health Care. 2015. Primary Health Care Department. Dietary Guidelines for the Brazilian population. Brasília: Ministry of Health of Brazil.

Cartea, M.E., Francisco, M., Soengas, P. and Velasco, P. 2010. Phenolic compounds in Brassica vegetables. Molecules 16(1): 251–280. doi:10.3390/molecules16010251.

Castro, A.I., Gomez-Arbelaez, D., Crujeiras, A.B., Granero, R., Aguera, Z., Jimenez-Murcia, S., Sajoux, I., Lopez-Jaramillo, P., Fernandez-Aranda, F. and Casanueva, F.F. 2018. Effect of a very low-calorie ketogenic diet on food and alcohol cravings, physical and sexual activity, sleep disturbances, and quality of life in obese patients. Nutrients 10(10): 1348. doi:10.3390/nu10101348.

Catry, E., Bindels, L.B., Tailleux, A., Lestavel, S., Neyrinck, A.M., Goossens, J.F., Lobysheva, I., Plovier, H., Essaghir, A., Demoulin, J.B., Bouzin, C., Pachikian, B.D., Cani, P.D., Staels, B., Dessy, C. and Delzenne, N.M. 2018. Targeting the gut microbiota with inulin-type fructans: preclinical demonstration of novel approach in the management of endothelial dysfunction. Gut. 67(2): 271–283. doi:10.1136/gutjnl-2016-313316.

Chambers, K.F., Day, P.E., Aboufarrag, H.T. and Kroon, P.A. 2019. Polyphenol effects on cholesterol metabolism via bile acid biosynthesis, CYP7A1: A review. Nutrients 11(11): 2588. doi:10.3390/nu11112588.

Chang, Y.C., Lee, T.S. and Chiang, A.N. 2012. Quercetin enhances ABCA1 expression and cholesterol efflux through a p38-dependent pathway in macrophages. J. Lipid Res. 53(9): 1840–1850. doi:10.1194/jlr.M024471.

Chang, Y.J., Pong, L.Y., Hassan, S.S. and Choo, W.S. 2020. Antiviral activity of betacyanins from red pitahaya (*Hylocereus polyrhizus*) and red spinach (*Amaranthus dubius*) against dengue virus type 2 (GenBank accession no. MH488959). Access Microbiol. 2(1): 1–6.

Choo, W.S. and Yong, W.K. 2011. Antioxidant properties of two species of Hylocereus fruits. Adv. Appl. Sci. Res. 2(3): 418–425.

Coelho, L.C., Dionísio, A.P., Lira, S.M., Santos, G.B.M., Zocolo, G.J., Silva, G.S., Holanda, M.O., Marques, C.G., Costa, J.T.G., Maciel, G.L. and Guedes, M.I.F. 2020. Características físico-químicas e efeito no estresse oxidativo da pitaia vermelha em camundongos *swiss* diabéticos. Res. Soc. Dev. 9(7): e899975035. doi:10.33448/rsd-v9i7.5035.

Cordeiro, M.H.M., Da Silva, J.M., Mizobutsi, E.H. and Da Mota, W.F. 2015. Caracterização física, química e nutricional da pitaia-rosa de polpa vermelha. Rev. Bras Frutic. 37(1). doi:10.1590/0100-2945-046/14.

Corrêa, T.A.F. and Rogero, M.M. 2019. Polyphenols regulating microRNAs and inflammation biomarkers in obesity. Nutrition 59: 150–157. doi:10.1016/j.nut.2018.08.010.

Costa, J.T.G., Lira, S.M., Holanda, M.O., Marques, C.G., Lima, C.L.S., Coelho, L.C., Santos, G.B.M., Dantas, J.B., Silva, R.J., Dionísio, A.P. and Guedes, M.I.F. 2020. Evidências do uso de frutas no tratamento complementar da Ansiedade e do Diabetes. Res. Soc. Dev. 9(9): e464997432. doi:10.33448/rsd-v9i9.7432.

Cupido, M., De-Nova, A., Guerrero-González, M.L., Pérez-Vázquez, F.J., Méndez-Rodríguez, K.B. and Delgado-Sánchez, P. 2022. GC-MS analysis of phytochemical compounds of *Opuntia megarrhiza* (Cactaceae), an endangered plant of Mexico. PeerJ Organic Chemistry 4: e5.

Denardin, C.C., Hirsch, G.E., Rocha, R.F., Vizzoto, M., Henriques, A.T., Moreira, J.C.F., Guma, F.T.C.R. and Emanuelli, T. 2015. Antioxidant capacity and bioactive compounds of four Brazilian native fruits. J. Food Drug Anal. 23(3): 387–398. doi:10.1016/j.jfda.2015.01.006.

Dias, P.S.M. 2016. Composição centesimal, atividade antioxidante, teor de compostos fenólicos e ecotoxicidade da polpa de frutos de pitaia branca (*Hylocereus undatus*) e pitaia vermelha (*Hylocereus polyrhizus*) [Monografia]. Juiz de Fora: Universidade Federal de Juiz de Fora.

Esquivel, P., Stintzing, F.C. and Carle, R. 2007. Phenolic compound profiles and their corresponding antioxidant capacity of purple pitaya (*Hylocereus* sp.) genotypes. Z Naturforsch C J. Biosci. 62c: 636–644. doi:10.1515/znc-2007-9-1003.

Faridah, A., Holinesti, R. and Syukri, D. 2015. Betalains from Red Pitaya Peel (*Hylocereus polyrhizus*): Extraction, spectrophotometric and HPLC-DAD identification, bioactivity and toxicity screening. Pak. J. Nutr. 14: 976–982.

Fathordoobady, F., Mirhosseini, H., Selamat, J. and Manap, M.Y.A. 2016. Effect of solvent type and ratio on betacyanins and antioxidant activity of extracts from *Hylocereus polyrhizus* flesh and peel by supercritical fluid extraction and solvent extraction. Food Chem. 202(1): 70–80. doi:10.1016/j.foodchem.2016.01.121.

Gadioli, I.L., Da Cunha, M. De S.B., De Carvalho, M.V.O., Costa, A.M. and Pineli, L.L. De O. 2018. A systematic review on phenolic compounds in Passiflora plants: Exploring biodiversity for food, nutrition, and popular medicine. Crit. Rev. Food Sci. Nutr. 58(5): 785–807. doi:10.1080/10408398.2016.1224805.

Gyuris, A., Szlávik, L., Minárovits, J., Vasas, A., Molnár, J. and Hohmann, J. 2009. Antiviral activities of extracts of *Euphorbia hirta* L. against HIV-1, HIV-2 and SIVmac251. *In Vivo* 23(3): 429–432.

Hidalgo, G.-I. and Almajano, M.P. 2017. Content, and radical scavenging determination: a review. Antioxidants 6(7): 1–27. doi:10.3390/antiox6010007.

Holanda, M.O., Lira, S.M., Silva, J.Y.G., Marques, C.G., Coelho, L.C., Lima, C.L.S., Costa, J.T.G., Silva, G.S., Santos, G.B.M., Zocolo, G.J., Dionísio, A.P. and Guedes, M.I.F. 2021. Intake of pitaya (*Hylocereus polyrhizus* (F.A.C. Weber) Britton & Rose) beneficially affects the cholesterolemic profile of dyslipidemic C57BL/6 mice. Food Biosci. 42: 101181. doi:10.1016/j.fbio.2021.101181.

Ismail, O.M., Abdel-Aziz, M.S., Ghareeb, M.A. and Hassan, R.Y.A. 2017. Exploring the biological activities of the *Hylocereus polyrhizus* extract. J. Innov. Pharm. Biol. Sci. 4(1): 1–6.

Jamilah, B., Shu, C.E., Kharidah, M., Dzulkifly, M.A. and Noranizan, A. 2011. Physico-chemical characteristics of red pitaya (*Hylocereus polyrhizus*) peel. Int. Food Res. J. 18(1): 279–286.

Jeronimo, M.C., Orsine, J.V.C. and Novaes, M.R.C.G. 2017. Nutritional pharmacological and toxicological characteristics of pitaya (*Hylocereus undatus*): A review of the literature. Afr. J. Pharm. Pharmacol. 11(27): 300–304. doi.org/10.5897/AJPP2016.4582.

Joshi, M. and Prabhakar, B. 2020. Phytoconstituents and pharmaco-therapeutic benefits of pitaya: A wonder fruit. J. Food Biochem. 44(7): e13260. doi:10.1111/jfbc.13260.

Kanchana, P., Devi, S.K.S.V., Pushpa Latha, P. and Spurthi, N. 2018. Phytochemical evaluation and pharmacological screening of antiparkinson's and laxative activities of Hylocereus undatus (white pitaya) in rodents. IOSR J. Pharm. 8(4): 78–92.

Khalili, R.M.A., Norhayati, A.H., Rokiah, M.Y., Asmah, R., Mohd Nasir, M.T. and Siti Muskinah, M. 2006. Proximate composition and selected mineral determination in organically grown red pitaya (Hylocereus sp.). J. Trop. Agric. Food Sci. 34(2): 269–276.

Khalili, R.M.A., Norhayati, A.H., Rokiah, M.Y., Asmah, R., Siti Muskinah, M. and Manaf, A.A. 2009. Hypocholesterolemic effect of red pitaya (Hylocereus sp.) on hypercholesterolemia induced rats. Int. Food Res. J. 16(3): 431–440.

Khalili, R.M.A., Abdullah, A.B.C. and Manaf, A.A. 2014. Isolation and characterization of oligosaccharides composition in organically grown red pitaya, white pitaya and papaya. Int. J. Pharm. Pharm. Sci. 6(2): 131–136.

Kim, H., Choi, H.-K., Moon, J.Y., Kim, Y.S., Mosaddik, A. and Cho, S.K. 2011. Comparative antioxidant and antiproliferative activities of red and white pitayas and their correlation with flavonoid and polyphenol content. J. Food Sci. 76(1): c38–c45. doi:10.1111/j.1750-3841.2010.01908.x.

Krajka-Kuzniak, V., Paluskczak, J., Szaefer, H. and Baer-Dubowska, W. 2013. Betanin, a beetroot component, induces nuclear fator erythroid-2-related fator 2-mediated expression of detoxifying/antioxidant enzymes in human liver cell lines. Br J. Nutr. 110(12): 2138–2149. doi:10.1017/S0007114513001645.

Kuang, X., Chiou, J., Lo, K. and Wen, C. 2021. Magnesium in joint health and osteoarthritis. Nutr. Res. 90: 24–35. doi:10.1016/j.nutres.2021.03.002.

Le Bellec, F., Vaillant, F. and Imbert, E. 2006. Pitahaya (*Hylocereus* spp.): a new fruit crop, a market with a future. Fruits 61(4): 237–250. doi:10.1051/fruits:2006021.

Lepinay, A.L., Larrieu, T., Joffre, C., Acar, N., Garate, I., Castanon, N., Ferreira, G., Langelier, B., Guesnet, P., Brétillon, L., Parnet, P., Layé, S. and Darnaudéry. 2015. Perinatal high-fat diet increases hippocampal vulnerability to the adverse effects of subsequent high-fat feeding. Psychoneuroendocrinology 53: 82–93. doi:10.1016/j.psyneuen.2014.12.008.

Lim, H., Min, D.S., Park, H. and Kim, H.P. 2018. Flavonoids interfere with NLRP3 inflammasome activation. Toxicol. Appl. Pharmacol. 355(1): 93–102. doi:10.1016/j.taap.2018.06.022.

Lim, H.K., Tan, C.P., Karim, R., Ariffin, A.A. and Bakar, J. 2010. Chemical composition and DSC thermal properties of two species of Hylocereus cacti seed oil: *Hylocereus undatus* and *Hylocereus polyrhizus*. Food Chem. 119(4): 1326–1331. doi:10.1016/j.foodchem.2009.09.002.

Lima, C.A. 2013. Caracterização, propagação e melhoramento genético da pitaya comercial e nativa do Cerrado [Tese]. Brasília: Universidade de Brasília.

Lira, S.M., Dionísio, A.P., Holanda, M.O., Marques, C.G., Silva, G.S., Coelho, L.C., Santos, G.B.M., Abreu, F.A.P., Magalhães, F.E.A., Rebouças, E.L., Guedes, J.A.C., Oliveira, D.F., Guedes, M.I.F. and Zocolo, G.J. 2020. Metabolic profile of pitaya (*Hylocereus polyrhizus* (F.A.C. Weber) Britton & Rose) by UPLC-QTOF-MSE and assessment of its toxicity and anxiolytic-like effect in adult zebrafish. Food Res. Int. 127: 108701. doi:10.1016/j.foodres.2019.108701.

Liu, F., Prabhakar, M., Ju, J., Long, H. and Zhou, H.-W. 2017. Effect of inulin-type fructans on blood lipid profile and glucose level: a systematic review and meta-analysis of randomized controlled trials. Eur. J. Clin. Nutr. 71(1): 9–20. doi.10.1038/ejcn.2016.156.

Luo, H., Cai, Y., Peng, Z., Liu, T. and Yang, S. 2014. Chemical composition and *in vitro* valuation of the cytotoxic and antioxidant activities of supercritical carbon dioxide extracts of pitaya (dragon fruit) peel. Chem. Cent. J. 8(1): 1–7. doi:10.1186/1752-153X-8-1.

Manihuruk, F.M., Suryati, T. and Arief, I.I. 2017. Effectiveness of the Red Dragon Fruit (*Hylocereus polyrhizus*) peel extract as the colorant, antioxidant, and antimicrobial on beef sausage. Media Peternakan. 40(1): 47–54. doi:10.5398/medpet.2017.40.1.47.

Martinez, R.M., Longhi-Balbinot, D.T., Zarpelon, A.C., Staurengo-Ferrari, L., Baracat, M.M., Georgetti, S.R., Sassonia, R.C., Verri, W.A. Jr and Casagrande, R. 2015. Anti-inflammatory activity of betalain-rich dye of *Beta vulgaris*: effect on edema, leukocyte recruitment, superoxide anion and cytokine production. Arch. Pharm. Res. 38(4): 494–504. doi:10.1007/s12272-014-0473-7.

Martin-Ventura, J.L., Rodrigues-Diez, R., Martinez-Lopez, D., Salaices, M., Blanco-Colio, L.M. and Briones, A.M. 2017. Oxidative stress in human atherothrombosis: sources, markers and therapeutic targets. Int. J. Mol. Sci. 18(11): 2315. doi:10.3390/ijms18112315.

Miao, E.A., Rajan, J.V. and Aderem, A. 2011. Caspase-1-induced pyroptotic cell death. Immunol. Rev. 243(1): 206–214. doi:10.1111/j.1600-065X.2011.01044.x.

Moo-Huchin, V.M., Estrada-Mota, I., Estrada-Léon, R., Cuevas-Glory, L., Ortiz-Vázquez, E., Vargas, Y., Vargas, M.L., Betancur-Ancona, D. and Sauri-Duch, E. 2014. Determination of some physicochemical characteristics, bioactive compounds and antioxidant activity of tropical fruits from Yucatan, Mexico. Food Chem. 152(1): 508–515. doi:10.1016/j.foodchem.2013.12.013.

Mozaffarian, D. 2016. Dietary and policy priorities for cardiovascular disease, diabetes, and obesity. Circulation 133(2): 187–225. doi:10.1161/CIRCULATIONAHA.115.018585.

Nurul, S.R. and Asmah, R. 2014. Variability in nutritional composition and phytochemical properties of red. Int. Food Res. J. 21(4): 1689–1697.

Omidizadeh, A., Yusof, R.M., Ismail, A., Roohinejad, S., Nateghi, L. and Bakar, M.Z.A. 2011. Cardioprotective compounds of red pitaya (*Hylocereus polyrhizus*) fruit. J. Food Agric. Environ. 9(3–4): 152–156.

Omidizadeh, A., Yusof, R.M., Roohinejad, S., Ismail, A., Bakar, M.Z.A. and Bekhit, A.E.D.A. 2014. Anti-diabetic activity of red pitaya (*Hylocereus polyrhizus*) fruit. RSC Adv. 4(108): 62978–62986. doi:10.1039/C4RA10789F.

Pandey, K.B. and Rizvi, S.I. 2009. Plant polyphenols as dietary antioxidants in human health and disease. Oxid. Med. Cell Longev. 2(5): 270–278. doi:10.4161/oxim.2.5.9498.

Partiff, V.J., Rubba, P., Bolton, C., Marotta, G., Hartog, M. and Mancini, M. 1994. A comparison of antioxidant status and free radical peroxidation of plasma lipoproteins in healthy young persons from Naples and Bristol. Eur. Heart J. 15(7): 871–876. doi:10.1093/oxfordjournals.eurheartj.a060603.

Poolsup, N., Suksomboon, N. and Paw, N.J. 2017. Effect of dragon fruit on glycemic control in prediabetes and type 2 diabetes: A systematic review and meta-analysis. PLoS One 12(9): e0184577. doi:10.1371/journal.pone.0184577.

Ramli, N.S., Brown, L., Ismail, P. and Rahmat, A. 2014. Effects of red pitaya juice supplementation on cardiovascular and hepatic changes in high-carbohydrate, high-fat diet-induced metabolic syndrome rats. BMC Compl. Alternative Med. 14(189). doi:10.1186/1472-6882-14-189.

Ramli, N.S., Ismail, O. and Rahmat, A. 2014. Influence of conventional and ultrasonic-assisted extraction on phenolic contents, betacyanin contents, and antioxidant capacity of red dragon fruit (*Hylocereus polyrhizus*). Sci. World J. 964731. doi:10.1155/2014/964731.

Ravichandran, G., Lakshmanan, D.K., Murugesan, S., Elangovan, A., Rajasekaran, N.S. and Thilagar, S. 2021. Attenuation of protein glycation by functional polyphenolics of dragon fruit (*Hylocereus polyrhizus*); an *in vitro* and *in silico* evaluation. Food Res. Int. 140(1): 110081. doi:10.1016/j.foodres.2020.110081.

Ren, K., Ting, J. and Guo-Jun, Z. 2018. Quercetin induces selective uptake of HDL-cholesterol via promoting SR-BI expression and activation of the PPARγ/LXRα pathway. Food & Function. 9: 624–635. doi:10.1039/C7FO01107E.

Rice-Evans, C.A., Miller, N.J. and Paganga, G. 1996. Structure-antioxidant activity relationships of flavonoids and phenolic acids. Free Radic Biol. Med. 20(7): 933–956. doi:10.1016/0891-5849(95)02227-9.

Rivera, A., Becerra-Martínez, E., Pacheco-Hernández, Y., Landeta-Cortés, G. and Villa-Ruano, N. 2020. Synergistic hypolipidemic and hypoglycemic effects of mixtures of *Lactobacillus nagelii*/betanin in a mouse model. Trop. J. Pharm. Res. 19(6): 1269–1276. doi:10.4314/tjpr.v19i6.23.

Roriz, C.L., Heleno, S.A., Alves, M.J., Oliveira, M.B.P.P., Pinela, J., Dias, M.I., Calhelha, R.C., Morales, P., Ferreira, I.C.F.R. and Barros, L. 2022. Red pitaya (*Hylocereus costaricensis*) peel as a source of valuable molecules: Extraction optimization to recover natural colouring agents. Food Chem. 372: 131344. https://doi.org/10.1016/j.foodchem.2021.131344.

Rudra, S.G., Nishad, J., Jakhar, N. and Kaur, C. 2015. Food industry waste: mine of nutraceuticals. Int. J. Sci. Environ. Technol. 4(1): 205–229.

Sagar, N.A., Pareek, S., Sharma, S., Yahia, E.M. and Lobo, M.G. 2018. Fruit and vegetable waste: bioactive compounds, their extraction, and possible utilization. Compr. Rev. Food Sci. F. 17(1): 512–531. doi:10.1111/1541-4337.12330.

Sani, H.A., Baharoom, A., Ahmad, M.A. and Ismail, I.I. 2009. Effectiveness of *Hylocereus polyrhizus* extract in decreasing serum lipids and liver MDA-TBAR level in hipercholesterolemic rats. Sains Malays. 38(1): 271–279.

Song, H., Chu, Q., Yan, F., Yang, Y., Han, W. and Zheng, X. 2016. Red pitaya betacyanins protects from diet-induced obesity, liver steatosis and insulin resistance in association with modulation of gut microbiota in mice. J. Gastroenterol. Hepatol. 31: 1462–1469. doi:10.1111/jgh.13278.

Song, J., Kwon, O., Chen, S., Daruwala, R., Eck, P., Park, J.B. and Levine, M. 2002. Flavonoid inhibition of sodium-dependent vitamin C transporter 1 (SVCT1) and glucose transporter isoform 2 (GLUT2), intestinal transporters for vitamin C and glucose. J. Biol. Chem. 277(18): 15252–15260. doi: 10.1074/jbc.M110496200.

Tahir, T., Bakri, S., Patellongi, I., Aman, M., Miskad, U.A., Maryunis, M., Yusuf, S., Rahayu, A.I., Syam, A.D. and Hasriyani. 2017. Evaluation of topical red dragon fruit extract effect (*Hylocereus Polyrhizus*) on tissue granulation and epithelialization in Diabetes Mellitus (DM) and Non-DM Wistar Rats: Pre Eliminary Study. Int. J. Sci. Basic Appl. Res. 32(1): 309–320.

Tahir, T., Febrianti, N., Wahyuni, S. and Rabia, Syam, Y. 2020. Evaluation of acute wound healing potential of red dragon fruit (*Hylocereus polyrhizus*) extract cream on type III collagen and Epidermal Growth Factor (EGF) levels: An animal study. Medicina Clinica Practica. 3(suppl 1): 100091. doi:10.1016/j.mcpsp.2020.100091.

Taira, J., Tsuchida, E., Katoh, M.C., Uehara, M. and Ogi, T. 2015. Antioxidant capacity of betacyanins as radical scavengers for peroxyl radical and nitric oxide. Food Chem. 166(1): 531–536. doi:10.1016/j.foodchem.2014.05.102.

Tenore, G.C., Novellino, E. and Basile, A. 2012. Nutraceutical potential and antioxidant benefits of red pitaya (*Hylocereus polyrhizus*) extracts. J. Funct. Foods 4(1): 129–136. doi:10.1016/j.jff.2011.09.003.

Tien, N.N.T., Le, N.L., Khoi, T.T. and Richel, A. 2022. Influence of location, weather condition, maturity, and plant disease on chemical profiles of dragon fruit (*Hylocereus* spp.) branches grown in Vietnam. Biomass Conv. Bioref. https://doi.org/10.1007/s13399-021-02146-w.

Tran, D.H., Yen, C.R. and Chen, Y.K.H. 2015. Effects of bagging on fruit characteristics and physical fruit protection in red pitaya (Hylocereus spp.). Biol. Agric Hortic. 31(3): 158–166. doi.org/10.1080/01448765.2014.991939.

Tsai, Y., Lin, C.G., Chen, W.L., Huang, Y.C., Chen, C.Y., Huang, K.F. and Yang, C.H. 2019. Evaluation of the antioxidant and wound-healing properties of extracts from different parts of hylocereus polyrhizus. Agronomy 9(1). doi:10.3390/agronomy9010027.

Twarda-Clapa, A., Olczak, A., Białkowska, A.M. and Koziołkiewicz, M. 2022. Advanced glycation end-products (AGEs): Formation, chemistry, classification, receptors, and diseases related to AGEs. Cells 11: 1312. https://doi.org/10.3390/cells11081312.

Villalobos-Gutiérrez, M.G., Schweiggert, R.M., Carle, R. and Esquivel, P. 2012. Chemical characterization of Central American pitaya (*Hylocereus* sp.) seeds and seed oil. CYTA J. Food. 10(1): 78–83. doi:10.1080/19476337.2011.580063.

Wahdaningsih, S., Wahyuono, S., Riyanto, S. and Murwanti, R. 2020. Terpenoid-lupeol of red dragon fruit (*Hylocereus polyrhizus*) and its immunomodulatory activity. Pak. J. Pharm. Sci. 33(2): 505–510. doi:10.36721/PJPS.2020.33.2.REG.505-510.1.

Wahdaningsih, S., Wahyuono, S., Riyanto, S. and Murwanti, R. 2021. Lymphocyte proliferation and nitric oxide-producing activities of lupeol Isolated from red dragon fruit (*Hylocereus polyrhizus*) extract. MCBS 5(1): 8–12. doi:10.21705/mcbs.v5i1.166.

Wang, Y., Chen, P., Tang, C., Wang, Y., Li, Y. and Zhang, H. 2014. Antinociceptive and anti-inflammatory activities of extract and two isolated flavonoids of *Carthamus tinctorius* L. J. Ethnopharmacol. 151(1): 944–950. doi:10.1016/j.jep.2013.12.003.

Werdiningsih, W. and Suhartati, S. 2018. Effects of Red Pitaya (Hylocereus polyrhizus) on lipid profile of male white rats (*Rattus norvegicus*) receiving high fat diet. Folia Med. Indones. 54(1): 16. doi:10.20473/fmi.v54i1.8046.

Wichienchot, S., Jatupornpipat, M. and Rastall, R.A. 2010. Oligosaccharides of pitaya (dragon fruit) flesh and their prebiotic properties. Food Chem. 120(1): 850–857. doi:10.1016/j.foodchem.2009.11.026.

Wu, L.C., Hsu, H.-W., Chen, Y.-C., Chiu, C.-C., Lin, Y.-I. and Ho, J.-A. 2006. Antioxidant and antiproliferative activities of red pitaya. Food Chem. 95: 319–327. doi.org/10.1016/j.foodchem.2005.01.002.

Wu, X., Wang, Y., Huang, X., Fan, C., Wang, G., Zhang, X., Zhang, Q.W. and Ye, W.C. 2011. Three new glycosides from *Hylocereus undatus*. J. Asian Nat. Prod. Res. 13(8). doi:10.1080/10286020.2011.586944.

Wybraniec, S., Platzner, I., Geresh, S., Gottlieb, H.E., Haimberg, M., Mogilnitzki, M. and Mizrahi. 2001. Betacyanins from vine cactus *Hylocereus polyrhizus*. Phytochemistry 58(8): 1209–1212. doi:10.1016/s0031-9422(01)00336-3.

Yi, Y., Zhang, Q.-W., Li, S.-L., Wang, Y., Ye, W.-C., Zhao, J. and Wang, Y.T. 2012. Simultaneous quantification of major flavonoids in "*Bawanghua*", the edible flower of Hylocereus undatus using pressurized liquid extraction and high performance liquid chromatography. Food Chem. 135(2): 528–533. doi:10.1016/j.foodchem.2012.05.010.

Yong, Y.Y., Dykes, G., Lee, S.M. and Choo, W.S. 2017. Comparative study of betacyanin profile and antimicrobial activity of Red Pitahaya (*Hylocereus polyrhizus*) and Red Spinach (*Amaranthus dubius*). Plant Foods Hum. Nutr. 72: 41–47. doi: 10.1007/s11130-016-0586-x.

Zaid, R.M., Mishra, P., Wahid, Z.A. and Sakinah, A.M.M. 2019. *Hylocereus polyrhizus* peel's high-methoxyl pectin: A potential source of hypolipidemic agent. Int. J. Biol. Macromol. 134: 361–367. doi:10.1016/j.ijbiomac.2019.03.143.

Zannis, V.I., Fotakis, P., Koukos, G., Kardassis, D., Ehnholm, C., Jauhiainen, M. and Chroni, A. 2015. HDL biogenesis, remodeling, and catabolism. *In*: von Eckardstein, A. and Kardassis, D. (eds.). High Density Lipoproteins. Handbook of Experimental Pharmacology, Springer, Cham. 224: 53–111. doi:10.1007/978-3-319-09665-0_2.

Chapter 17

Functional Properties of Plant By-products

Studies in Animal Models and Clinical Applications

Kamila Sabino Batista,[1] *Naís Lira Soares,*[1]
Juliana Gondim de Albuquerque,[1,2,3] *Renata Leite Tavares,*[1]
João Victor Gama de Albuquerque,[1] *Geyse Araújo Costa*[1] and
Jailane de Souza Aquino[1,*]

Introduction

The change in the lifestyle of modern society is characterized by a sedentary way of life and inadequate high-calorie food reflects the process of nutritional and epidemiological transition. This practice has triggered an increase in the incidence and prevalence of chronic Non-Communicable Diseases (NCDs) and their comorbidities such as obesity, metabolic syndrome, diabetes, hypertension and other cardiovascular diseases (Chrysochou, 2010; Ice et al., 2012). NCDs are the main causes of morbidity and mortality in the world and therefore curative medicine is often more prominent than preventive medicine, which raises costs in public health and causes harm to society (Wang, 2018). In contrast, consumers are more aware that their food choices can have consequences to their health and that maintaining

[1] Experimental Nutrition Laboratory, Department of Nutrition, Federal University of Paraíba (UFPB), Cidade Universitária, s/n-Castelo Branco III, João Pessoa 58051-900, Paraíba, Brazil.
[2] Post Graduate Program in Nutrition Sciences, Federal University of Pernambuco (UFPE), Cidade Universitária s/n, Recife 50670-901, Brazil.
[3] Post Graduate in Biotechnology, Division of Biological and Health Sciences, Universidad Autónoma Metropolitana (UAM), Ciudad de Mexico 09340, Mexico.
* Corresponding author: jailane.aquino@acadêmico.ufpb.br

a healthy lifestyle (Chrysochou, 2010) in association with physical activity is an essential aspect and has positive impacts on health (Chang and Lin, 2015).

The replacement of the westernized dietary pattern has been guided by health professionals and public policies with the inclusion of diverse plant sources in the diet and sometimes even giving rise to the spread of other eating practices such as veganism and vegetarianism (Nebl et al., 2019). The recommendation of the World Health Organization (WHO) and the Food and Agriculture Organization (FAO) is to consume at least 400 g of vegetables per day (excluding potatoes and other starchy tubers) as a strategy to prevent NCDs around the world (FAO/WHO, 2020).

Parallel to this scenario, the food agroindustry has also been following changes in people's lifestyles, including changes in the production, processing and commercialization profile. However, the expansion in this sector brings with it the generation of agro-industrial by-products, which are materials generated in the agricultural or industrial process which do not present a potential market and therefore it's end is disposal (Brasil, 2018). The disposal of this by-product represents between 40 and 50% of production, constitutes a potential pollutant for the environment, represents losses of raw materials and energy, and in turn requires significant investments in treatments that reduce the negative impacts to nature (Dilucia et al., 2020). On the other hand, using these by-products to produce fertilizers, biofuels and animal feed is a common practice, but it may not yet be sufficient for all the needs (Rao and Rathod, 2019). Once there is a destination for using these by-products which do not dispose them, then they can be called by-products according to the Brazilian legislation. Nevertheless, this name refers to animal feed and not for humans, which can be a confounding factor between the terms, residues, by-products and co-products (Brasil, 2018).

In this sense, studies using by-products from processing vegetables as ingredients of other foods or to produce nutraceuticals or phytotherapeutics have been carried out aiming to reduce the environmental impact and the development of technologies which add value to the products obtained and can improve the nutritional or technological properties (O'Shea et al., 2012; Gawlik-Dziki et al., 2015; Kabir et al., 2015; Felice et al., 2020). Among the main by-products generated in the vegetable production chain are the skins, peels, seeds and pomace (Majerska et al., 2019) which may in some cases represent a greater source of bioactive compounds (BACs) with beneficial health properties than the total content originally presented by the vegetable parts generally considered edible (Myoda et al., 2010; Fracassetti et al., 2013).

The by-products generally have recognized content of dietary fibres, phenolic compounds, carotenoids and vitamin C, in addition to several minerals which are related to antioxidant, anti-inflammatory, prebiotic, antimicrobial and anticarcinogenic properties, among others (Batista et al., 2018a). Depending on the plant source, they may also contain mono and polyunsaturated fatty acids which are important for the functioning of the organism such as oilseeds and oils extracted from them (Coimbra and Jorge, 2012). However, most studies are restricted to the occurrence, characterization and potential applications of BACs in *in vitro* studies (Coelho et al., 2019; Rebollo-Hernanz et al., 2019). Despite the important information that *in vitro* tests can provide, there are recognized limitations for this type of test. Thus, *in vivo* studies, which are non-clinical models with animals or clinical with humans, have been developed to evaluate the bioactivity of the by-products and will be presented below.

Non-clinical and clinical studies on the bioactivity of plant by-products

Vegetable by-products have a high bioactive compound content such as dietary fibres and phenolic compounds which demonstrate a positive modulatory effect on the composition of intestinal microbiota, since the presence of these nutrients increases the production of Short-Chain Fatty Acids (SCFAs), which are used as substrates for the growth of beneficial bacteria in the gastrointestinal tract, such as *Bifidobacterium* and *Lactobacillus*, acting as prebiotics (de Andrade et al., 2020). Ingestion of these prebiotics reduces the risk of developing several diseases such as colon cancer, irritable bowel syndrome, obesity and metabolic syndrome; the latter is defined as a set of metabolic disorders such as type II diabetes, dyslipidemia and hypertension, among other disorders (Farias et al., 2019), since the intestinal microbiota can act at the systemic level through several axes such as intestine-hepatic, intestine-lung, intestine-kidneys and intestine-brain (among others), with these being bidirectional ways.

Likewise, the presence of antioxidant substances in plant by-products present anti-inflammatory action, mainly due to the minimization of oxidative processes in the body (Tajaldini et al., 2020), such as reduced lipid peroxidation, damage to proteins and DNA, which helps in preventing inflammation (Kargutkar and Brijesh, 2016; Athaydes et al., 2019). The antioxidant and anti-inflammatory activities of plant by-products can be conferred by the synergistic action of their polyphenolic components, among them flavonoids and tannins, being attributed to their chemical composition (Kargutkar and Brijesh, 2016; Somensi et al., 2017; de Souza et al., 2017; Athaydes et al., 2019). Other nutrients such as vitamin C, carotenoids and some minerals, also have antioxidant action (Caimari et al., 2015; Patro et al., 2016; Alvarez-Suarez et al., 2017).

The antioxidant effect of flavonoids is related to the presence of hydroxyl groups (OH) in their structure, which allow them to be oxidized by free radicals, making them less reactive. Thus, the more hydroxyls present in the molecule, the greater its antioxidant power (Mahmoud et al., 2019). They also have a protective effect against free radicals by increasing the levels of antioxidant enzymes such as superoxide dismutase (SOD), catalase (CAT) and Glutathione Peroxidase (GPx) which are part of the body's antioxidant defence system (Kargutkar and Brijesh, 2016; Charradi et al., 2018).

On the other hand, the anti-inflammatory potential is related to the ability of these compounds to bind to cyclooxygenases (COXs), which catalyze the conversion of arachidonic acid into prostaglandins and thromboxanes (inflammatory mediators); thus, by binding COXs, they can decrease the inflammatory process (Mahmoud et al., 2019).

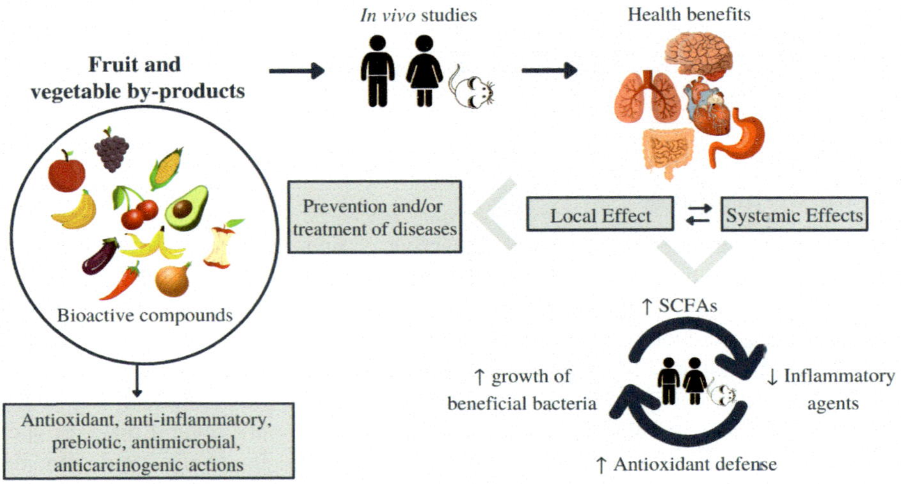

Figure 17.1. Main effects and mechanism of action of fruit and vegetable by-products in the body.

Most plant by-products with potent antioxidants and anti-inflammatory agents studied are presented as hydroalcoholic extracts due to their better technological characteristics, such as greater stability and solubility, in addition to easy access to humans (Somensi et al., 2017; de Souza et al., 2017). Figure 17.1 provides a schematic summary of the main effects of fruit and vegetable by-products.

Non-clinical studies using animal models which are healthy or affected by diseases have been useful in determining toxicity and mainly the dose-response of by-products as interventions for health promotion, which has an impact on the safety of their use. Some studies which assessed health claim properties from administering plant by-products are listed in Tables 17.1 and 17.2. For teaching purposes, the tables were divided into studies with fruit by-products (Table 17.1) and other vegetables (Table 17.2), according to the volume of publications cited.

The small number of studies that address the subject of administering by-products to human beings may reflect barriers and individual and socioeconomic prejudices. However, there is a need to translate the results obtained from basic research into clinical research, respecting all testing phases and also taking into account ethics and law, which characterizes translational science (Cohrs et al., 2014). In this sense, considering the levels of scientific evidence, the performance of randomized controlled clinical studies, as well as systematic reviews with the subject can also be foundations for an increase in the use and culture of consumption of fruit and vegetable by-products, vegetables and legumes.

Although still scarce, some clinical studies have been conducted to assess health claim properties from administering plant by-products. For didactic purposes, the tables were divided into studies with fruit by-products (Table 17.3) and vegetables by-products (Table 17.4), according to the volume of publications cited.

Table 17.1. Non-clinical studies involving the administration of fruit by-products.

Animal model	Managed by-products	Dose and time of administration	Study design	Main results	Reference
Male Wistar rats (100–150 g)	Methanolic extract of tamarind (*Tamarindus indica*) seeds	100 and 200 mg/kg for 15 d	Ulcer	Produces a dose-dependent cytoprotective effect (dose of 200 mg) in ulcer models induced by ibuprofen, alcohol and pyloric binding.	(Kalra et al., 2011)
Male Wistar rats (9–10 wk)	Proanthocyanidins from grape (*Vitis vinifera*) seeds	10, 20 and 40 mg, for 7 d	Recurrent colitis	Significantly reduced the severity of macroscopic lesion scores in a dose-dependent manner (mainly at a dose of 400 mg); significantly decreased the extent and severity of histological signs of cell damage (edema and infiltration of inflammatory cells); significant increase in colon antioxidant enzymes SOD and GPx (except in the group treated with a low dose) and increased GSH activity and content.	(Wang et al., 2011)
Male Sprague-Dawley rats (116 ± 4.5 g)	Grape (*Vitis vinifera*) seed extract powder	400 mg/kg for 10 d	Ulcerative colitis	The severity of the ulcerative colitis markers in the distal ileum and proximal colon decreased.	(Cheah et al., 2013)
Male Wistar rats (100–150 g)	Pineapple peel extract (*Ananas comosus*)	50, 100, 250 and 500 mg/kg, for 21 d	Arthritis	Dose of 500 mg: Anti-rheumatic activity; increased levels of antioxidant enzymes in the liver, kidney and spleen; decreased levels of inflammatory prostaglandins in the serum.	(Kargutkar and Brijesh, 2016)
Male Sprague-Dawley rats, free from specific pathogens (8 wk)	IDF of Pear (*Pyrus* spp.) by-product - seeds, residual pulp and skin	HD and HFD added 8% of the IDF of the pear by-product; *ad libitum* for 5 wk	Obesity-induced by a high-fat diet	The HFD diet resulted in: reduction in body weight and adipocyte size; improved glucose tolerance and glucose homeostasis; reduction in total cholesterol, LDL and leptin (but did not alter ghrelin); reduced liver expression of the FAS and SREBP-1c genes and increased CYP and AMPK; increased the bacterial genus *Akkermansia* in faeces. The HD diet resulted in: reduced blood ALT and increased *Bifidobacterium* in stool.	(Chang et al., 2017)
Male Wistar rats (220–240 g)	Aqueous extract of orange peel (*Citrus sinensis* L.)	100, 200 and 400 mg/kg, for 15 d	Peptic ulcer	In all doses: protects against gastric ulcer; anti-inflammatory properties; antioxidant properties.	(Selmi et al., 2017)

Table 17.1 contd. ...

...Table 17.1 contd.

Animal model	Managed by-products	Dose and time of administration	Study design	Main results	Reference
Female Wistar rats (200–250 g) and Swiss mice (25–30 g)	Hydroalcoholic extract of *Avocado-do-Mato* Persea major (Meisn.) L.E. Kopp (Lauraceae) bark	30, 100 and 300 mg/kg acute ulcer model (single dose); chronic ulcer model (twice a day, for 7 d)	Gastric ulcer	Acute ulcer: gastroprotective effects (doses of 30 and 300 mg). Chronic ulcer: gastric healing (300 mg dose). Decrease in inflammatory parameters (300 mg dose). Increased activity of antioxidant enzymes (dose of 300 mg).	(Somensi et al., 2017)
Male and female and Wistar rats (60 d , ~ 200 g) and male Swiss mice (60 d , ~ 30 g)	Gabiroba (*Campomanesia adamantium*) bark extract	100–1000 mg for 15 d	Neuropathic pain	Doses 100–300 mg: anti-inflammatory activity (male mice); action antihyperalgesic and antidepressant (rats).	(de Souza et al., 2017)
Male Wistar rats (20 ± 2 g)	Banana pulp and peel (*Musa sapientum* L.)	600 mg/kg for 2 wk	Stress	Anxiolytic effects. Effects like antidepressants. Increased memory function in animals treated with banana pulp and peel, compared to control animals.	(Samad et al., 2017)
Female Wistar rats (~ 22 wk)	By-products of acerola (*Malpighia emarginata* DC), cashews (*Anacardium occidentale* L.), and guava (*Psidium guajava* L.) lyophilized - residual pulp, seeds and peels.	400 mg/kg body weight, via gavage for 4 wk	Diet-induced dyslipidaemia	Fruit by-products, mainly acerola, increased serum HDL, faecal moisture, SCFA faecal excretion, organic acids in the faeces and *Bifidobacterium* spp. and *Lactobacillus* spp. in faeces (modulation of the microbiota); reduced serum lipids (TG, TC, LDL, and VLDL), body weight and faecal pH; less deposit of visceral and hepatic fats; protection against tissue damage in the intestine and liver.	(Batista et al., 2018a)
Male *Swiss* mice (7 wk)	By-product of jabuticaba (*Myrciaria jaboticaba*) - leather	Control diets and high in fat added 4% of the skin of jabuticaba; *ad libtum* for 10 wk	Fatty liver induced by a high-fat diet	Reduced body weight, accumulation of body and liver fat, IL-6 and TNF-α mRNA levels in epididymal adipose tissue, ALT and AST in the blood, PPAR-α mRNA in the liver, hepatic TBARS; increased blood HDL, faecal excretion of TG and cholesterol; increased the activity of SOD, CAT, GPx and GSH; higher faecal content of acetate and butyrate.	(Batista et al., 2018b)

Male Wistar rats (2 mon)	Grape (*Vitis vinífera*) seed powder	0.4, 4, 8 and 16 g/kg of feed for 8 wk	Healthy	Anti-inflammatory effect (dose of 16 g and dose of 8 g at a lower level); antioxidant effect (all doses, the dose of 16 g is more prominent); Doses of 8 and 16 g/kg induce mild adverse effects, such as diarrhoea.	(Charradi et al., 2018)
Spontaneously hypertensive male rats (10 wk)	Açaí (*Euterpe oleracea* Mart.) seed extract	200 mg/kg for 45 d	Hypertension and diabetes	Prevention of oxidative damage; preservation of renal morphology and function; anti-inflammatory action.	(Cordeiro et al., 2018)
Male Wistar rats (150–200 g)	Grape seed extract (*Vitis vinifera*)	25, 50 and 100 mg/kg for 12 wk	Liver cancer	Inhibited the formation of foci of altered hepatocytes (all doses); decreased oxidative stress (doses of 50 and 100 mg); anti-inflammatory activity (all doses).	(Hamza et al., 2018)
Female Wistar rats (180-200 g)	Soluble dietary fibres from the skin of yellow passion fruit (*Passiflora edulis f. Flavicarpa*)	Pre-treatment: single oral dose (0.1, 1 and 10 mg/kg body weight) or single intraperitoneal dose (up to 1 mg/kg body weight)	Ethanol-induced ulcer	All doses of soluble dietary fibres from the passion fruit skin administered by both routes decreased gastric lesions and prevented the fall in GSH levels and gastric mucus.	(Abboud et al., 2019)
Male Swiss mice (8 wk)	Avocado seed extract (*Persea americana* Mill.)	10, 35, 75 mg / kg, single dose	Gastric ulcer	Gastroprotective activity (in all doses); Increase in endogenous antioxidant enzymes (75 mg dose); Increased mucus production (doses of 10 and 75 mg); Decreased inflammatory process (doses of 35 and 75 mg).	(Athaydes et al., 2019)
Male ApoE knockout mice (C57BL/6 strain) (8 wk)	Peach seed oil (*Prunus perssica*)	2 or 5 g/kg for 4 wk	Atherosclerosis	All doses: anti-inflammatory response *in vitro*; improvement of the lipid profile *in vivo*.	(Hao et al., 2019)
Male Sprague-Dawley rats (180-220 g)	Total flavonoids of grapefruit peel (*Citrus changshanhuyo*)	25, 50 and 100 mg/kg, for 8 wk	Non-alcoholic fatty liver disease	High dose: Mitigated liver damage; suppressed systemic and intrahepatic inflammation; reversed the serum changes in the lipid profile and pro-inflammatory cytokines.	(Jiang et al., 2019)

Table 17.1 contd. ...

...Table 17.1 contd.

Animal model	**Managed by-products**	**Dose and time of administration**	**Study design**	**Main results**	**Reference**
Male C57BL/6J mouse (6 wk)	Dry extracts of polyphenol from bagasse and grape seed (*Vitis vinifera "Kyoho"*)	200 mg/kg of body weight per gavage for 7 d .	Treatment of damage caused by consuming a diet rich in fat and water plus an antibiotic cocktail.	The dry polyphenol extract from the grape seed decreased total cholesterol and increased the *Prevotella* (G-) genus in the intestinal microbiota. Both extracts increased the phylum *Verrucomicrobia* [mainly the genera *Akkermansia* (G-)] and also the genera *Alloprevotella* (G-) and *Unclassified Enterobacteriaceae* (G-); and reduced the phylum *Actinobacteria* (G +) and the genus *Streptococcus* (G +) in the intestinal microbiota. However, they did not restore the genera *Bifidobacterium* (G +), *Lactococcus* (G +), and *Lactobacillus* (G +).	(Lu et al., 2019)
Male Sprague-Dawley rats (200 g)	Extract rich in fibre (3.5 g/100 g) and polyphenols (50 mg GAE/d) obtained from palm oil processing (*Elaeis guineenses*)	OPP added in water (50 mg GAE/d) or the western diet (2000 mg GAE/kg), administration for 4 wk	Western diet consumption vs. large intestine health	OPP-D increased the mass of the cecal content; the concentration of total SCFA, acetate and propionate in the cecal content; the number of goblet cells producing mucus; and abundance of some bacteria cecal as *Faecalibacterium prausnitzii*, *Akkermansia muciniphila* and *Ruminococcus gnavus*. OPP-F increased the concentration of butyrate in the cecal content.	(Conlon et al., 2020)
Male Sprague-Dawley rats (120 g)	Powdered olive seed or nanoparticles	HFHF diet added with 5 or 10% powdered olive seed or nanoparticles administered for 8 wk	Endothelial dysfunction induced by HFHF	The tested doses, mainly the 10% dose caused a reduction in serum glucose, insulin levels, HOMA-IR, serum lipid profile (TG, TC, LDL and VLDL), lipase activity, IL-6, and serum CRP, serum MDA; increase in serum vasodilators (prostacyclin, nitric oxide and eNOS) and reduction in vasoconstrictor endothelin-1.	(Genedy et al., 2020)

Male Wistar Hannover rats (4 wk)	Skin, residual pulp, seeds and apple stems (*Malus domestica*)	AIN-210 diet rich in fat and supplemented with apple by-product containing 20% dietary fibre; offered *ad libitum* for 5 wk .	Damage caused by a high-fat diet.	The treated group showed an increase in HDL and a reduction in TG in the blood; reduction of lipids in the liver; increased faecal excretion of bile acids; reduction in faecal pH; increased faecal moisture; elevation of SCFA (acetic, butyric, and propionic) and branched SCFA (isobutyric and isovaleric) in the cecum.	(Mateos-Aparicio et al., 2020)
Male C57BL/6 mouse (without weight and age information)	Red raspberry seed	A high-fat diet with added polyphenols from red raspberry seed (0.1%) administered for 16 wk	Obesity-induced by a high-fat diet	The polyphenols of the red raspberry seed significantly increased the abundance of *Bifidobacterium*, an unclassified genus of *Clostridiales* and *Roseburia*; and decreased the abundance of *Ruminococcus* and *Mogibacteriaceae*.	(Xian et al., 2021)

ALT, Alanine aminotransferase; AMPK, AMP-activated protein kinase; AST, aspartate aminotransferase; CAT, catalase; CRP, C-reactive protein; CYP, cytochrome P450; eNOS, nitric oxide endothelial synthase; FAS, fatty acid synthase; G-, Gram-negative; G+, Gram-positive; GAE / d, gallic acid equivalents; GPx, glutathione peroxidase; GSH, glutathione; HD, High dietary fibre diet; HDL, high density lipoprotein; HFD, high-fat and dietary fibre diet; HFHF, high fat and high fructose; HOMA-IR, homeostatic model assessment of insulin resistance; IDF, Insoluble dietary fibre; IL, interleukin; LDL, low density lipoprotein; MDA, malondialdehyde; mRNA, messenger ribonucleic acid; OPP, oil palm preparation; OPP-D, oil palm preparation added to drinking water; OPP-F, oil palm preparation added to food; PPAR-α, Peroxisome proliferator-activated receptor alpha; SCFA, short-chain fatty acid; SOD, superoxide dismutase; SREBP-1c, Transcriptional factor sterol regulatory element-binding protein 1c; TBARS, thiobarbituric acid reactive substances; TC, total cholesterol; TG, triglycerides; TNF-α, tumour necrosis factor alpha; VLDL, very low-density lipoprotein.

Table 17.2. Non-clinical studies which involved the administration of vegetables by-products.

Animal model	Managed by-products	Dose and time of administration	Study design	Main results	Reference
Male Sprague-Dawley rats (100–120 g)	Soluble dietary fibre from soybean hulls	4% of the diet, for 4 wk	Hypercholesteraemic	Hypocholesterolaemic activity (decreased TG and LDL and increased HDL).	(Liu et al., 2016)
Male Wistar Hannover rats (~ 240 g)	Okara soybean by-product (*Glycine max* (L.) Merr.)	Diet high-fat containing 20% of Okara - 4 wk consumption *ad libitum*	Damage caused by consumption of a high-fat diet	Reduced body weight, urea and plasma TG; better absorption of calcium and magnesium.	(Pérez-López et al., 2018)
Male mice with accelerated senescence 8 and resistance to senescence 1 at 16 wk of age	Okara soybean by-product (*Glycine max* (L.) Merr.)	AIN-93M diet containing 7.5 or 15% okara - 26 wk of consumption *ad libitum*	Age-related cognitive impairment	Mice fed with 7.5% of okara had a longer latency in the memory test (Passive-Avoidance Task); greater NeuN intensity in the hippocampus; increased BDNF and the level of acetylcholine in the brain, and decreased TNF-α. The 15% okara diet reduced body weight, increased faecal weight, altered the composition of the cecal microbiota (*Clostridiales*, *Bacteriodales* and *Ciriobacteriales*), but did not alter the serum levels of lactic acid and butyric acid. Both diets caused an increase in the expression of the acetylcholine synthesizing enzyme and less time spent in the opposite of the target (Barnes maze test).	(Corpuz et al., 2019)
Male C57BL mice (3 to 4 wk)	Modified dietary fibres from okara (*Glycine max* (L.) Merr.)	A high-fat diet containing 10% (w/w) of modified dietary fibres - 12 wk consumption *ad libitum*	Damage caused by consumption of a high-fat diet	Reduced body weight and body fat; reduced TC, TG, ω-6 fatty acids, alanine, citrate, creatine and succinate in the blood.	(Dai et al., 2019)
Male Syrian golden hamsters (~ 47 g)	Artichoke by-product (*Cynaracardunculus* L. subsp. *Scolymus* (L.) Hayek)	A high-fat diet containing 20% fibre from the artichoke by-product - 3 wk consumption *ad libitum*	Damage caused by consumption of a high-fat diet	TG, TC, LDL and VLDL plasma levels were reduced; there was a reduction in total fats, TG, TC and esterified cholesterol in the liver; increased faecal excretion of total fats, TG, bile acids and proteins.	(Villanueva-Suárez et al., 2019)

Swiss disease-free albino mice (20–25 g) of both sexes	Dry seeds for the production of *Cucurbita moschata* extract with ethanolic solvents and ethyl acetate	*Cucurbita moschata* ethanolic extract (100 mg/kg or 200 mg/kg, powder) and *Cucurbita moschata* ethyl acetate extract (100 mg/kg or 200 mg/kg, powder) administered for 7 d	Anxiolytic and motor coordination activity	The ethanolic extract of *Cucurbita moschata* (200 mg/kg) showed the best result in both models compared to the drug Alprazolam and also a change in motor coordination was observed in contrast to the mice in the control group. Both extracts showed relevant anxiolytic activity.	(Arora et al., 2020)
Golden Hamster (*Mesocricetus auratus*) - 70–90 g of body weight	Bioactive peptides isolated from stems and leaves of broccoli (*Brassica oleracea var. Italica*) through hydrolysis with different enzymes (alkalase, trypsin, neutrase, papain and pepsin)	500 mg of different peptides/kg body weight - 9 d of supplementation	Damage caused by consumption of a high-fat diet	Broccoli peptides (hydrolysis with neutralization): greater antioxidant activity *in vitro*, reduces blood levels of TC. Broccoli peptides (hydrolysis with papain): reduces blood levels of TC and LDL.	(Chen et al., 2020)
Male Wistar rats (180–220 g)	Pumpkin seeds (*Cucurbita pepo* L.)	Diet added of 5 and 10% of raw or roasted pumpkin seed administered for 14 d	Nephrotoxicity caused by cisplatin	Both pumpkin seeds, mainly roasted seed, reduced serum urea, serum creatinine and potassium levels, renal 8-OHdG, renal MDA, kidney ROS, the renal activity of arginase II; increased serum albumin and sodium levels, CAT and GST kidney activity, NPSH, T-SHs and CAT, NO kidney levels.	(Oyetayo et al., 2020)
Male mice (8 wk)	Extracts of red rice bran or dark purple (purple-black)	Westernized high-fat diet with 10% red rice or dark purple administered for 12 wk	Diet-induced atherosclerosis	Both diets added reduced TC, TG and LDL/β2-oxidized pro-atherogenic glycoprotein I plasma levels, the size of atherosclerotic plaques in the aorta and in aortic (iNOS, TNF-α, IL-1β) and hepatic pro-inflammatory modulators (TNF-α, IL-1α, IL-1β).	(Tan et al., 2020)

...Table 17.2 contd.

Animal model	**Managed by-products**	**Dose and time of administration**	**Study design**	**Main results**	**Reference**
Male Wistar rats (without weight and age information)	Potato skin extract (*Solanum tuberosum* L.)	Oral administration of a single dose (50 mg/kg, 100 mg/kg, 200 mg/kg or 400 mg/kg)	Analgesic and anti-inflammatory activity	Doses of 100 mg/kg and 200 mg/kg reduced pain and doses of 100 mg/kg, 200 mg/kg, and 400 mg/kg reduced the volume of edema, but to a lesser extent than paracetamol and diclofenac, respectively.	(Wahyudi et al., 2020)
Male BALB/C mice (18–22 g)	Dietary fibre from soybean husks	Oral dose of insoluble dietary fibre (300 mg/kg) or soluble dietary fibre (300 mg/kg) administered for 42 d	Inflammatory bowel disease induced by sodium dextran sulphate	Both dietary fibre decreased weight loss, diarrhoea and faecal bleeding, serum TNF-α and levels of NF-κB in the colon. The soluble dietary fibre significantly increased the relative abundance of *Barnesiella, Lactobacillus, Ruminococcus,* and *Flavonifractor*.	(Yang et al., 2020)

8-OHdG, 8-hydroxy-21-deoxyguanosine; BDNF, brain-derived neurotrophic factor; CAT, catalase; GST, glutathione-S-transferase; HDL, high density lipoprotein; IL, interleukin; iNOS, inducible nitric oxide synthase; LDL, low density lipoprotein; MDA, malondialdehyde; NeuN, neuronal nuclei; NF-κB, nuclear factor kappa B; NO, nitric oxide; NPSH, non-protein thiol; ROS, oxygen-reactive species; TC, total cholesterol; TG, triglycerides; TNF-α, tumour necrosis factor alpha; T-SH, total thiol; VLDL, very low-density lipoprotein; ω-6, omega 6.

Table 17.3. Clinical studies which involved the administration of fruit by-products.

Population	Managed by-products	Dose and time of administration	Study design	Main results	Reference
Men and women (n = 68)	Grape seed extract	100 mg/day for four wk	Healthy	Improved endothelial function through antioxidant and anti-inflammatory effects in healthy individuals.	(Kominami et al. 2014)
Men and women 20–70 yr (n = 39)	Passion fruit seed	2 doses of 5 mg of purified piceatannol from the passion fruit seed	Eutrophic or overweight individuals	There was no difference in mood assessed by the profile of mood states and the visual analogue scale tests.	(Kitada et al. 2017)
Women 25–65 yr (n = 10)	Grape pomace	1400 mg/day for 21 d	Healthy	Reduced fasting glycaemia and increased production of SCFA.	(Gil-Sánchez et al. 2018)
Men and women 18–60 yr (n = 221)	A drink containing orange pomace	473 mL per day for three wk	Healthy	Laxative effect and provided modulation of the intestinal microbiota.	(Dennis-Wall et al. 2019)
Obese men and women (n = 40)	Grape seed extract	300 mg/day for 12 wk	Overweight or obese individuals	Reduction in body weight, BMI, waist circumference, neuropeptide-Y, TNF-α and increased CRP sensitivity.	(Parandoosh et al. 2020)

BMI, body mass index; CRP, C-reactive protein; SCFA, short-chain fatty acid; TNF-α, tumour necrosis factor alpha.

Table 17.4. Clinical studies which involved the administration of vegetable by-products.

Population	Managed by-products	Dose and time of administration	Study design	Main results	Reference
Men and women 18–75 yr (n = 44)	Carrot flour (pulp + peel)	2 doses (≈ 14.4 g) per d for four wk	Healthy	Increase in the concentration of total carotenoids in the plasma, with consequent stimulation of the proper functioning of the immune system.	(Castro et al., 2019)
Post- menopausal women aged 48–64 yr (n = 23)	Pumpkin seed oil	3 g/d for six wk	Healthy	Improved arterial haemodynamics in post-menopausal women and can therefore be effective in preventing and treating hypertension and cardiovascular diseases in this population.	(Wong et al., 2019)
Men and women 30–35 yr (n = 50)	Fenugreek seed	5 g, three times a day for eight wks	Type II diabetes	Reduced fasting glycaemia and systolic blood pressure. Improved liver and kidney function.	(Hadi et al., 2020)
Men and women 25–65 yr (n = 140)	Lettuce seed	20 mg/day for 12 wk	Dyslipidaemia	Improved the lipid profile compared to the placebo group. Showed an inhibitory effect on the elevation of liver enzymes induced by atorvastatin and possible liver toxicity.	(Moghadam et al., 2020)

Concluding remarks and perspectives

Both studies with animal models and clinical studies have leveraged advances in research aimed at assessing the exploitation of plant subproducts from demonstrating their functional potential. However, it is important to highlight that there are no ideal models, and the most suitable ones are those based on the objectives and hypotheses of the study and the association of complementary techniques, when necessary, as well as on the levels of scientific evidence.

By-products obtained from grapes and soybeans stand out among the most studied vegetable by-products in *in vivo* models, which demonstrates the need for further studies with by-products from other plant sources.

Additionally, combining healthy eating practices with the inclusion of food, nutraceutical and herbal products obtained from vegetable by-products can be a promising marketing strategy, since the raw material is of low cost and would not be discarded into the environment, which is aligned with the global sustainability trends and the sustainable development objectives proposed by the United Nations, in addition to promoting greater accessibility of these products to a greater part of the population, bringing several benefits such as health maintenance or disease prevention.

References

Abboud, K.Y., Da Luz, B.B, Dallazen, J.L, Werner, M.F.P., Cazarin, C.B.B., Júnior, M.R.M., Iacomini, M. and Cordeiro, L.M.C. 2019. Gastroprotective effect of soluble dietary fibres from yellow passion fruit (*Passiflora edulis* f. flavicarpa) peel against ethanol-induced ulcer in rats. J. Funct. Foods 54: 552–558. DOI: 10.1016/j.jff.2019.02.003.

Alvarez-Suarez, J.M., Giampieri, F., Gasparrini, M., Mazzoni, L., Santos-Buelga, C., González-Paramás, A.M., Forbes-Hernández, T.Y., Afrin, S., Páez-Watson, T., Quiles, J.L. and Battino, M. 2017. The protective effect of acerola (Malpighia emarginata) against oxidative damage in human dermal fibroblasts through the improvement of antioxidant enzyme activity and mitochondrial functionality. Food Funct. 8(9): 3250–3258. DOI: 10.1039/C7FO00859G.

Arora, I., Behl, T., Grover, M., Sachdeva, M., Pal, G. and Khan, N. 2020. Study of anxiolytic and motor co-ordination activity of Cucurbita moschata and its possible mechanism through GABA receptors. Obes. Med. 18: 100204. DOI: 10.1016/j.obmed.2020.100204.

Athaydes, B.R., Alves, G.M., de Assis, A.L.E.M., Gomes, J.V.D., Rodrigues, R.P., Campagnaro, B.P., Nogueira, B.V., Dâmaris, S., Kuster, R.M., Pereira, T.M.C., Kitagawa, R.R. and Gonçalves, R.D.C.R. 2019. Avocado seeds (*Persea americana* Mill.) prevents indomethacin-induced gastric ulcer in mice. Food Res. Int. 119: 751–760. DOI: 10.1016/j.foodres.2018.10.057.

Batista, Â.G., da Silva-Maia, J.K., Mendonça, M.C.P., Soares, E.S., Lima, G.C., Junior, S.B., Cruz-Höfling, M.A. and Júnior, M.R.M. 2018b. Jaboticaba berry peel intake increases short chain fatty acids production and prevent hepatic steatosis in mice fed high-fat diet. J. Funct. Foods 48: 266–274. DOI: 10.1016/j.jff.2018.07.020.

Batista, K.S., Alves, A.F., dos Santos Lima, M., da Silva, L.A., Lins, P.P., de Sousa, J.A.G., Silva, A.S., Toscano, L.T., Meireles, B.R.L.A., Cordeiro, A.M.T.M., Conceição, M.L., Souza, E.L. and Aquino, J.S. 2018a. Beneficial effects of consumption of acerola, cashew or guava processing by-products on intestinal health and lipid metabolism in dyslipidaemic female Wistar rats. Br. J. Nutr. 119(1): 30–41. DOI: 10.1017 / S0007114517003282.

Brasil. Instrução Normativa N° 81, de 19 de dezembro de 2018. [S.l.]: Ministério da Agricultura, Pecuária e Abastecimento, Gabinete do Ministro, 2018. Available in: <https://www.gov.br/agricultura/pt-br/assuntos/inspecao/produtos-vegetal/legislacao-1/biblioteca-de-normas-vinhos-e-bebidas/instrucao-normativa-no-81-de-19-de-dezembro-de-2018.pdf/view>.

Caimari, A., Puiggròs, F., Suárez, M., Crescenti, A., Laos, S., Ruiz, J.A., Alonso, V., Moragas, J., Del Bas, J.M. and Arola, L. 2015. The intake of a hazelnut skin extract improves the plasma lipid profile and reduces the lithocholic/deoxycholic bile acid faecal ratio, a risk factor for colon cancer, in hamsters fed a high-fat diet. Food Chem. 167: 138–144. DOI: 10.1016 / j.foodchem.2014.06.072.

Castro, M., Tatuszka, P., Cox, D.N., Bowen, J., Sanguansri, L., Augustin, M.A. and Stonehouse, W. 2019. Effects on plasma carotenoids and consumer acceptance of a functional carrot-based product to supplement vegetable intake: A randomized clinical trial. J. Funct. Foods 60: 103421. DOI: 10.1016/j.jff.2019.103421.

Chang, C.C.A. and Lin, Y.C. 2015. Physical activity and food consumption: The moderating role of individual dieting tendency. J. Health Psychol. 20(5): 490–499. DOI: 10.1177/1359105315573469.

Chang, S., Cui, X., Guo, M., Tian, Y., Xu, W., Huang, K. and Zhang, Y. 2017. Insoluble dietary fiber from pear pomace can prevent high-fat diet-induced obesity in rats mainly by improving the structure of the gut microbiota. J. Microbiol. Biotechnol. 27(4): 856–867. DOI: 10.4014 / jmb.1610.10058.

Charradi, K., Mahmoudi, M., Bedhiafi, T., Jebari, K., El May, M.V., Limam, F. and Aouani, E. 2018. Safety evaluation, anti-oxidative and anti-inflammatory effects of subchronically dietary supplemented high dosing grape seed powder (GSP) to healthy rat. Biomed. Pharmacother. 107: 534–546. DOI: 10.1016/j.biopha.2018.08.031.

Cheah, K.Y., Bastian, S.E., Acott, T.M., Abimosleh, S.M., Lymn, K.A. and Howarth, G.S. 2013. Grape seed extract reduces the severity of selected disease markers in the proximal colon of dextran sulphate sodium-induced colitis in rats. Dig. Dis. Sci. 58(4): 970–977. DOI: 10.1007/s10620-012-2464-1.

Chen, H., Xia, L.S., Zhang, X.D. and Wei, Z.J. 2020. Antioxidant and hypolipidemic potential of peptides from broccoli stems and leaves. Curr. Top. Nutraceutical Res. 18(1): 16–21. DOI: 10.37290/ctnr2641-452X.

Chrysochou, P. 2010. Food health branding: The role of marketing mix elements and public discourse in conveying a healthy brand image. J. Mark. Commun. 16(1-2): 69–85. DOI: 10.1080/13527260903342787.

Coelho, E.M., de Souza, M.E.A.O., Corrêa, L.C., Viana, A.C., de Azevêdo, L.C. and dos Santos Lima, M. 2019. Bioactive compounds and antioxidant activity of mango peel liqueurs (Mangifera indica L.) produced by different methods of maceration. Antioxidants 8(4): 102. DOI: 10.3390/antiox8040102.

Cohrs, R.J., Martin, T., Ghahramani, P., Bidaut, L., Higgins, P.J. and Shahzad, A. 2015. Translational medicine definition by the European Society for Translational Medicine. New Horiz. Transl. Med. 2(3): 86–88. DOI: 10.1016/j.nhtm.2014.12.002.

Coimbra, M.C. and Jorge, N. 2012. Fatty acids and bioactive compounds of the pulps and kernels of Brazilian palm species, guariroba (Syagrus oleraces), jerivá (Syagrus romanzoffiana) and macaúba (Acrocomia aculeata). J. Sci. Food Agric. 92(3): 679–684. DOI: 10.1002/jsfa.4630.

Conlon, M.A., Sambanthamurthi, R., Tan, Y.A., Sundram, K., Fairus, S. and Abeywardena, M.Y. 2020. Consumption of an oil palm fruit extract promotes large bowel health in rats. Nutrients 12(3): 644. DOI: 10.3390/nu12030644.

Cordeiro, V.D.S.C., de Bem, G.F., da Costa, C.A., Santos, I.B., de Carvalho, L.C.R.M., Ognibene, D.T., da Rocha, A.P.M., de Carvalho, J.J., de Moura, R.S. and Resende, A.C. 2018. Euterpe oleracea Mart. seed extract protects against renal injury in diabetic and spontaneously hypertensive rats: role of inflammation and oxidative stress. Eur. J. Nutr. 57(2): 817–832. DOI: 10.1007/s00394-016-1371-1.

Corpuz, H.M., Arimura, M., Chawalitpong, S., Miyazaki, K., Sawaguchi, M., Nakamura, S. and Katayama, S. 2019. Oral administration of Okara soybean by-product attenuates cognitive impairment in a mouse model of accelerated aging. Nutrients 11(12): 2939. DOI: 10.3390/nu11122939.

Dai, B., Huang, S. and Deng, Y. 2019. Modified insoluble dietary fibers in okara affect body composition, serum metabolic properties, and fatty acid profiles in mice fed high-fat diets: an NMR investigation. Food Res. Int. 116: 1239–1246. DOI: 10.1016/j.foodres.2018.10.011.

de Andrade, R.M.S., Silva, S., Costa, C.M.D.S.F., Veiga, M., Costa, E., Ferreira, M.S.L., Gonçalves, E.C.B.A. and Pintado, M.E. 2020. Potential prebiotic effect of fruit and vegetable byproducts flour using *in vitro* gastrointestinal digestion. Food Res. Int. 137: 109354. DOI: 10.1016/j.foodres.2020.109354.

de Souza, J.C., Piccinelli, A.C., Aquino, D.F., de Souza, V.V., Schmitz, W.O., Traesel, G.K., Cardoso, C.A.L., Kassuya, C.A.L. and Arena, A.C. 2017. Toxicological analysis and antihyperalgesic, antidepressant, and anti-inflammatory effects of Campomanesia adamantium fruit barks. Nutr. Neurosci. 20(1): 23–31. DOI: 10.1179/1476830514Y.0000000145.

Dennis-Wall, J.C., Burns, A.M., Solch, R.J., Ukhanova, M., Dahl, W.J., Christman, M.C., Boileau, T., Brauchla, M., Shin, J-E., Júnior, C.N., Mai, V. and Langkamp-Henken, B. 2019. A beverage containing orange pomace improves laxation and modulates the microbiome in healthy adults: a randomised, blinded, controlled trial. J. Funct. Foods 60: 103438. DOI: 10.1016/j.jff.2019.103438.

Dilucia, F., Lacivita, V., Conte, A. and Del Nobile, M.A. 2020. Sustainable use of fruit and vegetable by-products to enhance food packaging performance. Foods 9(7): 857. DOI: 10.3390/foods9070857.

FAO/WHO. Promoting fruit and vegetable consumption around the world. 2020. World Health Organization, Available in: <https://www.who.int/dietphysicalactivity/fruit/en/#:~:text=A%20recently%20published%20WHO%2FFAO,alleviation%20of%20several%20micronutrient%20deficiencies%2C>.

Farias, D.P., de Araujo, F.F., Neri-Numa, I.A. and Pastore, G.M. 2019. Prebiotics: Trends in food, health and technological applications. Trends Food Sci. Technol. 93: 23–35. DOI: 10.1016/j.tifs.2019.09.004.

Felice, F., Fabiano, A., De Leo, M., Piras, A.M., Beconcini, D., Cesare, M.M., Branca, A., Zambito, Y. and Di Stefano, R. 2020. Antioxidant effect of cocoa by-product and cherry polyphenol extracts: A comparative study. Antioxidants 9(2): 132. DOI: 10.3390/antiox9020132.

Fracassetti, D., Costa, C., Moulay, L. and Tomás-Barberán, F.A. 2013. Ellagic acid derivatives, ellagitannins, proanthocyanidins and other phenolics, vitamin C and antioxidant capacity of two powder products from camu-camu fruit (Myrciaria dubia). Food Chem. 139(1-4): 578–588. DOI: 10.1016/j.foodchem.2013.01.121.

Gawlik-Dziki, U., Kaszuba, K., Piwowarczyk, K., Świeca, M., Dziki, D. and Czyż, J. 2015. Onion skin—Raw material for the production of supplement that enhances the health-beneficial properties of wheat bread. Food Res. Int. 73: 97–106. DOI: 10.1016/j.foodres.2015.02.008.

Genedy, E.H.F., Mohamed, E.A.K., Sharaf, E.H.A.A., Shosha, N.N.H. and smail, S.H. 2020. Ameliorative Effect of olive seed or nano-olive seed powder against endothelial dysfunction induced by high fat-high fructose diet in rats. Int. J. Food Sci. Biotechnol. 5(4): 94. DOI: 10.11648/j.ijfsb.20200504.18.

Gil-Sánchez, I., Esteban-Fernández, A., de Llano, D.G., Sanz-Buenhombre, M., Guadarrana, A., Salazar, N., Gueimonde, M., de los Reyes-Gavilánc, C.G., Gómez, L.M., Bermejo, M.L.G., Bartolomé, B. and Moreno-Arribas, M.V. 2018. Supplementation with grape pomace in healthy women: Changes in biochemical parameters, gut microbiota and related metabolic biomarkers. J. Funct. Foods 45: 34–46. DOI: 10.1016/j.jff.2018.03.031.

Hadi, A., Arab, A., Hajianfar, H., Talaei, B., Miraghajani, M., Babajafari, S., Wolfgang, M. and Tavakoly, R. 2020. The effect of fenugreek seed supplementation on serum irisin levels, blood pressure, and liver and kidney function in patients with type 2 diabetes mellitus: A parallel randomized clinical trial. Complement. Ther. Med. 49: 102315. DOI: 10.1016/j.ctim.2020.102315.

Hamza, A.A., Heeba, G.H., Elwy, H.M., Murali, C., El-Awady, R. and Amin, A. 2018. Molecular characterization of the grape seeds extract's effect against chemically induced liver cancer: *In vivo* and *in vitro* analyses. Sci. Rep. 8(1): 1–16. DOI: 10.1038/s41598-018-19492-x.

Hao, E., Pang, G., Du, Z., Lai, Y.H., Chen, J.R., Xie, J., Zhou, K., Hou, X., Hsiao, C-D. and Deng, J. 2019. Peach kernel oil downregulates expression of tissue factor and reduces atherosclerosis in ApoE knockout mice. Int. J. Mol. Sci. 20(2): 405. DOI: 10.3390/ijms20020405.

Ice, G., Sadruddin, A.F.A. and Williams, S.R. 2012. Cultural and socioeconomic contexts and nutritional transitions: The anthropology of global overweight and obesity trends of older adults. SES Heal. Implic. 183–202.

Jiang, J., Yan, L., Shi, Z., Wang, L., Shan, L. and Efferth, T. 2019. Hepatoprotective and anti-inflammatory effects of total flavonoids of Qu Zhi Ke (peel of Citrus changshan-huyou) on non-alcoholic fatty liver disease in rats via modulation of NF-κB and MAPKs. Phytomedicine 64: 153082. DOI: 10.1016/j.phymed.2019.153082.

Kabir, F., Tow, W.W., Hamauzu, Y., Katayama, S., Tanaka, S. and Nakamura, S. 2015. Antioxidant and cytoprotective activities of extracts prepared from fruit and vegetable wastes and by-products. Food Chem. 167: 358–362. DOI: 10.1016/j.foodchem.2014.06.099.

Kalra, P., Sharma, S. and Suman, S.K. 2011. Antiulcer effect of the methanolic extract of Tamarindus indica seeds in different experimental models. J. Pharm. Bioallied Sci. 3(2): 236. DOI: 10.4103/0975-7406.80778.

Kargutkar, S. and Brijesh, S. 2016. Anti-rheumatic activity of Ananas comosus fruit peel extract in a complete Freund's adjuvant rat model. Pharm. Biol. 54(11): 2616–2622. DOI: 10.3109/13880209.2016.1173066.

Kitada, M., Ogura, Y., Maruki-Uchida, H., Sai, M., Suzuki, T., Kanasaki, K., Hara, Y., Seto, H., Kuroshima, Y., Monno, I. and Koya, D. 2017. The effect of piceatannol from passion fruit (Passiflora edulis) seeds on metabolic health in humans. Nutrients 9(10): 1142. DOI: 10.3390/nu9101142.

Kominami, M., Yasutake, Y., Ono, Y. and Shibata, H. 2014. Supplementation of olive and grape seed extracts improves vascular function in healthy humans: a randomized controlled study. Atherosclerosis 235(2): e80. DOI: 10.1016/j.atherosclerosis. 2014.05.207.

Liu, C., Lin, X.L., Wan, Z., Zou, Y., Cheng, F.F. and Yang, X.Q. 2016. The physicochemical properties, *in vitro* binding capacities and *in vivo* hypocholesterolemic activity of soluble dietary fiber extracted from soy hulls. Food Funct. 7(12): 4830–4840. DOI: 10.1039/C6FO01340F.

Lu, F., Liu, F., Zhou, Q., Hu, X. and Zhang, Y. 2019. Effects of grape pomace and seed polyphenol extracts on the recovery of gut microbiota after antibiotic treatment in high-fat diet-fed mice. Food Sci. Nutr. 7(9): 2897–2906. DOI: 10.1002/fsn3.1141.

Mahmoud, A.M., Bautista, R.J.H., Sandhu, M.A. and Hussein, O.E. 2019. Beneficial effects of citrus flavonoids on cardiovascular and metabolic health. Oxid. Med. Cell. Longev. 2019: 5484138. DOI: 10.1155/2019/5484138.

Majerska, J., Michalska, A. and Figiel, A. 2019. A review of new directions in managing fruit and vegetable processing by-products. Trends Food Sci. Technol. 88: 207–219. DOI: 10.1016/j.tifs.2019.03.021.

Mateos-Aparicio, I., De la Peña Armada, R., Pérez-Cózar, M.L., Rupérez, P., Redondo-Cuenca, A. and Villanueva-Suárez, M.J. 2020. Apple by-product dietary fibre exhibits potential prebiotic and hypolipidemic effectsin high-fat fed Wistar rats. Bioact. Carbohydr. Diet. Fibre. 23: 100219. DOI: 10.1016/j.bcdf.2020.100219.

Moghadam, M.H., Ghasemi, Z., Sepahi, S., Rahbarian, R., Mozaffari, H.M. and Mohajeri, S.A. 2020. Hypolipidemic effect of Lactuca sativa seed extract, an adjunctive treatment, in patients with hyperlipidemia: a randomized double-blind placebo-controlled pilot trial. J. Herb. Med. 23: 100373. DOI: 10.1016/j.hermed.2020.100373.

Myoda, T., Fujimura, S., Park, B., Nagashima, T., Nakagawa, J. and Nishizawa, M. 2010. Antioxidative and antimicrobial potential of residues of camu-camu juice production. J. Food Agric. Environ. 8(2): 304–307.

Nebl, J., Schuchardt, J.P., Wasserfurth, P., Haufe, S., Eigendorf, J., Tegtbur, U. and Hahn, A. 2019. Characterization, dietary habits and nutritional intake of omnivorous, lacto-ovo vegetarian and vegan runners—a pilot study. BMC Nutr. 5(1): 1–14. DOI: 10.1186/s40795-019-0313-8.

O'Shea, N., Arendt, E.K. and Gallagher, E. 2012. Dietary fibre and phytochemical characteristics of fruit and vegetable by-products and their recent applications as novel ingredients in food products. Innov. Food Sci. Emerg. Technol. 16: 1–10. DOI: 10.1016/j.ifset.2012.06.002.

Oyetayo, F.L., Akomolafe, S.F. and Osesanmi, T.J. 2020. Effect of dietary inclusion of pumpkin (Cucurbita pepo L.) seed on nephrotoxicity occasioned by cisplatin in experimental rats. J. Food Biochem. 44(10): e13439. DOI: 10.1111/jfbc.13439.

Parandoosh, M., Yousefi, R., Khorsandi, H., Nikpayam, O., Saidpour, A. and Babaei, H. 2020. The effects of grape seed extract (Vitis vinifera) supplement on inflammatory markers, neuropeptide Y, anthropometric measures, and appetite in obese or overweight individuals: A randomized clinical trial. Phytother. Res. 34(2): 379–387. DOI: 10.1002/ptr.6529.

Patro, G., Bhattamisra, S.K., Mohanty, B.K. and Sahoo, H.B. 2016. *In vitro* and *in vivo* antioxidant evaluation and estimation of total phenolic, flavonoidal content of Mimosa pudica L. Pharmacognosy Res. 8(1): 22. DOI: 10.4103/0974-8490.171099.

Pérez-López, E., Veses, A.M., Redondo, N., Tenorio-Sanz, M.D., Villanueva, M.J., Redondo-Cuenca, A., Marcos, A., Nova, E., Mateos-Aparicio, I. and Rupérez, P. 2018. Soybean Okara modulates gut microbiota in rats fed a high-fat diet. Bioact. Carbohydr. Diet. Fibre. 16: 100–107. DOI: 10.1016/j.bcdf.2018.09.002.

Rao, P. and Rathod, V. 2019. Valorization of food and agricultural waste: a step towards greener future. Chem. Rec. 19(9): 1858–1871. DOI: 10.1002/tcr.201800094.

Rebollo-Hernanz, M., Zhang, Q., Aguilera, Y., Martín-Cabrejas, M.A. and Gonzalez de Mejia, E. 2019. Relationship of the phytochemicals from coffee and cocoa by-products with their potential to modulate biomarkers of metabolic syndrome *in vitro*. Antioxidants 8(8): 279. DOI: 10.3390/antiox8080279.

Samad, N., Muneer, A., Zaman, A., Ayaz, M.M. and Ahmad, I. 2017. Banana fruit pulp and peel involved in antianxiety and antidepressant effects while invigorate memory performance in male mice: Possible role of potential antioxidants. Pak. J. Pharm. Sci. 30(3(Supl.)): 989–995.

Selmi, S., Rtibi, K., Grami, D., Sebai, H. and Marzouki, L. 2017. Protective effects of orange (Citrus sinensis L.) peel aqueous extract and hesperidin on oxidative stress and peptic ulcer induced by alcohol in rat. Lipids Health Dis. 16(1): 1–12. DOI: 10.1186/s12944-017-0546-y.

Somensi, L.B., Boeing, T., Cury, B.J., Steimbach, V.M.B., Niero, R., de Souza, L.M., da Silva, L.M. and de Andrade, S.F. 2017. Hydroalcoholic extract from bark of Persea major (Meisn.) LE Kopp (Lauraceae) exerts antiulcer effects in rodents by the strengthening of the gastric protective factors. J. Ethnopharmacol. 209: 294–304. DOI: 10.1016/j.jep.2017.08.007.

Tajaldini, M., Samadi, F., Khosravi, A., Ghasemnejad, A. and Asadi, J. 2020. Protective and anticancer effects of orange peel extract and naringin in doxorubicin treated esophageal cancer stem cell xenograft tumor mouse model. Biomed. Pharmacother. 121: 109594. DOI: 10.1016/j.biopha.2019.109594.

Tan, X.W., Kobayashi, K., Shen, L., Inagaki, J., Ide, M., San Hwang, S. and Matsuura, E. 2020. Antioxidative attributes of rice bran extracts in ameliorative effects of atherosclerosis-associated risk factors. Heliyon 6(12): e05743. DOI: 10.1016/j.heliyon.2020.e05743.

Villanueva-Suárez, M.J., Mateos-Aparicio, I., Pérez-Cózar, M.L., Yokoyama, W. and Redondo-Cuenca, A. 2019. Hypolipidemic effects of dietary fibre from an artichoke by-product in Syrian hamsters. J. Funct. Foods 56: 156–162. DOI: 10.1016/j.jff.2019.03.013.

Wahyudi, I.A., Ramadhan, F.R., Wijaya, R.I.K., Ardhani, R. and Utami, T.W. 2020. Analgesic, anti-inflammatory and anti-biofilm-forming activity of potato (*Solanum tuberosum* L.) peel extract. Indonesi. Jo. Cancer Chemoprevention 11(1): 30–35. DOI: 10.14499/indonesianjcanchemoprev11iss1pp30-35.

Wang, F. 2018. The roles of preventive and curative health care in economic development. PloS One 13(11): e0206808. DOI: 10.1371/journal.pone.0206808.

Wang, Y.H., Ge, B., Yang, X.L., Zhai, J., Yang, L.N., Wang, X-X., Shi, J-C. and Wu, Y.J. 2011. Proanthocyanidins from grape seeds modulates the nuclear factor-kappa B signal transduction pathways in rats with TNBS-induced recurrent ulcerative colitis. Int. Immunopharmacol. 11(10): 1620–1627. DOI: 10.1016/j.intimp.2011.05.024.

Wong, A., Viola, D., Bergen, D., Caulfield, E., Mehrabani, J. and Figueroa, A. 2019. The effects of pumpkin seed oil supplementation on arterial hemodynamics, stiffness and cardiac autonomic function in postmenopausal women. Complement. Ther. Clin. Pract. 37: 23–26. DOI: 10.1016/j.ctcp.2019.08.003.

Xian, Y., Fan, R., Shao, J., Toney, A.M., Chung, S. and Ramer-Tait, A.E. 2021. Polyphenolic fractions isolated from red raspberry whole fruit, pulp, and seed differentially alter the gut microbiota of mice with diet-induced obesity. J. Funct. Foods 76: 104288. DOI: 10.1016/j.jff.2020.104288.

Yang, L., Lin, Q., Han, L., Wang, Z., Luo, M., Kang, W., Liu, J, Wang, J., Ma, T. and Liu, H. 2020. Soy hull dietary fiber alleviates inflammation in BALB/C mice by modulating the gut microbiota and suppressing the TLR-4/NF-κB signaling pathway. Food Funct. 11(7): 5965–5975. DOI: 10.1039/d0fo01102a.

Chapter 18

The Benefits of Integral Use of Food in Waste Reduction, Nutrient Density and Human Health

Eliakim Aureliano da Silva,[1] *Tamiris Ramos Silva,*[1]
Maria Clara Feijó de Figueiredo,[1]
Beatriz Gonçalves Feitosa dos Santos,[1] *Lavínia Alves de Sousa,*[1]
Rosa Maria de Oliveira Santos,[1] *Fhanuel Silva Andrade,*[2]
Joilane Alves Pereira-Freire[1] and *Stella Regina Arcanjo Medeiros*[1,*]

Introduction

Global transformations in the social, economic and political fields are notable, a situation conducive to the emergence of demographic and nutritional models. In view of this fact, relevant changes in food patterns arise and scientific research points to the use of food waste resulting from domestic consumption and/or agro-industrial waste, reducing waste and increasing nutritional value. It is worth noting that food shortage is seen in the manifestation of social inequality, in food waste that prevails in all countries, nevertheless developing nations stand out negatively. culturally the usable parts of food and that which can be used, emphasizing food enrichment are routinely discarded, generating millions of tons of food waste (organic) throughout the year (Leão and Recine, 2011; FAO, 2011; Cardoso et al., 2015).

[1] Research Laboratory II, Federal University of Piaui, Campus Senator Helvídio Nunes de Barros - Junco, Picos - PI, 64607-670.
[2] Postgraduate Program in Food and Nutrition, Federal University of Piaui, Campus Ministro Petrônio Portella, Ininga, Teresina-PI, 64.049-550.
* Corresponding author: stellaarcanjo@ufpi.edu.br

The integral use of food in the fight against hunger and nutritional deficiencies is still an underexploited subject, despite promising experiences in which the integral use of some foods has a significant impact on the diet of underprivileged populations, and over time it is argued that it is of great value for all social classes to consume the food in its entirety when possible. But a large part of society is rooted in inappropriate habits; and despite scientific evidence, the waste of some food parts that could be consumed is still being seen (Vilhena and Silva, 2007; Nunes, 2009; Farias et al., 2016; Cardoso and Vieira, 2019).

In view of this, a continuous search for new food products from edible food parts that were earlier discarded is necessary, providing increased nutrient density, reducing food waste and the risks of chronic diseases, in addition to influencing food security with improvements in food, nutrition and health conditions of the Brazilian population.

The integral use of food, public policies and the reduction of food waste

To meet the constant increase in world population, food industries also had to adapt to this change, increasing their production and concomitant to this, the amount of waste and byproducts generated through these processes, which became a problem to environmental and social agencies, because this change contributes to problems such as climate change, environmental degradation and economic inequality (Caldeira et al., 2020; Jayesree et al., 2021).

According to the United Nations (UN) Food and Agriculture Organization (FAO, 2019), it is estimated that each year about 1.3 billion tons of food is lost in the world, which represents more than 30% of all food production for human consumption. Among the food groups, fruits and vegetables are the most wasted during processing, because due to their perishability and the need to increase their shelf life, they must go through some processing steps, such as cutting, peeling, boiling, canning, freezing, among others, generating a waste production of approximately 25 to 30% of the amount of food produced. These wastes usually comprise the inedible parts of the plants, such as leaves, seeds, stalks, roots, bagasse (FAO, 2019; Sagar et al., 2018).

Among the targets of the Sustainable Development Goals (SDGs) adopted by the United Nations, one of them is to halve global food waste and reduce food losses along the production and supply chain by 2030 in order to protect our planet's natural resources and climate for future generations (Target 12.3) (UNEP, 2015).

In Brazil, until the year 2020 there was no detailed law regarding the donations of surplus food produced for commercial purposes, however, in June 2020, Law No. 14.016 was enacted, in which the establishments focused on production and supply of food were authorized to donate the surpluses that were not commercialized and that were still suitable for human consumption. However, following some prerequisites, the expiration date of the food must be regarded, with effective conservation conditions, without compromising its integrity, health security and finally the food should still contain its nutritional properties (Brasil, 2020).

The large amount of food waste generated is alarming and discussed at length, since the impacts on the environment and sustainability are always observed at a

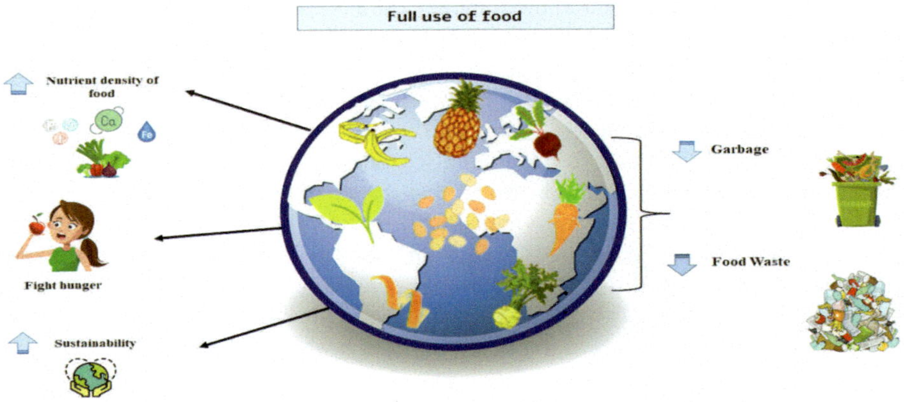

Figure 18.1. Demonstration of the benefits brought by the integral use of food as a global strategy (Source: prepared by the authors).

global level (Cedes, 2018; Gallo, 2018). The integral use of food has also come to be seen as an important global strategy for reducing waste and increasing the nutrient density of food preparations, helping to fight hunger, reducing waste and the amount of waste produced (Fig. 18.1).

Some European countries participating in the *EU FUSION* project (a project aimed at reducing food waste in Europe and using resources more efficiently) classify these commonly uneaten parts of food into: unavoidable, avoidable and possibly avoidable food waste. The unavoidable ones are fruit cores, peels and bones; the possibly avoidable ones include potato or carrot peels as these can be consumed. These countries consider it important to use these wastes to generate other high-value products, but they are wary of the risk of rebound effects, which may be an incentive to generate more waste (Zink and Geyer, 2017; Teigiserova et al., 2019).

However, the integral utilization of certain foods is one way to reduce the accumulation of waste and also improve the nutritional quality of food preparations, through the consumption of peels, seeds, stalks, which are usually considered waste by the population. Scientific research shows that in some foods the content of nutrients and bioactive compounds is higher in the parts that are discarded when compared to what is consumed and these compounds bring numerous benefits to human health, as shown in Table 18.1.

Just as functional foods bring health benefits, the edible parts of foods that are discarded also have compounds with functional claims and have been proven to have bioactive compounds, which provide benefits to human health (Table 18.1).

The integral use of food and the benefits to human health

The social nature of food is closely related to diet quality, health status and lifestyle, playing an important role in the quality of human life. Food practices involving daily aspects, such as what one wants to eat, choices, the way of preparing and consuming the food and even the place where the meal is taken, represent sociocultural aspects such as cultural identity, social condition and family memory, food and nutrition

Table 18.1. Non-conventional edible parts of vegetables and fruits, bioactive compounds present and their health benefits. Source: Own authorship (2021).

Non-conventional edible parts of vegetables and fruits	Bioactive compounds	Health benefits	References
Carrot peelings	Rich in b-carotene	Antioxidant function assisting in the elimination of free radicals.	Jayesree et al. (2021)
Apple pomace	Dihydrocalcones (floretin and florizin)	They are able to inhibit active glucose transport into cells by SGLT1 and SGLT2 in the small intestine, new possibility to treat diabetes.	Esparza et al. (2020)
Orange peel	D-Limonene	Antimicrobial action	Ferronatto and Rossi (2018)
Bark of the buriti fruit	Rich in minerals and bioactive compounds (chlorogenic and protocatechuic acid)	Antioxidant function assisting in the elimination of free radicals.	Pereira-Freire et al. (2018)
Peel of citrus fruit, peel of banana, pomegranate, mango and papaya	Pectin	Wound healing and tissue engineering	Oliveira et al. (2018); Khamsucharit et al. (2018); Yang et al. (2018)
Red grape marc	Anthocyanin	Reduction of oxidative stress, elimination of free radicals, assisting in reducing the risk of cancer and disease, as well as cholesterol. regulation	Teigiserova et al. (2019).
Broccoli stalks	Fibers	Fiber helps in maintaining proper serum glucose and cholesterol levels.	Silva et al. (2021)
Pumpkin seed	Polyunsaturated oils	Reduction of cardiovascular diseases and helminthic action	Vieira et al. (2021)
Olive pomace	Phenolic alcohol	Antioxidant function assists in the elimination of free radicals.	Fava et al. (2017)

education to promote choices more favorable to health in complex processes (Galesi-Pacheco et al., 2019).

Adequate nutrition depends, among other aspects, on people's knowledge about types and characteristics of foods that make them more or less healthy, the ease and proximity of shopping locations, preferences developed over the course of life and the presence of health problems (Medina et al., 2019).

The association between food and individuals' health/disease processes has been widely explored in literature, particularly emphasizing the influence that a dietary

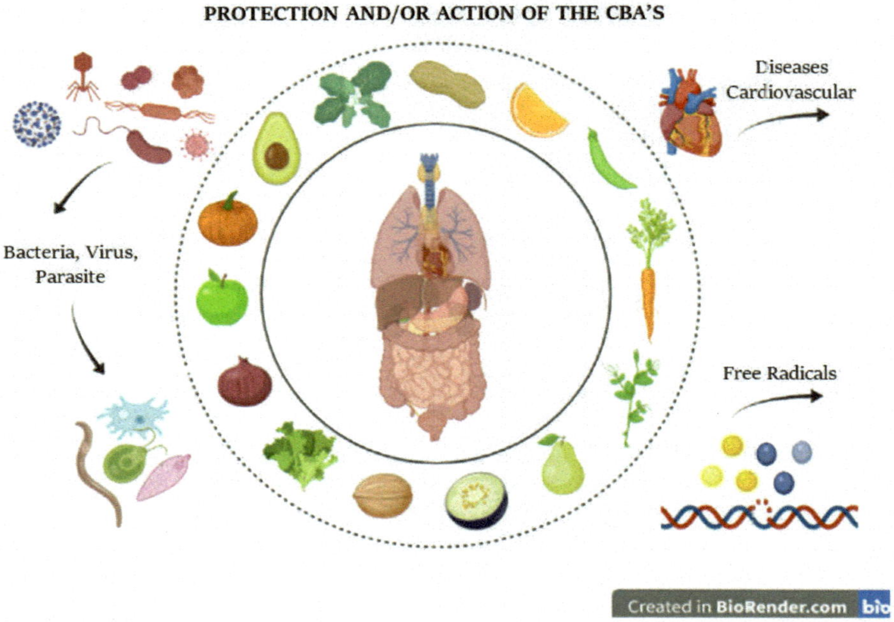

Figure 18.2. Benefits of functional foods and bioactive compounds (CBAs) to cardiovascular health, immunity and free radical inhibition (Source: prepared by the authors).

pattern considered healthy plays on the population's well-being and quality of life. Based on this premise, several strategies in the field of nutrition, aiming to promote healthy habits, are based on the debate about the importance of eating healthily as a primary factor that defines food choices (Batista and Moreira, 2020).

Functional foods are those foods or ingredients that produce beneficial health effects beyond their basic nutritional functions, including the metabolic or physiological role that these nutrients or non-nutrients has in growth, development, maintenance and other functions in the human body that influence the reduction or control of chronic degenerative diseases (Cañas and Braibante, 2019) (Fig. 18.2).

The protection that these foods provide to human health occurs through their constituents, the bioactive compounds, which have a mechanism of action and complex synergistic effects. This optimizing action may be subject to interference from several variables such as nutritional status, mode and amount of food consumed, gastrointestinal integrity, genetics and microbiota bacteria (Fig. 18.2). The influence of the genetic factor stands out, since research proves that polymorphisms caused by bioactive compounds in specific genes cause metabolic and functional changes responsible for benefiting the organism. Also modulating the expression of genes involved in physiometabolic responses of complex chronic pathologies such as cardiovascular diseases, obesity and type 2 diabetes (Milenkovic et al., 2011; Chuengsamarn et al., 2012; Berná et al., 2014; Konings et al., 2014).

Although research on the usually discarded parts of food is still scarce, some results show that fruit peels have higher amounts of nutrients than the pulp, thus being an alternative nutritional source and a strategy to reduce food waste (Bramont et al., 2018; Weurich, 2016).

In a study conducted by Karnjanawipagul et al. (2010) when analyzing the concentrations of b-carotene in carrot pulp, approximately 14.59 mg of the compound was found for 100 g of vegetable, while in the peel approximately 204.5 mg was found, a value much higher than the parts often consumed (Mamma and Christakopoulos, 2014). The b-carotene has provitamin A activity, important for eye health and also has antioxidant action, capable of reducing the amount of free radicals present in the body, reducing the risk of degenerative diseases and also protects the skin from damage caused by UV radiation (Marinova and Ribavora, 2007).

The carotenoids produced and stored in pumpkin, especially b-carotene, have an active property, thus ensuring the maintenance, growth and development of the epithelium, essential for correct night vision and immune function. Regarding the fatty acids present in the pumpkin seed, the most studied are fatty acids: oleic, linoleic and palmitic acids (Montesano, 2018; Vale et al., 2019).

Regarding broccoli, it is common to observe that stalks and leaves are discarded as organic waste or used for animal feed or completely discarded by the consumer during the preparation process. However, the usable residues can contain up to a number of fibers averagely six times higher and minerals twice as much as the broccoli flower itself, which is currently still the only part consumed and valued by the population (Coutinho, 2021).

The actions of pectin in human health are associated with the reduction of cholesterol (reduction of 6 to 7% of LDL cholesterol) and blood glucose (decreased glycemic response and insulin secretion). It also has the characteristic of a gel-forming element and can be used when a product is not able to generate this element in an adequate amount. According to the RDC No 45/2010, pectin is a food additive that has the technological function of "thickener". Still, its use is based on *quantum satis,* that is, proportion/quantity needed to achieve the desired aspect, provided that its use does not cause harmful practice of fraud and that its addition in a product is approved for the product in question (ANVISA, 2010; Moura, 2016).

In the industry, many residues are discarded after the extraction of juices and pulps, parts such as peels, seeds, however, these residues can undergo several processes for obtaining substances with high nutritional value, such as essential oil. The utilization of oils is still exploited in small quantities, due to more ingenious extraction processes, however the orange peel and pumpkin seeds are positive examples of the effects obtained from using these components (Farhat et al., 2011).

Table 18.1 displays some examples of usually discarded food parts and the bioactive compounds present in them, it is possible to observe that all of them have beneficial effects to health and even functional claims registered by the National Health Surveillance Agency (ANVISA), as is the case of fibers, present in most fruit peels and broccoli stalks in which fibers assist the bowel function and help maintain adequate levels of serum glucose and cholesterol, its consumption should be associated with a balanced diet and healthy lifestyle habits (Silva and Brito, 2015; ANVISA, 2019).

Still observing the aforementioned table and the most current literature, one can mention that fruits have stood out among other foods in the integral utilization of food, since several studies show that their regular consumption and their parts reduce the incidence of certain pathologies, besides providing a huge diversity of bioactive compounds and the technological potential of their residues (Vizzoto, 2012).

An alternative to the use of these parts is the preparation of bran and flour that allows the preparation of other products, such as breads, drinks and desserts, with better nutritional characteristics. Thus, the integral use of food, in addition to promoting health benefits through the compounds present, still considers the idea of sustainability. The use of these usually discarded food parts besides being accessible and low-cost offers many nutrients, reducing malnutrition and hunger, especially in low-income families (Camargo, 2016).

Conclusions

The integral utilization of food is a field that tends to grow more and more, especially in the face of the social, economic and environmental crises that countries have been facing, and can be seen as the beginning of the reflection of man's awareness to positive changes in the environment. Considering that essential use of food prevents malnutrition and hunger, it reduces waste and can contribute to feeding the population, especially the low-income population. Thus, EIA does not only promote health benefits through the compounds present, but also regards the idea of sustainability.

In this chapter, it became clear that an excellent alternative for the integral use of food is the preparation of bran and flour, which leads to the explanation of other products, such as breads, drinks and desserts, with better nutritional characteristics. It was also possible to share the nutritional content and inedible parts of some foods, as well as to inform about the nutritional and economic potential of Non-Conventional Food Plants - NCFPs, which during harvests are lost due to the lack of technological scientific studies and the cultural devaluation of native flora.

It is known that only scientific research is able to answer and solve some relevant questions, as well as different research options in the development of products and contemporary studies of functional foods, or even part of them, that are discarded, as well as the valorization/utilization of non-conventional food plants and fruits, whose potential have not been explored satisfactorily in various areas of science and technology.

References

ANVISA. Agência Nacional de Vigilância Sanitária. Alegações de propriedade funcional aprovadas, 2019. Disponível em: Alegações de propriedade funcional aprovadas_Anvisa.pdf — Português (Brasil) (www.gov.br). Acesso em: 11 de setembro de 2021.

Batista, L.D. and Moreira, P.V.L. 2020. Decisões Alimentares de Beneficiárias do Programa Bolsa Família: a Renda é o Único Determinante do "Comer Saudável"? Revista Brasileira de Ciências da Saúde. 24(1): 91–104, Disponível em:< https://periodicos.ufpb.br/ojs2/index.php/rbcs/article/view/47695/29836>

Berná, G. et al. 2014. Nutrigenetics and nutrigenomics insights into diabetes etiopathogenesis. Nutrients 6: 5338–69.

Bramont, W.B., Leal, I.L., Umsza-Guez, M.A., Guedes, A.S., Alves, S.C.O., Reis, J.H.O., Barbosa, J.D.V. and Machado, B.A.S. 2018. Comparação da Composição Centesimal, Mineral e Fitoquímica de Polpas e Cascas de Dez Diferentes Frutas. Rev. Virtual Quim. 10(4), no prelo. Data de publicação na Web: 24 de julho de 2018.

BRASIL. Lei n°14.016, de 23 de junho de 2020. Dispões sobre o combate ao desperdício de alimentos e a doação de excedentes de alimentos para o consumo humano. Disponível em: L14016 (planalto.gov.br).

Caldeira, C., Vlysidis, A., Fiore, G., Laurentiis, V., Vignali, G. and Sala, S. 2020. Sustainability of food waste biorefinery: a review on valorisation pathways, techno-economic constraints, and environmental assessment. Bioresource Technology. doi: https://doi.org/10.1016/j.biortech.2020.123575.

Camargo, T.L. 2016. APROVEITAMENTO INTEGRAL DE ALIMENTOS: menos resíduos e mais saúde. Porto Alegre. 93 f. Trabalho de Conclusão de Curso (Graduação em Nutrição) - Universidade Federal do Rio Grande do Sul, Faculdade de Medicina.

Cañas, G.J.S. and Braibante, M.E.F. 2019. A Química dos Alimentos Funcionais. Quím. nova esc. – São Paulo. 41(3): 216–223, ago, 2019. Disponível em: https://www.researchgate.net/publication/334639335_A_Quimica_dos_Alimentos_Funcionais.

Cardoso, A.M.A. and Vieira, T.A. 2019. Práticas de redução do desperdício de alimentos: o caso de um projeto social em Santarém, Pará. Multitemas. 24(58).

Cardoso, F.T. et al. 2015. Aproveitamento integral de Alimentos e o seu impacto na Saúde. Sustentabilidade em Debate. Brasília, 6,3: 131–143.

CEDES – Centro de Estudos e Debates Estratégicos. 2018. Consultoria Legislativa da Câmara dos Deputados. Perdas e desperdício de alimentos – estratégias para redução. Série de cadernos de trabalhos e debates 3. Brasília, DF.

Chuengsamarn, S. et al. 2012. Curcumin extract for prevention of type 2 diabetes. Diabetes Care 35: 2121–7.

Coutinho, A.J.M. 2021. Cupcake adicionado de farinha de resíduos de brócolis: análise físico-química e sensorial entre crianças. Multitemas, 5–20.

ESPARZA, I. et al. 2020. Fruit and vegetable waste management: Conventional and emerging approaches. Journal of Environmental Management 265: e.110510.

FAO, Food and Agriculture Organization of the United Nations. 2021. Global food losses and food waste – Extent, causes and prevention, 2011. Disponível em: < http://www.fao.org/3/mb060e/mb060e00.htm > Acesso em: 15 de abr. 2021.

FAO, IFAD, UNICEF, WFP and WHO. 2019. The Stateof Food Security and Nutrition in the World 2019. Safe guarding again steconomics lowdown sand down turns. Rome, FAO. Disponível em: http://www.fao.org/3/ca5162en/ca5162en.pdf. Acesso em: 06 de julho de 2020.

Farias, P.K.S., Souza, S.D.O., Santana, I.M.O., Prates, R.P., Gusmão, A.C.M. and Soares, P.D.F. 2016. Desenvolvimento e análise sensorial de diferentes tipos de hambúrgueres funcionais utilizando o reaproveitamento de alimentos. Cad. Ciênc. Agra. 8(3): 07–14.

Farhat, A. et al. 2011. Microwave steam diffusion for extraction of essential oil from orange peel: Kinetic data, extract's global yield and mechanism. Food and Bioproducts Processing 125: 255–261.

Fava, G. et al. 2017. Hydroxytyrosol recovery from olive mill wastewater: process optimization and development of a pilot plant. CLEAN- Soil, Air, Water. v.45, e.4.

Ferronatto, A.N. and Rossi, R.C. 2018. Extração e aplicação do óleo essencial da casca da laranja como um ingrediente natural. Estudos Tecnológicos em Engenharia 12: 78–93.

Galesi-Pacheco, L.F. et al. 2019. Food practices and their meanings in the daily routine of a university. Rev. Nutr. 32: e180208. Disponível em: https://www.scielo.br/pdf/rn/v32/1415-5273-rn-32-e180208.pdf.

Gallo, J.M.A.S. 2018. Avaliação da percepção de participantes do Banco de Alimentos da Companhia de Entrepostos e Armazéns Gerais de São Paulo sobre o processo de irradiação de alimento. Divulgação do tratamento de alimentos por radiação ionizante: desenvolvimento de questionário e análise sensorial. Tese (doutorado) – IPEN – Instituto de Pesquisas Energéticas e Nucleares. Autarquia associada a Universidade de São Paulo – USP, São Paulo, 245 f.

Jayesree, N. et al. 2021. Valorisation of carrot peel wasteby water-induced hydrocolloidal complexation for extraction of carote and pectin. Chemosphere 272: e.129919.

Karnjanawipagul, P., Nittayanuntawech, W., Rojsanga, P. and Suntornsuk, L. 2010. Analysis of -Carotene in carrot by spectrophotometry. Mahidol University Journal of Pharmaceutical Science 37: 8–16.

Khamsucharit, P. et al. 2018. Characterization of pectin extracted from banana peels of different varieties. Food Sci. Biotechnol. 27(3): 623–629.

Konings, E. et al. 2014. The effects of 30 days resveratrol supplementation on adipose tissue morphology and gene expression patterns in obese men. Int. J. Obes (Lond) 38: 470–3.

Leão, M.M. and Recine, E. 2011. O direito humano à alimentação adequada. pp. 471–488. *In*: Taddei, J.A., Lang, R.M.F., Longosilva, G. and Toloni, M.H.A. (eds.). Nutrição em Saúde Pública. São Paulo: Rubio.

Mamma, D. and Christakopoulos, P. 2014. Biotransformation of citrus by-products into value added products. Waste and Biomass Valorization 5(4): 529–549.

Marinova, D. and Ribarova, F. 2007. HPLC determination of carotenoids in Bulgarian berries. Journal of Food Composition and Analysis, Davis 20(5): 370–374.

Medina, L.P.B. et al. 2013. Desigualdades sociais no perfil de consumo de alimentos da população brasileira. Pesquisa Nacional de Saúde. Rev. Bras. Epidemiol. vol.22 supl.2 Rio de Janeiro 2019. Disponível em: https://www.scielo.br/pdf/rbepid/v22s2/1980-5497-rbepid-22-s2-e190011-supl-2.pdf

Milenkovic, D. et al. 2011. Hesperidin displays relevant role in the nutrigenomic effect of orange juice on blood leukocytes in human volunteers: a randomized controlled cross-over study. PLoS One 6, ISSN 1932-6203.

Montesano, D. 2018. Chemical and nutritional characterization of seed oil from Cucurbita máxima L. (var. Berrettina). Pumpkin Foods 7: 1–14.

Moura, F.A. de. 2015. Propriedades físico-químicas e efeito prebiótico de pectina hidrolisada obtida de resíduos agroindustriais. 2015. 77 fls. Tese (Doutorado em Ciência e Tecnologia de Alimentos), Universidade Federal de Santa Maria, Rio Grande do Sul - RS.

Nunes, J.T. 2009. Aproveitamento integral dos alimentos: qualidade nutricional e aceitabilidade das preparações. [Monografia]. Brasília: Centro de Excelência em Turismo - CET, Universidade Federal de Brasília.

Oliveira, A.N. et al. 2018. Optimization of pectin extraction from Ubá mango peel through surface response methodology. Int. J. Biol. Macromol. 113: 395–402.

Sagar, N.A., Pareek, S., Sharma, S., Yahia, E.M. and Lobo, M.G. 2018. Fruit and vegetable waste: bioactive compounds, their extraction, and possible utilization. Compr. Rev. Food Sci. Food Saf. 17: 512–531.

Silva, K.T.V. et al. 2021. Elaboração de uma quiche de pupunha com brócolis e frango. Saber Científico. 4: 52–57.

Silva, M.P.L. and Brito, T.S. 2015. Aproveitamento integral dos alimentos e edução nutricional em uma creche do Rio de Janeiro: estudo de caso. 2015. TCC (Bacharelado) Universidade Federal Fluminense, Bacharelado em Nutrição, Niterói.

Teigiserova, D.A., Hamelin, L. and Thomsen, M. 2019. Review of high-value food waste and food residues biorefineries with focus on unavoidable wastes from processind. Resources, Conservation & Recycling 149: 413–426.

UNEP, 2015. Transforming Our World: the 2030 Agenda for Sustainable Development. https://doi.org/10.1007/s13398-014-0173-7.2.

Vale, C.P. et al. 2019. Composição e propriedades da semente de abóbora. Fag Journal of Health (FJH) 1: 79–90.

Vilhena, M.O. and Silva, M.C. 2007. Aproveitamento integral de alimentos orgânicos: arte culinária verde. In: 2º Jornada Nacional da Produção Científica em Educação Profissional e Tecnológica, 2007, São Luís/MA.

Vieira, K.H. et al. 2021. Caracterização da farinha de semente de abóbora obtida por secagem em micro-ondas e estufa. Jornal Brasileiro de Desenvolvimento. 7: 22267–22283.

Vizzotto, M. 2012. Propriedades funcionais das pequenas frutas. Informe Agropecuário 33: 84–88.

Weyrich, O.D. 2016. Cascas de frutas: estudo das propriedades nutricionais e tecnológicas. Dissertação (Mestrado) – Universidade Federal de Goiás, Escola de Agronomia (EA), Programa de Pós-Graduação em Ciência e Tecnologia de Alimentos, Cidade de Goiás.

Yang, X. et al. 2018. Pomegranate peel pectin can be used as an effective emulsifier. Food Hydrocolloids 85(April): 30–38.

Zink, T. and Geyer, R. 2017. Circular economy rebound. J. Ind. Ecol. 21: 593–602.

Index

A

active packaging 64, 65, 68
Agro-industrial waste 1, 3, 4
Anacardium occidentale 105, 114
Annonaceous 188
anti-inflammatory activity 258, 259, 264
antimicrobial activity 120, 121, 126, 128
antioxidant 119–121, 125–129, 256
Avocado seed 202–205, 207, 208, 210, 211

B

Basil 119–122, 125–127
bioactive compounds 202–204, 207–211, 233, 235, 236, 238, 239, 254
Biofilms 61
Biogenic amines 163, 164
Biological activities 233, 238, 239, 246
Biomedical 4
Bioprospecting 1–6
biotechnological applications 137, 185, 214, 216
Biotechnology 118–120, 127–129
Biotransformation 147–149
by-product 133, 136, 137, 140

C

Cacti 87, 91–93, 95
Caruru 33
Cashew tree 105, 106, 108, 114
chemical composition 122–125, 129, 185–188, 216, 217, 227
Chicory 31, 32
CNL 107–109, 111–114
coatings 72–83
Co-Products 1, 3, 4
Cubiu 29, 30
Culinary 220, 224, 228

E

Edible coating 73–76, 79, 83

F

Family Annonaceae 189
Fermentation 147–150, 154
Flavonoids 166, 168, 170
Food 2, 3, 6
food safety 84
Food waste 144–146, 148–154
functional food 201, 202, 208–211, 232, 233

G

gastrointestinal disorders 167

H

Hylocereus polyrhizus 232, 233
hypercholesterolemia 202, 207
hyperglycemia 202, 207
hypocholesterolemic potencial 51

I

Integral use of food 275, 276, 278
Intellectual property 10, 13, 15
isothiocyanates 168

M

Mauritia flexuosa L. 41, 42, 47, 50
Maxixe 30
Microbial bioprocesses 147, 149
Moringa 15–18
Moringa oleifera 133–135, 137
mycotoxins 176

N

Nutritional aspects 185, 189
Nutritional composition 233, 234
nutritional potential 42
Nutritional Profile 95
nutritional value 272, 277

O

Ocimum spp. 119

P

Pharmacological properties 185, 191, 193
Platonia insignis Mart. 215–219, 221, 223, 228
Polymers 58–61, 65, 67, 68
Pomegranate 10, 18

S

seed cake 133, 136–141
Soursop 11–13, 18
sustainability 3–5, 66, 273, 278

T

Taro 26–29
Technological Applications 99

U

Umbu 13–15, 18

W

waste reduction 272

Y

Yam 21–24, 26

Editors' Biography

Dr. Joilane Alves Pereira Freire is a Professor at the Department of Nutrition, Federal University of Piaui, Picos-Piaui, Brazil. Her research area includes food biotechnology, tests of toxicity of natural products and natural resources with potential applications in the development of new food ingredients. Leader of the Research Group on Bioprospecting, Biotechnology and Innovation of natural products (BBIPN). She has published 34 research papers in national and international journals and edited/authored 11 books and 1 patent.

Dr. Stella Regina A. Medeiros is a Professor at the Department of Nutrition, Federal University of Piaui, Picos-Piaui, Brazil. Her research area includes food biotechnology, natural resources with potential applications in the development of functional foods. She has published more than 40 research papers in national and international journals. In addition, has edited/authored more than 31 books/chapters from eminent national publishers and also 10 patent applications.

Dr. Jailane de Souza Aquino is a Professor at the Department of Nutrition, Federal University of Paraíba, João Pessoa-Paraíba, Brazil. She is the Head Researcher of Experimental Nutrition Laboratory, Department of Nutrition, UFPB, Brazil. She is a member of the IFMSA - Score Research Exchange, acting as an advisor to medical students on exchange in Brazil. She is a member of the International Society for Developmental origins of health and disease - DOHaD. Leader of the Research Group on Food, Bioactives and Nutrition: on health and disease-from pregnancy to senescence (GPABiN). She has published more than 75 research papers in national and international journals. In addition, she has edited/authored 6 books and 4 patents. She has CNPq Research Productivity Scholarship - level 2.

Dr. Chistiane M. Feitosa is a Professor at the Department of Chemistry, Federal University of Piaui, Teresina-Piaui, Brazil. Her research area includes pharmacology, natural products and their secondary metabolites with the potential for the treatment of neurodegenerative diseases and antimicrobials. She has published more than 55 research papers in national and international journals. In addition, she has edited/authored more than 12 books/chapters from eminent publishers like CRC Press/Taylor and Francis, and also 5 patents.

Dr. Mahendra Rai is presently a visiting Professor at the Department of Microbiology, Nicolaus Copernicus University, Torun, Poland. Formerly he was Professor and Head of the Department of Biotechnology, SGB Amravati University, Maharashtra state, India. He was a visiting scientist at the University of Geneva, Debrecen University, Hungary; University of Campinas, Brazil; VSB Technical University of Ostrava, Czech Republic and National University of Rosario, Argentina and the University of Sao Paulo. He has published more than 425 research papers in national and international journals. In addition, he has edited/authored more than 69 books and 6 patents. Recently, he has been featured in Stanford's list of the top 2% of scientists under Nanoscience and Nanotechnology.